Lecture Notes
in Business Information Processing

354

More information about this series at http://www.springer.com/series/7911

Witold Abramowicz · Rafael Corchuelo (Eds.)

Business Information Systems

22nd International Conference, BIS 2019
Seville, Spain, June 26–28, 2019
Proceedings, Part II

 Springer

Editors
Witold Abramowicz 🆔
Poznań University of Economics
and Business
Poznań, Poland

Rafael Corchuelo 🆔
ETSI Informática
University of Seville
Seville, Spain

ISSN 1865-1348   ISSN 1865-1356   (electronic)
Lecture Notes in Business Information Processing
ISBN 978-3-030-20481-5   ISBN 978-3-030-20482-2   (eBook)
https://doi.org/10.1007/978-3-030-20482-2

This Springer imprint is published by the registered company Springer Nature Switzerland AG
The registered company address is: Gewerbestrasse 11, 6330 Cham, Switzerland

Preface

During the 22 years of the International Conference on Business Information Systems, it has grown to be a well-renowned event of the scientific community. Every year the conference joins international researchers for scientific discussions on modeling, development, implementation, and application of business information systems based on innovative ideas and computational intelligence methods. The 22nd edition of the BIS conference was jointly organized by the University of Seville, Spain, and Poznań University of Economics and Business, Department of Information Systems, Poland, and was held in Seville, Spain.

The exponential increase in the amount of data that is generated every day and an ever-growing interest in exploiting this data in an intelligent way has led to a situation in which companies need to use big data solutions in a smart way. It is no longer sufficient to focus solely on data storage and data analysis. A more interdisciplinary approach allowing one to extract valuable knowledge from data is required for companies to make profits, to be more competitive, and to survive in the even more dynamic and fast-changing environment. Therefore, the concept of data science has emerged and gained the attention of scientists and business analysts alike.

Data science is the profession of the present and the future, as it seeks to provide meaningful information from processing, analyzing, and interpreting vast amounts of complex and heterogeneous data. It combines different fields of work, such as mathematics, statistics, economics, and information systems and uses various scientific and practical methods, tools, and systems. The key objective is to extract valuable information and infer knowledge from data that then may be used for multiple purposes, starting from decision-making, through product development, up to trend analysis and forecasting. The extracted knowledge allows also for a better understanding of actual phenomena and can be applied to improve business processes. Therefore, enterprises in different domains want to benefit from data science, which entails technological, industrial, and economic advances for our entire society. Following this trend, the focus of the BIS conference has also migrated toward data science.

The BIS 2019 conference fostered the multidisciplinary discussion about data science from both scientific and practical sides, and its impact on current enterprises. Thus, the theme of BIS 2019 was "Data Science for Business Information Systems." Our goal was to inspire researchers to share theoretical and practical knowledge of the different aspects related to data science, and to help them transform their ideas into the innovations of tomorrow.

The first part of the BIS 2018 proceedings is dedicated to Big Data, Data Science, and Artificial Intelligence. This is followed by other research directions that were discussed during the conference, including ICT Project Management, Smart Infrastructures, and Social Media and Web-based Systems. Finally, the proceedings end with Applications, Evaluations, and Experiences of the newest research trends in various domains.

The Program Committee of BIS 2018 consisted of 78 members who carefully evaluated all the submitted papers. Based on their extensive reviews, 67 papers were selected.

We would like to thank everyone who helped to build an active community around the BIS conference. First of all, we want to express our appreciation to the reviewers for taking the time and effort to provide insightful comments. We wish to thank all the keynote speakers who delivered enlightening and interesting speeches. Last but not least, we would like to thank all the authors who submitted their papers as well as all the participants of BIS 2019.

June 2019   Witold Abramowicz

Organization

BIS 2019 was organized by the University of Seville and Poznań University of Economics and Business, Department of Information Systems.

Program Committee

Witold Abramowicz (Co-chair)	Poznań University of Economics and Business, Poland
Rafael Corchuelo (Co-chair)	University of Seville, Spain
Rainer Alt	Leipzig University, Germany
Dimitris Apostolou	University of Piraeus, Greece
Timothy Arndt	Cleveland State University, USA
Sören Auer	TIB Leibniz Information Center Science and Technology and University of Hannover, Germany
Eduard Babkin	LITIS Laboratory, INSA Rouen; TAPRADESS Laboratory, State University – Higher School of Economics (Nizhny Novgorod), Russia
Morad Benyoucef	University of Ottawa, Canada
Matthias Book	University of Iceland, Iceland
Dominik Bork	University of Vienna, Austria
Alfonso Briones	BISITE Research Group, Spain
François Charoy	Université de Lorraine – LORIA – Inria, France
Juan Manuel Corchado Rodríguez	University of Salamanca, Spain
Beata Czarnacka-Chrobot	Warsaw School of Economics, Poland
Christophe Debruyne	Trinity College Dublin, Ireland
Renata De Paris	Pontifical Catholic University of Rio Grande do Sul, Brazil
Yuemin Ding	Tianjin University of Technology, China
Suzanne Embury	The University of Manchester, UK
Jose Emilio Labra Gayo	Universidad de Oviedo, Spain
Werner Esswein	Technische Universität Dresden, Germany
Charahzed Labba	Université de Lorraine, France
Agata Filipowska	Poznań University of Economics and Business, Poland
Ugo Fiore	Federico II University, Italy
Adrian Florea	'Lucian Blaga' University of Sibiu, Romania
Johann-Christoph Freytag	Humboldt Universität Berlin, Germany
Naoki Fukuta	Shizuoka University, Japan
Claudio Geyer	UFRGS, Brazil
Jaap Gordijn	Vrije Universiteit Amsterdam, The Netherlands

Milena Stróżyna	Poznań University of Economics and Business, Poland
York Sure-Vetter	Karlsruhe Institute of Technology, Germany
Herve Verjus	Universite Savoie Mont Blanc – LISTIC, France
Krzysztof Węcel	Poznań University of Economics and Business, Poland
Hans Weigand	Tilburg University, The Netherlands
Benjamin Weinert	University of Oldenburg, Germany
Mathias Weske	University of Potsdam, Germany
Anna Wingkvist	Linnaeus University, Sweden
Julie Yu-Chih Liu	Yuan Ze University, Taiwan

Organizing Committee

Milena Stróżyna (Chair)	Poznań University of Economics and Business, Poland
Barbara Gołębiewska	Poznań University of Economics and Business, Poland
Inmaculada Hernández	University of Seville, Spain
Patricia Jiménez	University of Seville, Spain
Piotr Kałużny	Poznań University of Economics and Business, Poland
Elżbieta Lewańska	Poznań University of Economics and Business, Poland
Włodzimierz Lewoniewski	Poznań University of Economics and Business, Poland

Additional Reviewers

Simone Agostinelli
Asif Akram
Christian Anschuetz
Kimon Batoulis
Ilze Birzniece
Victoria Döller
Katharina Ebner
Filip Fatz
Lauren S. Ferro
Umberto Fiaccadori
Dominik Filipiak

Szczepan Górtowski
Lars Heling
Olivia Hornung
Amin Jalali
Maciej Jonczyk
Piotr Kałużny
Adam Konarski
Izabella Krzemińska
Vimal Kunnummel
Meriem Laifa
Pepe Lopis

Ronald Lumper
Pavel Malyzhenkov
Sławomir Mazurowski
Odd Steen
Piotr Stolarski
Ewelina Szczekocka
Marcin Szmydt
Jakub Szulc
Patrick Wiener

Contents – Part II

Contents – Part I

Artificial Intelligence

ICT Project Management

Smart Infrastructure

Social Media and Web-Based Systems

Trends in CyberTurfing in the Era of Big Data

Hsiao-Wei Hu[1(✉)], Chia-Ning Wu[1(✉)], and Yun Tseng[2]

[1] School of Big Data Management, Soochow University, Taipei, Taiwan
camihu@gmail.com, grace9796@gmail.com
[2] Deloitte & Touche, Taiwan, Taipei, Taiwan
christitseng@deloitte.com.tw

Abstract. Previous research on CyberTurfing has been scattered and fragmented in terms of methods and terminology. This paper presents a review of published research related to CyberTurfing in the era of big data. Our objectives were to identify the important terms in this domain and extract essential knowledge from previous studies. We also sought to gain an overview of the trends in CyberTurfing to provide guidance for subsequent research in this field.

Keywords: CyberTurfing · AstroTurfing · Fake review · Fake account · Term rectification · Opinion manipulation

1 Introduction

Product information is increasingly being accessed on virtual platforms (e.g., Facebook, LinkedIn, Amazon, TripAdvisor, Yelp) rather than conventional channels (e.g., TV commercials). According to the Marketing Intelligence & Consulting Institute [1], the channels most commonly used to obtain product-related information are portal websites and Facebook. Social network sites (SNS) provide a forum on which to share experiences with other customers. Bloggers and KOLs (key opinion leaders) also have considerable influence, as shown in Fig. 1.

Previous research [2–4] has shown that customers are accustomed to checking reviews on-line before making purchasing decisions. This has made electronic word-of-mouth (eWOM) an important target for companies seeking to enhance social marketing.

Regardless of whether is positive or negative, eWOM has a greater influence on the propagation of information than does traditional WOM [5]. Most potential customers are willing to believe that eWOM is shared without any bias with regard to malice or profit [6]. Positive e-WOM can help to develop the visibility and/or reputation of a company [7]. Negative eWOM can destroy a competitor. In the era of big data, the analysis of user-generated content is essential to eWOM marketing strategies.

Changes in consumption habits and marketing trends have rendered celebrity endorsements obsolete, and the shift from brick-and-mortar establishments to virtual spaces has had a profound effect on conventional business models. The internet has opened up a plethora of business opportunities; however, there is a dark side to these developments as well. Fake reviews, fake accounts, and spamming are a growing problem in virtual marketplaces. In 2018, the data analytics company, Cambridge

W. Abramowicz and R. Corchuelo (Eds.): BIS 2019, LNBIP 354, pp. 3–13, 2019.
https://doi.org/10.1007/978-3-030-20482-2_1

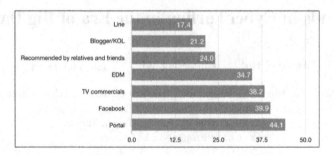

Fig. 1. Channels for people to obtain information before shopping

Analytica, revealed that it had collected personally identifiable information from as many as 87 million Facebook users, and used this information to influence the US presidential election via psychographic modeling.

In March of 2018, scholars from MIT, Harvard, Yale and Columbia University jointly published a research paper in Science, called "The science of fake news [8]". They described efforts to detect false news items and other types of fakery. The term CyberTurfing (CT) is growing in popularity as a catchword covering a range of topics including spam, faked content, and the operations that underlie these phenomena. One common example of CT is the hiring of a large number of writers to compose positive comments and coders to disseminate them to target audiences. The terminology is new; however, this issue can be traced back to the birth of the internet in the 1990s [9].

The issue of CT has received relatively little attention in the literature. In this study, we conducted a review of previous publications pertaining to the nature of CT and the trends in its development. We adopted a big data approach to identify 14,750 papers recently published in journals and conferences. Our objective was to gain an overview of existing research on CT. In this literature review, we focused on current trends as well as the terminology most commonly used in articles addressing the various aspects of CT.

The term Astroturfing refers to the practice of masking the sponsors of a message or organization to make it appear as though it originates from and is supported by grassroots participants. CT is essentially the online equivalent of AstroTurfing, particularly on social network sites (SNS) [10]. In [11] and [12], the authors defined CT as the artificial advocacy of a product, service, or political viewpoint, aimed at giving the appearance of a grassroots movement.

The emergence of the internet has made it possible to expand the scope to AstroTurfing in the form of CT. By connecting millions of people, the internet facilitates the transmission of information with no constraints in terms of borders, authentication mechanisms, or traceability. However, this has also made it far easier to create false identities and fabricate falsehoods [13, 14].

The objective underlying all forms of CT is to give a false impression of widespread support for or against a given agenda. The CT scenarios that writers/spammers present their target audiences differ according to their motives. One example is the presentation of advertisements on webpages or group forums aimed at luring users into clicking on links that lead to webpages prepared by the writers/spammers.

A group or organization hires people as writers of fake content and coders for its dissemination (hereafter referred to as spammers).

1. Spammers register multiple accounts on multiple platforms.
2. Spammers use the accounts collaboratively to publish comments and/or responses aimed at misleading the public (e.g., consumers, voters) in order to gain competitive advantage.

CT can be divided into three basic aspects: fake behavior, fake accounts, and fake content (Fig. 2).

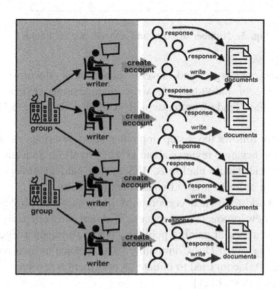

Fig. 2. Example scenario of CT

2   CT: Related Examples

In April 2013, Samsung was revealed that its subsidiary in Taiwan hired a consulting company called Pentai to conduct e-WOM marketing on the internet. That company hired students who pretended to be Samsung mobile phone users sharing positive experiences in order to gain competitive advantage over their competitors [15].

A local web forum (known as Mobile01) is considered the best 3C forum for Alexa in Taiwan [16]. Members share their experience with electronic products on this forum, and most of them believe in the authenticity of the posts and comments.

According to [17], the three smartphone brands that are most commonly discussed are hTC, Samsung, and Sony (Fig. 3). hTC is seen as a direct competitor of Samsung in Taiwan, due to similarities between the brands in terms of marketing position and brand awareness.

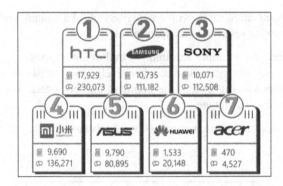

Fig. 3. Number of posts and reviews per 3C brands

TaiwanSamsungLeaks.org listed several confidential documents from Pentai describing the hiring of individuals to write disingenuous reviews on the forum [18, 19]. Taiwan's Fair Trade Commission (FTC) fined Samsung $340,000 for misleading advertising.

This case closely follows the 3-step implementation described earlier. 1. Samsung hired people to write negative reviews about HTC (a rival). 2. Samsung paid people to open multiple on social media platforms. 3. Samsung paid spammers to publish misleading comments in order to mislead consumers for profit.

For instance, world-renowned corporations have used CT strategies to produce vaporware around new products or services. Walmart was allegedly behind a fake YouTube video post aimed at gaining publicity. It has also been alleged that Walmart developed a fake blog called Our Community - Our Choice to gain publicity for new store openings [20]. Ask.com tried to initiate an information revolution against Google on the London Underground [21].

3   Methods

3.1   Systematic Literature Review

There has been considerable research into the detection of CT, including work on spammer detection, authorship identification, plagiarism detection, cyborgs, and clickbait. In this paper, we provide a systematic review of the literature to identify, evaluate, and interpret research relevant to CT based on a research method that is reliable, accurate, and applicable to auditing.

3.2   Sources

Automated searches were conducted on the electronic database, Semantic Scholar, to obtain papers relevant to this topic. To expand the keywords related to CT, we first defined a set of basic keywords based on the three aspects mentioned in the previous section (behavior, accounts, content), and "Overall", as shown in Table 1.

Table 1. Terms used in online search

Aspects	Terms
Behavior	Clickbait
Account	Fake Accounts
	Spammer
Content	Fake Reviews
	Fake News
	Spam Reviews
	Opinion Spam Detection
Overall	AstroTurfing
	CT

The field of CT is changing rapidly; therefore, we recursively expanded the terms and classified them according to aspects, as shown in Table 2.

Table 2. Terms used in online search

Aspects	Terms
Behavior	Clickbait
	Software Forensics
	Plagiarism
	Smear Campaign
Account	Fake Account
	Authorship Identification
	Spammer
	Authorship Attribution
	Cyborg
	Stylometry
Content	Deceptive Reviews
	Deception Reviews
	Fake Reviews
	Fake News
	Online Review Spam
	Spam Review
	Opinion Spam Detection
	Flogging
	Splogging
Overall	AstroTurfing
	AstroTurfer
	CT
	CyberTurfer

The final keywords in Table 2 were then used in a search on Semantic Scholar to find representative journal articles. The search resulted in 14,750 papers published in English since 1990.

3.3   Selection of Articles

Two criteria were used in the selection of articles representative of the published literature addressing our research questions.

The first criterion was whether the article discussed to the detection of CT via fake news, reviews, or opinions. This criterion also helped to identify papers describing the techniques or algorithms used to identify CT.

The second criterion was whether the article discussed actual instances of CT. Papers in this category could provide insight into the manner by which organizations manipulate consumers for profit. Papers dealing with algorithms and other technical issues were excluded from this category.

4   Findings

In the age of technological advances, online platforms provide easy access points for the dissemination of false information.

We discovered that the number of keywords and categories pertaining to CT has been increasing over time. We sought to determine the extent to which disinformation is disseminated, how rapidly it is spread, and the target audiences. We first categorized the existing articles according to the four approaches used in the detection of CT, as follows:

- Behavior: detection of activities by spammers
- Account: detection of fake accounts
- Content: detection of content aimed at misleading consumers or belonging to a fake account
- Overall: detection of all three aspects of CT (behavior, accounts, and content)

In Fig. 4, the X-axis indicates annual growth in the number of papers pertaining to CT. The Y-axis indicates the number of keywords used by researchers in articles on this topic. The size of the donut indicates the number of papers pertaining to CT, after normalization. Different colors denote different aspects.

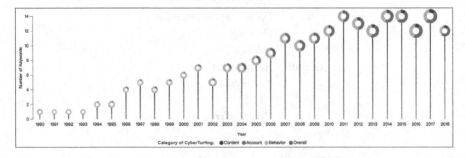

Fig. 4. Trend of relevant articles on CT with different aspects

We found that before 2003, most of the research on this topic focused on the behavior aspect of CT. The other aspects have received more attention in recent years. Since 2000, there has been a general increase in the number of papers pertaining to CT, as shown in Fig. 5.

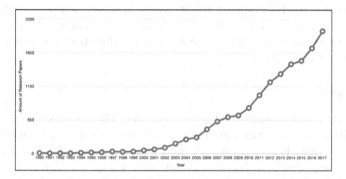

Fig. 5. Number of published articles on CT

4.1 Behavior

Our searches identified many terms pertaining to CT behavior. The ease with which information can be copied and pasted on the internet has prompted considerable interest in the issue of plagiarism [22]. The display of ads that can be accessed by clicking on a link (Clickbait) have also attracted interest [23].

As shown in Fig. 6, the issue of clickbait has recently been growing in popularity; however, the issue of plagiarism has always and continues to attract the most attention.

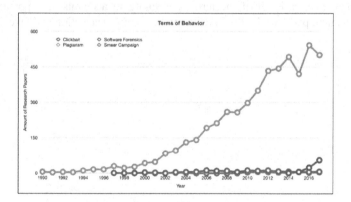

Fig. 6. Trends in the publishing of articles on CT behavior

The Video Game Voters Network (VGVN) is a grassroots organization of voting-age gamers who organize, defend against threats, and take action in support of computer and video games. Since its creation in 2006, more than 500,000 activists have joined. In 2009, the gaming industry sought to gain the support of gamers in countering efforts by lawmakers to implement controls on gaming content [24]. VGVN members sent thousands of letters to Capitol Hill and state legislative offices, which resulted in the overturning of an anti-video gaming bill in Utah [25, 26]. In this case, it is obvious that the members who highly engaged in video game industry made their voices heard in government and advocating for policy issues. However, considering the simple process of signing up, what incentive is there to drop out or join in of the organization. It is crucial to detect the participant of speaking up activities is spammers or those who really matter about the issue.

4.2 Accounts

CyberTurfers try to increase their influence by creating a large number of fake accounts or using robots to disseminate misinformation. It has been reported that in 2010, as many as 1.5 million Facebook accounts were available for purchase for use in misinformation campaigns [27]. Numerous applications have been developed to assist in the creation of maintenance of fake accounts or the compromising of real accounts in order to increase the number of subscribers and/or the number of votes for a particular post [28]. The platform SMOService uses the accounts of real users and robots to add large numbers of likes or dislikes in order to boost a YouTube video or Instagram post [29]. The strategy preferred on Twitter is to make a story appear as if it were spread by many users. The ranking of Twitter posts is based on the number of reposts, quotes, and likes. Twitter cyborgs are often used to retweet and reference Twitter posts. In other cases, Twitter cyborgs post the same message with slight changes in the text, which means that the same text can be used in all tweets. By increasing traffic, they are able to increase their influence [30]. Above, it's all related to practical application of fake accounts to reach influence diffusion.

As shown in Fig. 7, all of the terms pertaining to accounts are appearing with greater frequency, particularly those dealing with spammer detection and fake accounts.

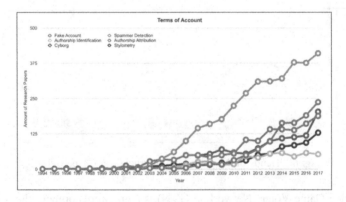

Fig. 7. Trends in the publishing of articles on CT accounts

4.3   Content

Social networks and community platforms have made it possible for people with similar interests to connect. Advancements in analysis techniques have also made it possible to target specific audiences based on the collection of user data [31]. Most of the large websites that depend on user recommendations (e.g., Amazon, Booking.com, Yelp) have been accused of deploying fake reviews [32]. In 2006, Sony launched a PS2 marketing campaign called "All I want for Xmas", which was meant to look like a fan blog, but was actually just a crass attempt to sway teenagers by creating false video-and-blogging content [33].

The lifestyle and food writer Oobah Bulter used to write fake reviews on TripAdvisor for a price of £10 with the aim of enhancing the ranking of restaurants. However, it was later revealed that he had never even visited most of the establishments. The situation shifted from nefarious to ridiculous in April 2017, when Bulter reviewed a fake restaurant called the Shed at Dulwich. He listed the street name, phone number, website, and menu as well as a few photos of the dishes, some of which were not food at all. The posting of fake reviews pushed the restaurant to the top of the TripAdvisor rankings. Bulter even received calls for reservations, before revealing that the entire story was a hoax [34]. As mentioned cases above, non-existent could be echoed by those who follow with interests; however, the fact is that they are misled by deliberate fake content.

As shown in Fig. 8, the issue of fake content did not attract the attention of researchers until 2005; however, it has been gaining attention in recent years.

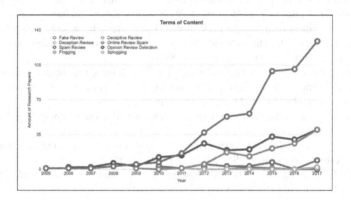

Fig. 8. Trends in the publishing of articles on CT content

4.4   Overall

In 1985, US Senator Lloyd Bentsen coined the term AstroTurfing, in reference to a "mountain of cards and letters" sent to his office to promote the interests of the insurance industry [35]. Today, misinformation is disseminated using anonymous online reviews, stories, or websites [36]. The term refers to the methods used by organizations when seeking to give potential customers the impression that ordinary online users are recommending their products and services [37].

The shift from the political perspective to online markets demonstrates the multi-dimensionality of this phenomenon.

5 Conclusions

This paper presents a comprehensive review of publications dealing with the detection of CT. The number of papers in this domain increased from just 6 in 1990 to 4,106 in 2017, giving a basis for future studies concerning CyberTurfung.

In the past, this topic was addressed by researchers from a variety of fields using a variety of methods; however, it appears that the topic of CT is being taken more seriously today. This review paper is meant to provide researchers seeking to enter this field with an overview of the developments and trends in CT. We believe that there is considerable room for further research on this topic.

The contributions of the paper are as follows: We first identified the terms that are important in this domain. Second, we sought to extract essential knowledge from previous studies. Third, also sought to gain an overview of the trends in CT to provide guidance in subsequent research in this field.

References

1. Chang, H.-C.: Taiwan online shopping consumer survey (2017)
2. Bonabeau, E.: The perils of the imitation age. Harvard Bus. Rev. **82**(6), 45–54 (2004)
3. Kamakura, W.A., Basuroy, S., Boatwright, P.: Is silence golden? an inquiry into the meaning of silence in professional product evaluations. Quant. Mark. Econ. **4**(2), 119–141 (2006)
4. Senecal, S., Nantel, J.: The influence of online product recommendations on consumers' online choices. J. Retail. **80**(2), 159–169 (2004)
5. Hudson, S., Huang, L., Roth, M.S., Madden, T.J.: The influence of social media interactions on consumer–brand relationships: a three-country study of brand perceptions and marketing behaviors. Int. J. Res. Mark. **33**(1), 27–41 (2016)
6. Hanson, W.A., Kalyanam, K.: Principles of Internet Marketing. South-Western College Pub., Cincinnati (2000)
7. Chen, Y., Xie, J.: Online consumer review: word-of-mouth as a new element of marketing communication mix. Manage. Sci. **54**(3), 477–491 (2008)
8. Shao, C., Ciampaglia, G.L., Varol, O., Flammini, A., Menczer, F.: The spread of fake news by social bots. arXiv preprint arXiv:1707.07592 (2017)
9. Wikipedia. https://en.wikipedia.org/wiki/Internet
10. Leiser, M.: AstroTurfing, 'CT' and other online persuasion campaigns. Eur. J. Law Technol. **7**(1), 1–27 (2016)
11. Heggde, G., Shainesh, G. (eds.) Social Media Marketing: Emerging Concepts and Applications. Springer (2018)
12. Jacobs, J.: Faking it how to kill a business through astroturfing on social media. Keeping Good Co. **64**(9), 567 (2012)
13. Zhang, J., Carpenter, D., Ko, M.: Online astroturfing: a theoretical perspective (2013)
14. Mackie, G.: Astroturfing infotopia. Theoria **56**(119), 30–56 (2009)
15. Wang, C.C., Day, M.Y., Lin, Y.R.: A real case analytics on social network of opinion spammers. In: 2016 IEEE 17th International Conference on Information Reuse and Integration (IRI), pp. 623–630. IEEE, July 2016

16. Mobile01. https://zh.wikipedia.org/wiki/Mobile01
17. W. Mobile01, 04 June 2016. 論壇上, 究竟有多少業配文呢?抓出寫手, 就靠社群帳號數據分析!. 論壇上, 究竟有多少業配文呢?抓出寫手, 就靠/. http://group.dailyview.tw/2016/06/03/mobile01
18. Chen, Y.R., Chen, H.H.: Opinion spam detection in web forum: a real case study. In: Proceedings of the 24th International Conference on World Wide Web, pp. 173–183. International World Wide Web Conferences Steering Committee, May 2015
19. Chen, Y.R., Chen, H.H.: Opinion spammer detection in web forum. In: Proceedings of the 38th International ACM SIGIR Conference on Research and Development in Information Retrieval, pp. 759–762. ACM, August 2015
20. Ciarallo, J.: Wal-Mart Busted for Astroturfing, Again. Adweek, 27 January 2010
21. Aaron, P.: Ask.Com's 'Revolt' Risks Costly Clicks. Wall Street J., 5 April 2007. http://online.wsj.com/article/SB117572581285960181.html?mod=googlewsj. Accessed 19 Apr 2009
22. Culwin, F., Lancaster, T.: A review of electronic services for plagiarism detection in student submissions. In: LTSN-ICS 1st Annual Conference, pp. 23–25, August, 2000
23. Potthast, M., Köpsel, S., Stein, B., Hagen, M.: Clickbait Detection. In: Ferro, N., et al. (eds.) ECIR 2016. LNCS, vol. 9626, pp. 810–817. Springer, Cham (2016). https://doi.org/10.1007/978-3-319-30671-1_72
24. Tigner, R.: Online astroturfing and the European union's unfair commercial practices directive. Commun. Coll. Week, 13 (2009)
25. More Than 250,000 Gamers Join Video Game Votes Network. (n.d.). http://www.theesa.com/article/250000-gamers-join-video-game-votes-network/
26. Cook, D.M., Waugh, B., Abdipanah, M., Hashemi, O., Rahman, S.A.: Twitter deception and influence: issues of identity, slacktivism, and puppetry. J. Inf. Warfare 13(1), 58–71 (2014)
27. Richmond, R.: For Sale: Fake and Stolen Facebook Accounts, 02 May 2010. https://www.nytimes.com/2010/05/03/technology/internet/03facebook.html
28. Viswanath, B., et al.: Towards detecting anomalous user behavior in online social networks. In: USENIX Security Symposium, pp. 223–238, August 2014
29. Gu, L., Kropotov, V., Yarochkin, F.: The fake news machine
30. Chu, Z., Gianvecchio, S., Wang, H., Jajodia, S.: Who is tweeting on Twitter: human, bot, or cyborg? In: Proceedings of the 26th Annual Computer Security Applications Conference, pp. 21–30. ACM, December 2010
31. Hamilton, M., Kaltcheva, V.D., Rohm, A.J.: Social media and value creation: the role of interaction satisfaction and interaction immersion. J. Inter. Mark. 36, 121–133 (2016)
32. Mukherjee, A., Liu, B., Glance, N.: Spotting fake reviewer groups in consumer reviews. In: Proceedings of the 21st International Conference on World Wide Web, pp. 191–200. ACM, April, 2012
33. Kaplan, A.M., Haenlein, M.: Two hearts in three-quarter time: how to waltz the social media/viral marketing dance. Bus. Horiz. 54(3), 253–263 (2011)
34. Butler, O.: I Made My Shed the Top Rated Restaurant On TripAdvisor, 06 December 2017. https://www.vice.com/en_uk/article/434gqw/i-made-my-shed-the-top-rated-restaurant-on-tripadvisor
35. Wade, A.: Good and bad reviews: the ethical debate over 'astroturfing'. The Guardian, 9 January 2011
36. Gallagher, K.: Astroturfing: 21 st century false advertising. ASA Institute for Risk & Innovation (2014)
37. Haikarainen, J.: Astroturfing as a global phenomenon (2014)

Keyword-Driven Depressive Tendency Model for Social Media Posts

Hsiao-Wei Hu(✉), Kai-Shyang Hsu, Connie Lee(✉), Hung-Lin Hu,
Cheng-Yen Hsu, Wen-Han Yang, Ling-yun Wang,
and Ting-An Chen

School of Big Data Management, Soochow University, Taipei, Taiwan
camihu@gmail.com, im.connilee@gmail.com,
andy86715@gmail.com, kevin19961009@gmail.com,
tim.wh.yang@gmail.com, anny110929@gmail.com,
ivy9670119@gmail.com, casehsu@scu.com.tw

Abstract. People are increasingly sharing posts on social media (e.g., Facebook, Twitter, Instagram) that include references to their moods/feelings pertaining to their daily lives. In this study, we used sentiment analysis to explore social media messages for hidden indicators of depression. In cooperation with domain experts, we defined a tendency towards depression as evidenced in social media messages based on DSM-5, a standard classification of mental disorders widely used in the U.S. We also developed three data engineering procedures for the extraction of keywords from posts presenting a depressive tendency. Finally, we created a keyword-driven depressive tendency model by which to detect indications of depression in posts on a major social media platform in Taiwan (PTT). The performance of the proposed model was evaluated using three keyword extraction procedures. The DSM-5-based procedure with manual filtering resulted in the highest accuracy (0.74).

Keywords: Big data · Social media · Depression · NLP · Sentiment analysis

1 Introduction

People are increasingly sharing posts on social media (e.g., Facebook, Twitter, Instagram) that include references to their moods/feelings pertaining to their daily lives.

Social media is restructuring communication and interactions between individuals, communities, and businesses. Researchers are engaged in investigating with the profound impact that the rapid evolution of social media exerts on user-generated content, due to its effects on purchasing behavior as well as the way it shapes the perceptions and emotional well-being of users [1].

The World Health Organization [2] has predicted that the prevalence of depression will increase over the next 20 years. Depressive disorders can affect one's general health and habits, including sleep patterns and eating behavior. They can also affect one's interpersonal relationships as well as work and school life. A number of useful self-diagnosis tools have been developed, DSM-IV-TR is one the well-known as shown

© Springer Nature Switzerland AG 2019
W. Abramowicz and R. Corchuelo (Eds.): BIS 2019, LNBIP 354, pp. 14–22, 2019.
https://doi.org/10.1007/978-3-030-20482-2_2

Table 1. DSM-5 diagnostic criteria pertaining to major depressive disorders

Criteria	Description	Manual Labeling
1	Depressed, sad, hopeless, discouraged most of the day	☐
2	Loss of interest or pleasure in previously enjoyed activities	☐
3	Impaired ability to think, concentrate, or make decisions	☐
4	Increased or reduced appetite	☐
5	Loss or gain in weight	☐
6	Common sleep disturbance (insomnia/hypersomnia)	☐
7	Psychomotor changes include agitation or retardation	☐
8	Decreased energy, tiredness and fatigue	☐
9	A sense of worthlessness or guilt, frequent thoughts of death, suicidal ideation, or suicide attempts	☐

in Table 1, such as the Chinese depression inventories, developed by the John Tung Foundation [3].

In one survey by the John Tung Foundation, it was revealed that approximately one million people in Taiwan suffer from depression. A 2005 survey revealed that nearly one in four students is affected by depression that may require professional assistance [4].

In this study, we worked with domain experts on the labeling and scoring of elements in articles dealing with depression. We then applied text-mining technology to analyze messages indicative of depression. A tendency towards depression in web posts was defined using the DSM-5, a standard classification scheme of mental disorders widely used by mental health professionals in the U.S. We developed four data engineering methods to enable the automatic extraction of terms related to a tendency towards depression. We believe that depressive expression is related to emotion with native languages, so it is crucial to analyze depressive tendency in Chinese. Therefore, we also developed a keyword-driven depressive tendency model to detect indications of depression in Chinese articles posted on a major social media platform in Taiwan, PTT.

PTT is a web forum in which threads are categorized according to topic. It includes a Prozac board for individuals suffering from depression, which functions as a platform on which to express one's self and provide mutual comfort. As shown in Fig. 1, PTT is a culturally-specific social media outlet.

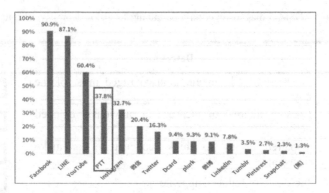

Fig. 1. Social media platforms in Taiwan listed in order of the number of accounts

The remainder of this paper is organized as follows: Sect. 2 summarizes previous works on methods for dealing with the exhibition of depressive tendencies in web posts. Section 3 describes the proposed model and methodology. Section 3 presents experiment results. Section 4 concludes the paper.

2   Related Work

In this study, major depressive disorders were identified using the DSM-IV-TR based on nine criteria. In this study, we developed a system to assist domain experts in labeling posts according to whether they present indications of depression, as outlined in Table 1. Domain experts input (in the text field) a sentence or corpus from a given post that they feel pertains to one of the diagnostic criteria, such as physiological information (e.g., headache, depressed, sad, hopeless, decreased energy).

2.1   Sentiment Analysis

Sentiment analysis of text involves the extraction of information pertaining to the opinions, sentiments, and emotions conveyed by writers within a topic of interest. This process is often equated to opinion mining; however, it also encompasses emotion mining. Opinion mining uses natural language processing (NLP) and machine learning (ML) to determine the attitude of a writer towards a given subject. Emotion mining uses similar procedures; however, the focus is on the detection and classification of the emotional response of the writer toward events or topics [5].

In [5], the authors provide a clear framework by which to perform sentiment analysis and/or opinion mining. They categorized sentiment analysis into opinion mining (expression of opinions) and emotion mining (the articulation of emotions). Opinion mining deals with the intellectual assessment of a given issue, which can be expressed as positive, negative, or neutral. Emotion mining deals with the emotional state of the writer at the time of writing a piece of text. Figure 2 illustrates the categorization of sentiment analysis, including subtasks.

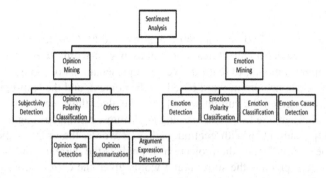

Fig. 2. Taxonomy of tasks involved in sentiment analysis [5]

In this study, we sought to identify signals of depressive tendencies in social media posts using sentiment analysis based on the framework proposed in [5]. Our focus was on emotion mining and the underlying data engineering methods.

Nearly all existing emotion-mining schemes rely on lexical information, which provides a priori information concerning the type and strength of emotions carried by a word or phrase.

Table 2 presents some lexicons useful in emotion-mining tasks and their characteristics.

Table 2. Summary of emotion-related lexicons [5]

Name	Author	Year	Size (words)	Set of emotions
Wordnet affect	Strapparava [6]	2004	4,787	hierarchy of emotions
LIWC	Pennebaker [7]	2007	5,000	affective or not, positive, negative, anxiety, anger, sadness
NRC	Mohammad [8]	2010	14,182	anger, fear, anticipation, trust, surprise, sadness, joy, disgust
NRC hashtag	Mohammad [9]	2013	32,400	anger, fear, anticipation, trust, surprise, sadness, joy, disgust
CBET	Gholipour Shahraki [5]	2015	24,000	anger, fear, joy, love, sadness, surprise, thankfulness, disgust, guilt

Most previous work on textual emotion mining has dealt with English language forms. Despite previous efforts [10] to detect emotions in the messages posted on Weibo (a Chinese microblog similar to Twitter), there remains a great deal of work to be done in this area. To the best of our knowledge, no emotion lexicons have been developed for the identification of depression. In this study, we developed three data engineering methods for the extraction of keywords from posts presenting a tendency toward depression (based on DSM-5 indicators) with the aim of compiling a depression-related lexicon. We believe this lexical framework represents a valuable tool for subsequent research in this field.

2.2   Word2vec

Word2vec is a tool based on deep learning and released by Google in 2013. This tool adopts two main model architectures, continuous bag-of-words (CBOW) model and continuous skipgram model, to learn the vector representations of words. The CBOW architecture predicts the current word based on the context, and the skip-gram predicts surrounding words given the current word [11]. The algorithms are described in detail in Mikolov et al. There are two main learning algorithms, continuous bag-of-words and continuous skip-gram. [12] With continuous bag-of-words, the order of the words in the history does not influence the projection. It predicts the current word based on the context. Skip-gram predicts the surrounding words given the current word. Unlike the standard bag-of-words model, continuous bag-of-words uses a distributed representation of the context. It's also important to state that the weight matrix between input and the projection layer is shared for all word positions [11].

3   Methods and Experiment Results

In this study, we applied techniques used in natural language processing and emotion mining in order to apply the knowledge of experts in establishing lexicons pertaining to depression. We then developed a keyword-driven depressive tendency model to enable the detection of signals of depression in comments posted on a major social media platform Taiwan (commonly called PTT).

PTT is a bulletin board system covering a wide range of topics, including gossip, sports, politics, literature, travel, online shopping, and even the military. PTT has more than 1.5 million registered users, who generate more than 20,000 articles every day. During peak times, more than 150,000 members are online simultaneously.

In this study, we collected approximately 19,000 posts on the PTT Prozac board for the period from October, 2012 to January, 2017 (http://www.ptt.cc/man/prozac/). An API tool was used to request raw data from PTT to be stored in our database. We then conducted data preprocessing in Python, wherein it was divided into training data

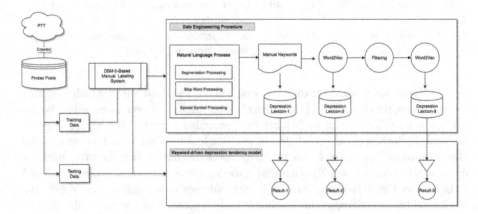

Fig. 3. Framework used in keyword-driven depressive tendency model

and testing data, as outlined in Fig. 3. Three data engineering methods were developed for the extraction of depression-related keywords, which were then labeled as follows: Depression Lexicon 1, Depression Lexicon 2, and Depression Lexicon 3.

We sought to facilitate the labeling procedure through establishment of a DSM-5-based manual labeling system. A screenshot of this system is presented in Table 1.

We built each lexicon as follows:

The first procedure involved obtaining 100 samples from 19,000 articles using jieba (a package of python) for word segmentation via NLP to obtain keywords for Depression Lexicon 1. We then used 26 keywords from Depression Lexicon 1 to identify indications of depression in 500 test samples; this resulted in accuracy of 66%. When using 25 keywords in conjunction with physiological information (e.g., headache, depressed, sad, hopeless, decreased energy), the accuracy dropped to 64%, as shown in Fig. 4.

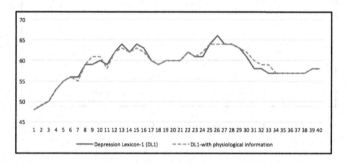

Fig. 4. Result obtained using Depression Lexicon 1

The second procedure was similar to the first, except that word2vec was used to expand the keywords to establish Depression Lexicon 2. Using 31 keywords from Depression Lexicon 2 to identify indications of depression in 500 test samples, we achieved accuracy of 61%. When using 15 keywords in conjunction with physiological information, the accuracy increased to 63%, as shown in Fig. 5.

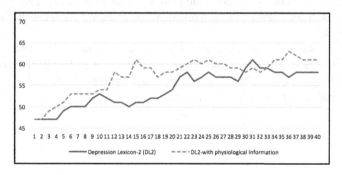

Fig. 5. Results using Depression Lexicon 2

The third procedure was the same as the second procedure but with 1,990 original keywords. We manually ameliorated the previous erroneous word segmentations and set stop words before re-running jieba for word segmentation. We annotated 469 keywords corresponding to the indicators from DSM-5 after determining whether the article indicated a state of depression. We then used word2vec to establish Depression Lexicon 3. Using 4 keywords, we achieved accuracy of 74%, as shown in Fig. 7. When using 4 keywords in conjunction with physiological information, the accuracy dropped slightly to 73%, as shown in Fig. 6.

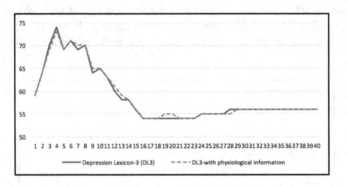

Fig. 6. Results obtained using Depression Lexicon 3

Note that the proposed DSM-5-based Manual Labeling System includes an information dashboard for use by domain experts in assessing the posting behavior of a client and identifying important messages. A screenshot of the proposed system is presented in Fig. 7.

As shown in Fig. 7, the pie-charts at the top of the dashboard indicates the month in which the largest number of posts were published (left side) and the weekdays in which the user was most active (right side).

The middle section of the interface indicates patterns of moods shifting along a timeline. Different moods are indicated by different colors; for example, the green blocks represent life-sharing, whereas the yellow blocks indicate expressions of mood. This timeline is meant to assist social workers in tracking patterns in the shifting moods of their clients.

The bottom section of the interface indicates the frequency with which the client published posts year by year. In the figure, we can see that the client was not very active at the beginning of 2015, began expressing himself/herself in June, and continued this level of activity until the end of the year.

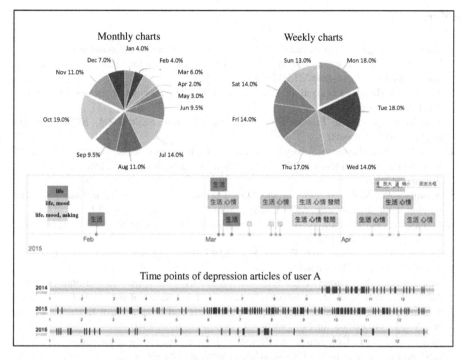

Fig. 7. Screenshot of depressive tendency information dashboard

4   Conclusions

Our primary objective in this study was to establish a depression lexicon using big-data analysis techniques. The proposed depression lexicon complements a decision support system by detecting indications of depression in posts on social media. This can be used to assist social workers in tracking the emotional fluctuations exhibited by their clients. We used sentiment analysis to explore social media messages for hidden indicators of depression. In cooperation with domain experts, we defined a tendency towards depression as evidenced by social media messages based on DSM-5, a standard classification of mental disorders widely used in the U.S. We also developed three data engineering procedures for the extraction of keywords from posts presenting a depressive tendency. Finally, we created a keyword-driven depressive tendency model by which to detect indications of depression in posts on a major social media platform in Taiwan (PTT).

The performance of the proposed model was evaluated using three keyword extraction procedures. The DSM-5-based procedure with manual filtering resulted in the highest accuracy (0.74). Considerable work remains to create a depressive tendency detection model that does not rely heavily on human input. Thus, future work in this area will focus on reducing dependence on manual labeling tasks.

References

1. Trapp, R.: Digital Transformation Drives Business For Social MediaManagers (2016). https://www.forbes.com/sites/rogertrapp/2016/10/12/digital-transformation-drives-business-for-social-media-managers/#521517bf3547
2. The World Health Organization: Depression (2018). http://www.who.int/news-room/fact-sheets/detail/depression
3. John Tung Foundation: Taiwanese depression self-diagnose Scale (2004). https://www.jtf.org.tw/overblue/taiwan1/
4. John Tung Foundation: A Survey of the Correlation between College Students' Life Stress and Depresstion Tendency (2005). https://www.jtf.org.tw/psyche/melancholia/survey.asp?This=66&Page=1
5. Yadollahi, A., Shahraki, A.G., Zaiane, O.R.: Current state of text sentiment analysis from opinion to emotion mining. ACM Comput. Surv. (CSUR) **50**(2), 25 (2017)
6. Strapparava, C., Valitutti, A.: Wordnet affect: an affective extension of wordnet. In: Lrec, vol. 4, pp. 1083–1086 (2004)
7. Pennebaker, J.W., Booth, R.J., Francis, M.E.: LIWC2007: linguistic inquiry and word count, Austin, Texas (2007). liwc.net
8. Mohammad, S.M., Turney, P.D.: Emotions evoked by common words and phrases: using Mechanical Turk to create an emotion lexicon. In: Proceedings of the NAACL HLT 2010 Workshop on Computational Approaches to Analysis and Generation of Emotion in Text, pp. 26–34. Association for Computational Linguistics, June, 2010
9. Mohammad, S.M., Kiritchenko, S., Zhu, X.: NRC-Canada: Building the state-of-the-art in sentiment analysis of tweets (2013). arXiv preprint arXiv:1308.6242
10. Li, W., Xu, H.: Text-based emotion classification using emotion cause extraction. Expert Syst. Appl. **41**(4), 1742–1749 (2014)
11. Mikolov, T., Chen, K., Corrado, G., Dean, J.: Efficient estimation of word representations in vector space (2013). arXiv preprint arXiv:1301.3781
12. word2vec, Google Code Archive (2013). https://code.google.com/archive/p/word2vec/

Exploring Interactions in Social Networks for Influence Discovery

Monika Ewa Rakoczy[1]([✉]), Amel Bouzeghoub[1], Katarzyna Wegrzyn-Wolska[2], and Alda Lopes Gancarski[1]

[1] SAMOVAR, CNRS, Telecom SudParis, 9 Rue Charles Fourier, 91000 Evry, France
monika.ewa.rakoczy@gmail.com,
{amel.bouzeghoub,alda.gancarski}@telecom-sudparis.eu
[2] Efrei Paris, 30 Avenue de la Republique, 94800 Villejuif, France
katarzyna.wegrzyn@groupe-efrei.fr

Abstract. Today's social networks allow users to react to new contents such as images, posts and messages in numerous ways. For example, a user, impressed by another user's post, might react to it by liking it and then sharing it forward to her friends. Therefore, a successful estimation of the influence between users requires models to be expressive enough to fully describe various reactions. In this article, we aim to utilize those direct reactive activities, in order to calculate users impact on others. Hence, we propose a flexible method that considers type, quality, quantity and time of reactions and, as a result, the method assesses the influence dependencies within the social network. The experiments conducted using two different real-world datasets of Facebook and Pinterest show the adequacy and flexibility of the proposed model that is adaptive to data having different features.

Keywords: Influence · Influencers · Social scoring · Social network analysis

1 Introduction

Nowadays, due to the increasing number of people using social network sites, the number of careers created using networking sites such as Youtube has risen as well. The majority of such businesses made via social networking are involved with creators generating content for their audiences, i.e. youtubers, bloggers, instagrammers, earning fame and influence, as well as revenue from advertisement. However, one of the concerns of both creators and advertising companies that collaborate with Internet creators is how to measure, prove and sustain their influence on audiences. To illustrate the problem, let us consider the network that consists of different users - proactive and reactive ones, having different interests, and having various numbers of connections to others. Now, we want to target the social campaign to people, using limited resources. The question is

W. Abramowicz and R. Corchuelo (Eds.): BIS 2019, LNBIP 354, pp. 23–37, 2019.
https://doi.org/10.1007/978-3-030-20482-2_3

how to select creators to share the event, which not only have (possibly large) audiences, but also the audience that would be involved in the campaign?

The topic of *Influence Maximization* [4] is not new in the literature, however, the proposed methods focus more on the spread of the information in the network already assumed to be indicating influence. Hence, such works suppose to have users influencing others, basing on "friend" relations on Facebook, sharing posts or even calling each other. While these approaches focus on expanding the reach of information within the network (i.e. the possibility of user *seeing* some content), none of them actually guarantees that we have actual influence between users (that is engaging with the message). However, while only the pair-wise relation between two users seems to be insufficient to immediately implicate the impact, analyzing and evaluating the engagement and reactions concerning a particular user seem to be good basis for evaluating the influence.

In this paper, we present a simple, general model, which we called ARIM (**Action-Reaction** Influence **M**odel), for evaluating influence between users using one (any) on-line social network platform. In the model, we base on the users' proactive and reactive behaviors, that can be found on basically any social networking site. We concentrate on data flexibility aspect of the model, so that it can be used with different datasets. Our focus is to create a model which uses data features that are possible to obtain for research, and in which each of the properties is connected to its overall expressiveness. Our contributions are as follows:

- We analyze characteristics of different social network sites, establish the key influence-related terms and present the simple social network sites interactions schema (Sect. 2).
- We propose a simple, general influence model ARIM, that focuses on three influence aspects important for model expressiveness namely *intensity* of users' reactions, *spread* and *engagement* and *time dependency* (Sect. 2).
- We built a framework based on the proposed model and test it using two real-world datasets, that allow us to validate ARIM (Sect. 3).

2   Description of Action-Reaction Influence Model

In this section, we firstly define the basic notions and the core of our idea. Then, we describe three components of ARIM model, explain their importance for the model expressiveness, and provide descriptions of methods for their calculation.

The aim of our work is to create a general model for evaluating influence, that is independent of the data platform on which it is applied on. On top of that, we focus on the simplicity of the model on both abstraction and data levels, for two reasons: to be compatible with the intuitive understanding of influence, and to be able to operate on the minimum features expressive enough to evaluate influence.

We define the influence from the perspective of Actions and Reactions on social network sites, which goes with the intuitive understanding of influence, and the way we tend to evaluate influence in real world. The core of the **Action-Reaction Influence Model (ARIM)** consists in the simple schema of information exchange on social network sites, depicted in Fig. 1. We base on the situation in which one user is performing an Action by generating the content, i.e positing photo or post. The second user, being (a part of) audience, is performing Reaction[1]) to the content. We follow the categorization of the relation from social networks that distinguishes upvotes, comments, etc. (see Sect. 2.1). We aim to measure the influence between Action generating user – the Subject of Influence – and the reacting user – the Object of Influence. In the following, we formally define the notions of influence and other related to it.

Preliminaries 1. *Let us assume a social network $SN =< E, L, T >$, where E is the set of entities, such as users, companies, or conferences, L is the set of links being Actions and Reactions (defined below), and T is a considered time interval.*

Definition 1 Action. *An action $a_i \in A$ performed by a source entity $e_i \in E$ at time $t_i \in T$ is represented as a triple $a_i = \{p_i, e_i, t_i\}$, where p_i is the type of the action a_i. Action a generates reactions on the other entities as an effect. A is the set of all actions, each can be of one of the types $\{PostingText, PostingPhoto, PostingVideo\}$.*

Definition 2 Reaction. *A reaction $r_j \in R$ is performed by an entity $e_j \in E$ at time t_j. The reaction is after and as a response to an action $a_i = \{p_i, e_i, t\}$ $(e_i \neq e_j)$, and is represented as a quadruple $r_j = \{q_j, e_j, t_j, s\}$, where q_j is type of reaction, $t_j \in T$ and $t_j > t_i$.*

R is the set of all reactions, each can be one of type $\{Upvoting, Commenting, Sharing\}$.

Definition 3 Influence. *The relation of influence between two entities e_s and e_t $(e_s, e_t \in E)$ occurs iff e_s performed at time t_i at least one action $a \in A$, on which e_t performed one or more reactions $\{r_1, .., r_n\} \in R$ in a particular considered interval of time T, where the start t_j of the time interval T is after action time t_i $(t_j > t_i)$. We say that e_s **influences** e_t. The entity e_s is a Subject of Influence or an **influencer**. The entity e_t is an Object of Influence, or an **influencee**.*

Proposition: The relation of influence is asymmetric, time-dependent and oriented. It involves actions generating content (proactive) and reactions (reacting on content).

Remark: Importantly, we do not assume that the Subject and Object of Influence have a relation between each other, i.e. follow, friendship, etc. Obviously, the actions of the Subject of Influence must be visible by the Object of Influence

---

[1] In the Action-Reaction schema, we refer to an Action as a self-activity of user u, while Reactions symbolize activities overtaken by other users in response to the user u Action.

in order for the Object to be able to react. However, this assumption implies only the visibility of the actions performed by the Subject, and represents more flexible approach in terms of the data requirements and applicability to various social network platforms without a notion of direct relation between users.

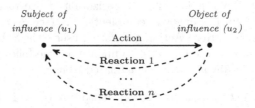

Fig. 1. The Action-Reaction schema. The subject generates content visible to the Object who reacts to the content, possibly by using multiple types of reactions several times.

The presented general influence definitions and schema aim to cover three degrees of expressiveness, namely: intensity of influence depending on the reactions, influence spread and engagement and influence time dependency, which we discuss and detail in what follows.

2.1 Intensity of Influence Depending on Reactions

The fact that our model is based on the Action-Reaction schema (Fig. 1) implies the flexibility in terms of the number of features the data can include. Obviously, by minimum, the data is required to have the users and reactions of one kind (e.g. comments). Considering the existing to-date potential sources of data, we enlisted most of major social network sites in order to gather all the similar features categorized to Action and Reaction. The results are presented in Table 1. From this comparison, it can be seen that most omnipresent reactions available to the users to perform upon the generated content are *upvotes* (also named *likes*, *claps*, *hearts*, etc. – in this work, all one-click reactions that imply appreciation of the content will be called by the general name *upvote*), and comments, and then shares (the action of sharing the content originally posted by other user). In this article, without loss of generality, we concentrate on these three reactions due to their ubiquity.

In particular, we want to focus on the implications that different types of reactions have on the influence strength. Indeed, upvotes, comments and shares have various functions and meanings within social networks. According to [10], the upvote is treated as a lightweight reaction, easy to perform to acknowledge the posted content, similar to "wordless nod". In comparison, comments are regarded as "more satisfying to receivers" [10]. Moreover, because of the quickness and easiness of upvoting, it may also be regarded as less meaningful than other reactions. Clearly, reacting using comment involves both effort (writing comment content in comparison to "default" value of upvote) and time

Table 1. Social Network Aspects depending on the type and/or site

Social network type	Action	Reaction
Social network sites		
Facebook	posts, photos	comments, likes ($\approx$upvotes), shares, mentions
LinkedIn	posts, updates	comments, likes ($\approx$upvotes), shares
G+	posts, photos	comments, +1 ($\approx$upvotes), shares
Content-sharing		
Microblogging		
Twitter	tweets ($\approx$posts)	replies ($\approx$comments), likes ($\approx$upvotes), retweets ($\approx$shares), mentions
Weibo	posts	replies ($\approx$comments), likes ($\approx$upvotes), retweets ($\approx$shares), mentions
Blogging		
Medium	posts	comments, claps ($\approx$upvotes), shares to outside platform, e.g. twitter/fb, bookmark
Politico	posts	comments, shares to outside platform, e.g. twitter/fb/g+
Creative content		
Youtube	videos	comments, likes ($\approx$upvotes), dislikes, shares, views
Instagram	photos, stories	comments, hearts ($\approx$upvotes)
Collaborative sites		
Yelp	reviews	upvote review, upvote profile

(writing and answering versus just one-click). Furthermore, a study [1] has shown that the majority of people share content due to its value – 94% of subjects share "valuable or entertaining content with others", 84% share to support causes and issues they care about. This means that when we see content that is highly impacting we are more willing to share it. This phenomenon is much different to the "casual" upvoting. It was also observed that receiving more complex reactions from acquaintances is corresponding to increase of relationship strength and closeness, as opposed to getting upvotes where no such association was noticed [2].

Taking all the above into consideration, we assumed the hierarchy of the reactions, in which the reaction of sharing is regarded as better descriptor of influence than comments and comments implying higher influence than upvotes. To illustrate, let us consider the following example: user u_1 created the post. User u_2 liked the post, while user u_3 commented on the post. The hierarchy aims to evaluate influence according to the importance of the reactions, in this case evaluating the influence of user u_3 higher than user u_2. Moreover, for each generated influencer content, we can specify several combinations of reactions of the influencee – user can *only* upvote, comment or share the content, can upvote and comment, upvote and share, comment and share, and obviously, all three at the same time, meaning the user is upvoting, commenting and sharing the same content.

We do notice that one might argue that the importance of combinations could be modeled by using the linear combination of weights and vector of reactions. However, we aim to put different importance to each of the reactions (and their possible combinations) by not only using weights vector but also by utilizing non-linear multiplications corresponding to reaction combination, so that the importance of the existence of two or three reactions at the same time can be stressed, and have greater value on the final score. In order to achieve this, we need to model reactions with the use of a non-linear function.

Considering all of the above, first, in order to deal with distinction of different types of reactions, we propose a *ReactionsIntensity* function that combines all reactions done by particular reacting user e_t to a particular action a in time interval T, using the following formula:

$$
\begin{aligned}
ReactionsIntensity(a, e_t, T) = {} & w_1 * |R_u(e_t, T)| + w_2 * |R_c(e_t, T)| + \\
& w_3 * |R_s(e_t, T)| + w_4 * |R_u(e_t, T)| * |R_c(e_t, T)| + \\
& w_5 * |R_u(e_t, T)| * |R_s(e_t, T)| + w_6 * |R_c(e_t, T)| * |R_s(e_t, T)| + \\
& w_7 * |R_u(e_t, T)| * |R_c(e_t, T)| * |R_s(e_t, T)|
\end{aligned}
\tag{1}
$$

In Formula 1, $R_u(e_t, T)$, $R_c(e_t, T)$, and $R_s(e_t, T)$ are sets returned by function $R_{type}(e_i, T)$. This function returns the set of reactions of one type performed by entity e_i within time interval T. As mentioned previously, we assume type being one of $\{Upvoting, Commenting, Sharing\}$, which results in three functions – $R_u(e_i, T), R_c(e_i, T), R_s(e_i, T)$ – returning the set of all reactions performed by entity e_i of type *upvoting*, *commenting* or *sharing*, respectively. Weights $w_1, .., w_7$ in Formula 1 signify the degree of the importance of each reaction combination. The weights should be tuned to emphasize each reaction combination (lower weight – less emphasis). The number of possible reactions can obviously be higher or lower depending on the types of reactions in the dataset. The function includes the multiplication of each pair of the reactions, so that in the case in which there is a combination of reactions the overall value of the measure increases. Inclusion of these combinations – cases when two or three reactions were done by one reacting user concurrently – reflects the fact that Reaction Strength is considered per Reacting user.

Above-mentioned *ReactionsIntensity* function returns the information about the intensity of reactions per particular reacting entity. Using this function, we can specify *ActionAvgIntensity* function which specifies how much appraisal on average action a performed by entity e_s received, in the form of reactions, from its audience within time interval T. In order to do that, we first define function $Audience(e_i, a, T)$ that returns the set of all entities (i.e. audience) that reacted to action a performed by entity e_i within time interval T. Accordingly, we can define *ActionAvgIntensity* function as:

$$ActionAvgIntensity(a, T) = \frac{\sum_{e_i \in Audience(e_s, a, T)} ReactionsIntensity(a, e_i, T)}{|Audience(e_s, a, T)|} \qquad (2)$$

where $ReactionsIntensity(a, e_i, T)$ is the defined above *ReactionsIntensity* function, and $Audience(e_s, a, T)$ is the set returned by *Audience* function of all entities that reacted to action a performed by entity e_s within time interval T.

2.2 Spread and Engagement

The reaction schema presented in Fig. 1 is a simple, but important generalization of interactions on social networks sites. However, further consideration involves the case of multiple reactions to the same user's content.

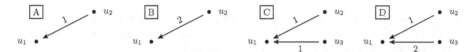

Fig. 2. Examples of different scenarios of users u_2 and u_3 reacting of the content generated by user u_1. The arrow symbolizes the reaction (e.g. comment) and the number signifies the number of reactions

Let us imagine a user u_1 performs an action – creates a post. Users u_2, u_3 can react differently, as pictured in Fig. 2. In two cases (Fig. 2A and B), only user u_2 reacted to the post. Intuitively, the comparison of these two situations leads to the conclusion that the influence of user u_1 should be higher in the Fig. 2B, as user u_2 has stronger response in this case. In other two cases (Fig. 2C and D), both users u_2 and u_3 reacted. We can say that the overall influence of user u_1 in Fig. 2D should be higher, because of more reactions obtained. While it is fairly easy to compare these pairs of situations, the A and B or C and D, the issue starts to be complicated when we want to order influence from highest to lowest in all the presented situations. Intuitively, without any particular model we can state that influence in case A will be the lowest, while influence in case D the highest. However, it is not that obvious in case of B and C. Is the fact that user u gets reaction from two separate users important enough to evaluate influence of u in situation B higher? Or is the strength of particular reaction more important than the spread of the audience?

In order to tackle this problem, we propose to differentiate two components of influence, namely influence spread and engagement. Intuitively, in the example above, user u_1 in situation C would have higher spread, while in situation B would obtain higher audience engagement value.

Definition 4 *Spread.* *Property of influence relation determining the number of audience members per action performed by an influencer i.e. the number of users affected by influencer action.*

The idea behind the spread is to calculate the active audience (meaning the one that is reacting to the initial activity), in order to determine the actual overall broadcast range. We define spread as the number of users that made reaction at least once, i.e. general audience cardinality. Thus, using Formulas presented in Sect. 2.1, the formula for spread calculation is as follows:

$$Spread(e_s, T) = \sum_{a_i \in A} |Audience(e_s, a_i, T)| \tag{3}$$

where A is the set of all actions performed by entity e_s within time interval T and $Audience(e_s, A, T)$ is the set, returned by function $Audience$ defined in Sect. 2.1, of entities who reacted to multiple actions (defined by set A) performed by entity e_s within time interval T.

Definition 5 *Engagement.* *Property of influence relation determining the strength of the audience reactions per action performed by an influencer.*

The engagement notion aims to conceptualize how powerful is the user's influence, therefore evaluates the overall involvement of already active users reacting to the content. Consequently, we propose to calculate engagement using the following formula:

$$Engagement(e_s, T) = \frac{\sum_{a_i \in AllActions(e_s, T)} ActionAvgIntensity(a_i, T)}{|AllActions(e_s, T)|} \tag{4}$$

where $AllActions(e_s, T)$ is the set returned by the function $AllActions$, which returns all actions performed by entity e_s within time interval T.

2.3   Time Dependency

The third crucial factor we consider as an influence component is time. As it was mentioned in the influence definition, we acknowledge that influence is occurring in time, hence it is necessary to examine and include different aspects of time in the influence estimation process. Here, we present the approach to include time dependency within ARIM.

Gaining the influence is a long-term and continuous process. Therefore, in many cases, considering the whole available dataset gives the best estimation of users influence to-date. However, while considering the whole data time period for calculation, including time aspects connected to actions occurring within this time is still valid.

To illustrate, let us consider two situations, in which users u_1 and u_2 are posting content for time period (t_1, t_{10}), and content is visible to the same number of users (audience). In the first situation, user u_1 is posting in each of the time points $t_1, ..., t_{10}$. In the second, user u_2 has posted only two times, in t_1 and t_{10}. Additionally, both users u_1, u_2 received during this time the same amount and type of reactions. While evaluating influence, the question is whether user u_1 posting much more ("constantly") should have equal value of the influence (as we count only the reactions that both users received), or should user u_2 indeed have higher influence? We propose to tackle this issue by introducing the component that favors lower frequency of performing actions. This is in accordance with the fact that we tend to appreciate more users with less posts but with higher quality and possible maximum gain for reactions. The presented idea aims to incorporate time-dependency aspect into ARIM without the need to divide the data according to the time.

Consequently, we introduce the abovementioned time component, $ActionFreq$, defined as:

$$ActionFreq(e_s, T) = e^{\frac{1}{|A|}} \tag{5}$$

where A is the set of all actions performed by entity e_s within time interval T. Using $ActionFreq$ function, we penalize the entities with larger number of actions, e.g. for two entities with equal *spread* and *engagement*, we favor the entity that performs less actions.

2.4 Influence Calculation

Finally, all the previous influence components are combined together, in order to calculate final influence value:

$$Influence(e_s, T) = Engagement(e_s, T) \times Spread(e_s, T) \times ActionFreq(e_s, T) \tag{6}$$

The formula calculating the final value of influence is using the whole available data, in order to achieve the best estimate. Formula 6 binds all the measures presented before using the multiplication. Thanks to this, the contribution of spread and engagement to the final influence score depends on their value. Therefore, the user with high engagement and low spread will have the same influence as the user with low engagement and high spread, assuming that they have equal amount of posts. This is important, as being the better spreader does not straightforwardly implies being more influential (as shown in e.g. [5]). Additionally, the fc component penalizes the users with bigger number of posts, which means that for two users with equal spread and engagement, we prefer the user that posts less.

3 Evaluation

This section describes experiments conducted in order to validate and check the proposed model ARIM.

In order to evaluate our proposed approach, we performed the experiments using two real datasets. The first one consists of data from Facebook [11] containing information about posts and their comments (without the text content) with precise information about the time of each of action/reaction. The second dataset includes data from Pinterest [13], that contains repins (shares) and likes (upvotes). Table 2 presents the basic statistics about both used datasets.

We conducted two sets of experiments, resulting in general influence score using ARIM model, utilizing datasets of Facebook and Pinterest.

The implementation was done using PostgreSQL ver 9.6 and R language ver 3.3.1.

Table 2. Statistics about used Facebook [11] and Pinterest [13] datasets

	Parameter	Number
Facebook	Number of acting users	1 067 026
	Number of users that reacted	23 426 682
	Number of posts	25 937 525
	Number of comments	104 364 591
	Time span of data	15/10/14–11/02/15
Pinterest	Number of acting users	1 307 527
	Number of users that reacted	8 314 067
	Number of posts	2 362 006
	Number of shares	37 087 685
	Number of upvotes	19 332 254
	Time span of data	03/01/13–21/01/13

Facebook. The experiments performed on Facebook dataset, due to the nature of the data that contained one type of relation - comments, were done using all weights equal to 1. Table 3 shows the Top 10 ranking of the most influential users. It can be observed that the user in the first place, while having low (relatively to other top 10 users) engagement rate, is having exceptionally high spread. At the same time, this person has created 96 posts, which is also relatively low. Despite the fact that the audience is not very reactive, i.e. commenting only once, and not entering into discussions, the user is considered very influential due to the user's huge spread for very few actions. Complementary to Table 3, Fig. 3 presents a comparison of Top 3 users in terms of influence score, engagement and spread

Table 3. Top 10 Influential users from Facebook [11]

#	Engagement	Spread	#Actions	Influence score
1	1.039	66181	96	69478
2	1.216	19793	549	24116
3	1.208	18093	148	22012
4	1.204	17030	103	20701
5	1.071	17817	200	19183
6	1.097	17040	941	18717
7	1.092	16087	263	17637
8	1.413	11086	67	15899
9	1.053	14185	998	14953
10	1.066	12678	34	13924

rate, and number of posts. The high spread rate of top 1 user relative to two other top users can be easily noticed.

Another interesting case can be observed on the 8th position in the ranking, with the person having the lowest spread rate in the ranking, significantly lower than both users on 9th and 10th position. Interestingly though, this person engagement rate is very high (highest value in the ranking), with additionally small number of posts. Therefore, this user higher place can be explained with the fact that ARIM is not only focused on both engagement and spread equally, but also it favors the smaller number of posts (see Sect. 2). Hence, the user on the 8th position in the ranking surpasses the next user (9th position) that although having high spread, he/she has also lower engagement and very high (998) number of posts (the biggest number of posts in the whole ranking).

Pinterest. The second set of experiments was conducted on Pinterest database, containing two types of reactions, namely shares (called on the site "repins") and upvotes ("likes"). In order to show the difference of the results when considering shares to be of more significant value than upvotes, we used ARIM model with: (1) equal weights for all reactions, $w_1 = w_2 = w_3 = 1$ (see Eq. 1 in Sect. 2), and (2) higher weight for shares, $w_1 = w_3 = 1, w_2 = 2$. Tables 4 and 5 present the obtained results from both runs. Additionally, the complementary Table 6 presents the detailed information about the users, containing the aggregated sums of upvotes and shares (2th and 3th column) and the number of times that both of the reactions occurred simultaneously (for the same post and from the same user). The information about the latter number (4th column) is important, as the Eq. 1 (Sect. 2) in ARIM model includes an addition component in the formula regarding cases of simultaneous occurrence of the reactions (as was described in Sect. 2.1).

Going back to the Top 10 rankings (Tables 4 and 5), it can be seen that for both experiment executions, the first three positions are unchanged. This can be explained by a very high spread value of each of the users, which is the predominant component for their high influence. Figure 4 shows in detail the ratios of spread, engagement and post number for each of the top 3 users.

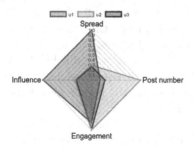

Fig. 3. Detailed comparison of top three users from influence score using Facebook dataset

Table 4. Top 10 Influential users from Pinterest dataset obtained using ARIM with equal weights (all reactions considered equal)

#	UID	Engagement	Spread	#Actions	Influence score
1	2777	1.314	23386	1282	30743
2	20703	1.249	19777	566	24747
3	2367	1.367	13512	1025	18487
4	5656	1.314	9843	535	12958
5	4000	1.286	9908	360	12778
6	1731	1.442	8553	328	12372
7	5074	1.389	8876	465	12358
8	820	1.262	9735	615	12304
9	4968	1.301	9013	569	11742
10	993	1.344	8580	387	11559

Table 5. Top 10 Influential users from Pinterest dataset obtained using ARIM with weight emphasis on share reactions

#	UID	Engagement	Spread	#Actions	Influence score
1	2777	2.263	23386	1282	52961
2	20703	1.935	19777	566	38329
3	2367	2.283	13512	1025	30877
4	820	2.224	9735	615	21690
5	4000	2.133	9908	360	21196
6	5656	2.133	9843	535	21032
7	4968	2.262	9013	569	20422
8	1731	2.360	8553	328	20245
9	5074	2.256	8876	465	20067
10	993	2.258	8580	387	19427

Moreover, from Table 6 we can see that all of the top 3 users have a very high "combination" number, which means that they were apprised by other users simultaneously using upvotes and shares.

The use of the emphasis on the shares can be clearly seen by the example of user 820. In the first rank with equal weights, this person position is low (8th). However, the stress by using the higher weight for share reactions results in the increase of the position of user 820 to 4th, jumping ahead of users like 5656 who, while having higher spread value, have less shares in total. Similarly, user 4968, who also has a high number of shares, is promoted from 9th to 7th position. Interestingly, in the case of second rank (Table 5), the emphasis on the shares also resulted in the top user 2777 having much higher influence score

Table 6. Additional information about users gained reactions from two ranks using Pinterest dataset

UID	Shares sum	Upvotes sum	#Shares and upvotes oncurrently
20703	15944	11467	1066
820	12617	4184	219
993	9251	4159	583
5074	8897	4816	694
2777	28920	10863	1358
1731	9242	4665	829
2367	19404	9028	1203
5656	10695	5786	885
4000	10729	5245	618
4968	10709	3621	466

in comparison to all other users. Indeed, the gap between users 2777 (1st) and 20703 (2nd) significantly rose (two fold), while the gap between users on second and third position (20703 and 2367) stayed similar. This is due to the fact that user 2777 has significantly higher share sum (28920 versus 15944 and 19404 for users 20703 and 2367 respectively), along with the high spread rate.

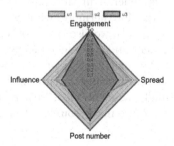

Fig. 4. Detailed comparison of top three users: 2777 (u1 – red), 20703 (u2 – green), 2367 (u3 – blue) from both influence scores using Pinterest dataset (Color figure online)

4   Related Work

Much of the current research is focused on the problem of Influence Maximization, in which the aim is to find the group of nodes in the social network for which the information spread will be maximal. The systems, e.g. [3,4], targeting influence maximization problem are focused on selecting a subset of users basing on the graph, where users are already connected via influence relations. In comparison, in this work we target evaluating the influence between the users and establishing a rank for them.

There are several works proposing various influence metrics for ranking influential users in social networks. Many of them are platform specific, for example operating on Twitter [6], or job portals [8], as opposed to this work, which aims to propose the general approach that can be tuned to any social network site. Few of the well-known, well-established influence metrics base their model only on the typology of the network, i.e. PageRank [7], or centrality measures [12]. However, these methods do not consider deeper network characteristics, such as differences between types of reactions between the users, or the aspect of time dependency.

To the best of our knowledge, the work of Rao et al. [9,11] is the closest to our approach. However, there are several key differences between the methods. Rao et al. system called Klout focuses on being multi-platform, while our ARIM model is build for single platform, targeting a platform-specific influencers' analysis. This means that ARIM not only needs less data, but also it can be more helpful to use it in practice, i.e. for small advertiser companies. On top of that, probably due to the company privacy, none of the articles published by Klout reveals enough details about the metric to really have insight into their work, e.g. in their article they mention using both weights and more than 3 thousand features to calculate the final score, however, they do not include any details about them.

In comparison to above-mentioned *state-of-the-art* methods, our approach includes several important differences that make ARIM better than other approaches. Firstly, ARIM is general, in the sense that it can be applied to various types of social networks. Furthermore, while we do consider topology of the influence (in the form of spread), we also target other influence aspects, such as reaction intensity, engagement and time-dependency. Moreover, our proposed model is also flexible as for the consideration of different reaction types, whose degree of importance can be easily tuned according to particular application requirements and needs. Finally, as mentioned, we targeted single-platform use, thanks to which it does not require many sources of data to work on.

5   Conclusion

In this paper, we presented the Action Reaction Influence Model that evaluates influence for each user basing on the proactive (actions) and reactive (reactions) behaviors of social network participants. The model targets three aspects of influence, namely intensity of influence, spread and engagement in the context of influence value, and time. Importantly, the model is flexible in terms of features that may or may not be available for particular social network site. Moreover, it also embraces the fact that different reactions types should be differently considered and should have various significance for the final influence score of a particular user. We have performed experiments on two real-world datasets including data from two well-known social network sites, namely Facebook and Pinterest. The results present interesting discoveries about the users influence and indicate the adequacy of the proposed ARIM model. They also show how different emphasis on various types of reactions can change the overall influence rank.

As future work, we intend to deepen further analysis of influence trends. In particular, we want to focus on users that, while not being at the highest positions in influence ranks, show the potential to gain the influence with time.

References

1. The psychology of sharing. why do people share online? The New York Times Customer Insight Group (2011). https://www.bostonwebdesigners.net/wp-content/uploads/POS_PUBLIC0819-1.pdf. Accessed 18 May 2018
2. Burke, M., Kraut, R.E.: Growing closer on facebook: changes in tie strength through social network site use. In: Proceedings of the SIGCHI Conference on Human Factors in Computing Systems, pp. 4187–4196. ACM (2014)
3. Chen, W., Lin, T., Tan, Z., Zhao, M., Zhou, X.: Robust influence maximization. In: Proceedings of the 22nd ACM SIGKDD International Conference on Knowledge Discovery and Data Mining, pp. 795–804. ACM (2016)
4. Kempe, D., Kleinberg, J.M., Tardos, É.: Maximizing the spread of influence through a social network. Theory Comput. **11**, 105–147 (2015)
5. Kitsak, M., et al.: Identification of influential spreaders in complex networks. arXiv preprint arXiv:1001.5285 (2010)
6. Laflin, P., Mantzaris, A.V., Ainley, F., Otley, A., Grindrod, P., Higham, D.J.: Discovering and validating influence in a dynamic online social network. Soc. Netw. Anal. Min. **3**(4), 1311–1323 (2013)
7. Page, L., Brin, S., Motwani, R., Winograd, T.: The pagerank citation ranking: bringing order to the web. Technical report, Stanford InfoLab (1999)
8. Rames, A., Rodriguez, M., Getoor, L.: Multi-relational influence models for online professional networks. In: Proceedings of the International Conference on Web Intelligence, pp. 291–298. ACM (2017)
9. Rao, A., Spasojevic, N., Li, Z., DSouza, T.: Klout score: measuring influence across multiple social networks. In: 2015 IEEE International Conference on Big Data (Big Data), pp. 2282–2289. IEEE (2015)
10. Scissors, L., Burke, M., Wengrovitz, S.: What's in a like?: attitudes and behaviors around receiving likes on facebook. In: Proceedings of the 19th ACM Conference on Computer-Supported Cooperative Work & Social Computing, pp. 1501–1510. ACM (2016)
11. Spasojevic, N., Li, Z., Rao, A., Bhattacharyya, P.: When-to-post on social networks. In: Proceedings of the 21st ACM SIGKDD International Conference on Knowledge Discovery and Data Mining, pp. 2127–2136. ACM (2015)
12. Zafarani, R., Abbasi, M.A., Liu, H.: Social Media Mining: An Introduction. Cambridge University Press, New York (2014)
13. Zhong, C., Shah, S., Sundaravadivelan, K., Sastry, N.: Sharing the loves: understanding the how and why of online content curation. In: 7th International AAAI Conference on Weblogs and Social Media (ICWSM13), Boston, US, July 2013

A Literature Review on Application Areas of Social Media Analytics

Kirsten Liere-Netheler, León Gilhaus(⊠), Kristin Vogelsang,
and Uwe Hoppe

University of Osnabrück, Katharinenstr. 1, 49069 Osnabrück, Germany
lgilhaus@uni-osnabrueck.de

Abstract. The use of social media is part of everyday life in both private and professional environments. Social media is used for communication, data exchange and the distribution of news and advertisements. Social Media Analytics (SMA) help to collect and interpret unstructured data. The measurement of user behavior serves to form opinions and evaluate the influence of individual actors. This results in a multitude of application areas for SMA. On the basis of a literature search, our aim is to determine the main application areas and summarize the current state of research. We describe these areas, show current findings from the literature and uncover gaps in research. The main application areas of SMA investigated in research are healthcare, tourism and natural disaster control.

Keywords: Social media analytics · Application areas · Review

1 Introduction

Social media (SM) have become indispensable in today's digital age. Facebook, Twitter and Co. are the main media for private communication, advertising and the dissemination of news. The term SM encompasses various interactive, collaborative, web-based applications and platforms [1]. They have become particularly popular for private persons. The main drivers of this development were technological simplicity, low costs and the increasing use of mobile devices [2]. The higher use of SM and the associated shift of communication to partially (public) virtual spaces also increases relevance in the entrepreneurial and organizational sector [3, 4]. The implementation of new technologies in the area of corporate communications and public relations promotes cooperation and the exchange of knowledge among employees within the company and maintains contact with customers, business partners and other stakeholders outside the company [5].

The added value of SM is not only a support of communication, but useful information can be drawn and analyzed from the generated data [2]. However, different data formats of the platforms lead to a more difficult analysis, because (semi-) structured and unstructured data sets are created [6]. For this reason, the research area Social Media Analytics (SMA) has gained enormous importance in recent years. SMA comprises the development and evaluation of "[...] informatics tools and frameworks to collect, monitor, analyze, summarize, and visualize social media data, usually driven

W. Abramowicz and R. Corchuelo (Eds.): BIS 2019, LNBIP 354, pp. 38–49, 2019.
https://doi.org/10.1007/978-3-030-20482-2_4

by specific requirements from a target application." [7, p. 14]. The measurement of user behavior and participation serves to record moods and opinions and to assess the influence of individual actors [8, 9]. The fundamental difference between SMA and traditional business analytics methods is that it uses real-time data rather than exclusively structured and historical data to gain insights into current issues while supporting effective decision making [10]. SMA are used in a variety of scenarios and are used and developed independently in various disciplines. The aim of this article is therefore to provide an overview of previously researched areas of application of SMA. The research question is therefore: *Which are the main application areas of social media analytics from a research perspective?*

For this purpose, a structured literature review will be conducted. In the following, an overview of existing reviews on SMA is given to differentiate our approach from these. Moreover, the procedure will be shown in order to ensure traceability. Afterward, the results are described and discussed in the form of the application areas of SMA.

2 Reviews on Social Media Analytics

In addition to a large number of empirical studies, a few reviews on SMA already exist in literature. Benefits that can be generated by the analysis of SM data as well as challenges can be found [11]. A review by Rathore et al. (2017) aims at SMA-supported decision making. The authors present analytical approaches that are used in different scenarios for better decision making. In addition, they take up methods that are rarely used but have a high potential for better decision making [12]. Moreover, reviews on explicit application areas can be found in literature. A review on SMA in healthcare shows the usefulness of data analysis in this area [13]. On the other hand Wang and Ye (2018) discuss the possibilities of using SMA as a means of managing disaster situations. For the categorization of the literature, they propose a framework that refers to the four most relevant dimensions in disaster management [14].

Furthermore, there are reviews which show how the challenges in the field of SMA can be overcome through the application of Big Data technologies. Sahatiya (2018) identifies analysis techniques and tools from the area of Big Data [15]. A collection of the challenges faced by researchers in this field of studies can be found in the article by Stieglitz et al. (2018). The results highlight that the primary challenges arise in the discovery, collection and preparation of SM data. They use their results to extend an existing framework model for SMA [16]. A similar goal is pursued by Sebei et al. (2018). In an initial analysis they find that there is a lack of clearly defined processing steps for SM data analysis. They take this as an opportunity to identify technologies in the advanced research field of Big Data that can be used for the analysis of SM data [17].

The existing reviews show the topicality, but above all the complexity, of the research needs in the field of SMA. There are publications that refer to individual application areas (healthcare, natural disaster) or that point out certain methods in specific application areas. However, to the best of our knowledge, there is a lack of a general overview of the application areas of SMA.

3 Method

The aim of this article is to provide a systematic overview of application areas of SMA that are discussed in research. In a first step, relevant contributions in the field of business informatics were sought [18]. Table 1 provides an overview of search terms, databases and search fields used. In addition to the databases used, the search was extended by the Basket of Eight of the Association for Information Systems (AIS) as well as conference proceedings of the European Conference on Information Systems (ECIS), International Conference on Information Systems (ICIS), Hawaii International Conference On System Sciences (HICSS) and Americas Conference on Information Systems (AMCIS). Furthermore, contributions from related research disciplines were examined. The systematic literature search was followed by a forward and backward search. The time span was set for the years 2008 to 2018 because the term SMA has been represented in Google Trends since 2008 [19].

Table 1. Searching criteria

Search terms	Databases	Search fields
Social Media Analytics OR Social Big Data Analytics	Science Direct AIS Electronic Library IEEE Xplore Web of Science Springer Link	Titels, Topics, Abstracts

A total of 1,782 articles were found in the selected databases, whereby some duplicates could be shortened directly. Articles without a clear reference to SMA and non-scientific documents or working reports were excluded. We only looked for articles clearly highlighting an application domain. This leads to 65 relevant publications. In order to be able to classify the articles, research topics, results, sources, authors, and the year of publication were first considered in order to carry out a pre-identification of the areas. Furthermore, an evaluation of the abstracts and keywords was carried out so that areas could be formulated. These were unified within the research team in several discussion rounds. Information on application domains, SM platforms and SMA analysis methods are the base for the following comparative analysis.

4 Application Areas

In total 8 application areas were identified which are currently very present in research. These could further be divided in public domains as well as private domains. Healthcare, natural disaster control and politics are application areas of public domains. Tourism, finance, industry, media and services were identified as private domains. A large part of the literature describes use cases in marketing activities like reputation management or competitor analysis. Table 2 provides an overview of the publications assigned to each area. The first articles found were published in 2009, after which the number increased

Table 2. Application areas

Domain	Application area	Publications	No.
Public domains	Healthcare	[20–35]	15
	Natural disaster control	[36–47]	12
	Politics	[48–55]	8
	Total articles in public domains		*35*
Private domains	Tourism	[27, 56–63]	9
	Finance	[64–71]	8
	Services	[72–76]	6
	Industry	[77–81]	5
	Media	[82, 83]	2
	Total articles in private domains		*30*
Total			**65**

steadily up to a maximum of 13 articles in 2016. 10 publications were found in 2017. The authors Stieglitz and Yang stand out with multiple publications.

Even though SMA pursues different analysis goals in the different domains, there is a high degree of overlap with regard to the methods used for data analysis. Text Mining, in the form of Sentiment Analysis and Topic Modeling, is the most commonly used analytical approach. Social Network Analysis (SNA) is often used to determine the influence of users and their connections. For analysis, automatic methods such as Natural Language Processing (NLP) and Machine Learning are increasingly being implemented. Data from the SM platforms Twitter and Facebook are often used. Researchers probably primarily used data from both SM platforms due to the uncomplicated data retrieval or because these are most promising due to a high number of users. In the following, the main topics of the application areas are presented.

4.1 Public Domains

Healthcare. In the health sector, SM platforms are widely used, but the quality of the information varies with the chosen channel. For example, 40% of shared content is fault [20]. The use of applications that warn users of potential misinformation is recommended [32]. The acceptance of contributions is mainly influenced by quality, emotionality and credibility [27]. Case studies already show the benefit of SMA for the forecast of disease outbreaks in infectious diseases [21, 30]. In addition, text mining methods can be used to develop early warning systems, e.g. for potential addiction risks [26]. SMA research in the field of drug safety is of high interest. In particular, the monitoring of adverse drug reactions (ADR) is often studied in research. Via SM, patients can exchange their experiences in the use of certain drugs [22, 31]. Akhtyamova et al. [23] give an overview of methods and techniques for the use of SMA for ADR, which call for an intensification of research in the field of NLP. The fact that patients' terminology does not correspond to medical jargon reduces the functionality of mechanical procedures [20].

Natural Disaster Control. In the field of disaster control, SMA can support the timely collection and analysis of information. Natural events such as earthquakes are recognized and localized in real time by analyzing data from SM [36, 37]. Information diffusion in times of crisis can also be supported [38]. A stronger use of SM as a communication instrument is therefore demanded [39]. Machine learning and algorithms for word processing on platforms can capture public opinion [40]. Crisis-related messages are in most cases formulated in objective, informal language [41] although emotional messages spread faster than non-emotional ones [42]. User behavior in crisis situations with regard to false reports is basically no different from that in normal situations [43]. An important field in this area is situational awareness during crisis [44, 45]. A framework for the extraction of important and disaster-relevant information to increase situational awareness is presented in research [46].

Politics. The research focuses on the analysis of political moods and the possibilities of election result prediction by SMA. Sentiment Analysis [48–50] is used above all. The number of tweets over a party correlates with the election result. A dominance of specific users is recognizable, since 40% of the tweets come from only 4% of the users [51]. Different data tracking approaches and explicit analysis methods are combined to identify characteristic terms in the political contributions and to identify them in previously unknown posts [52]. In this way parties can target selected groups of voters with appropriately coded posts [53]. In addition to articles on voting behavior, the automatic identification of racist text content is also important [54]. The influence of misinformation (fake news) from social bots or paid SM users is meaningful [84]. Here, SMA research faces the challenge of developing systems for identifying misinformation.

4.2 Private Domains

Tourism. SMA can primarily assist in obtaining information about customers and using it for marketing purposes in the tourism sector. Doing so, customer-generated information about popular travel destinations and attitudes to specific travel policies can be determined [56, 57]. The external impact of a destination perceived by tourists can be identified using SMA [58]. Furthermore, a connection between customer experience and satisfaction is established in online reviews [59]. Using a text mining analysis of reviews of hotels on online evaluation platforms (TripAdvisor, Expedia and Yelp). Xiang et al. [85] found significant differences in the presentation of hotels depending on the platform. SM data, in addition to the analysis of customer preferences, are helpful in investigating the movement of tourists and understanding their travel preferences [60, 61].

Finance. In the financial sector, SMA is used for mood analysis in order to forecast the development of stock market prices [64]. The aim is to support decision-makers in valuing financial investments. A correlation between sentiment on SM and stock market indices was found [65]. Emotional tweets have a negative effect on price performance [66]. Valid stock price forecasts are made possible by SMA in about 75% of cases [67].

Services. Besides the identified areas of application, some articles were summarized under services. These include for example sports talent management [72], small companies like pizzerias [73, 74], as well as large companies like an airline [75].

Case studies show the possibility for companies to better monitor their competitive environment and achieve competitive advantages [76]. Moreover, text mining of praise and complaint contributions on SM platforms improves customer loyalty as well as development and innovation processes [73]. Customer reactions before and after the introduction of a new product can also be analyzed with the help of SMA [74]. Observation and analysis of Twitter data are used to identify both positive and negative customer perceptions [75].

Industry. For industrial companies SMA can enable the development of innovative product ideas. Market opportunities can be identified based on the analysis of customer-generated data on evaluation portals [77]. Furthermore, suggestions for product improvement can be derived from social contributions [78]. Case studies were carried out regarding communication via SM after the Volkswagen "Dieselgate" emissions scandal became known. While customers mainly identified irony as a communication strategy [79], the company acted more passively which had a negative impact on its reputation [80].

Media. We identified only two articles dealing with the media industry. These examine the usefulness of SMA for TV marketing campaigns. SMA is used to investigate the impact of a TV channel's SM activities on the audience ratings of a TV program. A strong engagement in SM correlates positively with the ratings [82]. SNA and sentiment analysis can also be used to determine viewers' opinions about television programs and the commercials shown [83].

5 Discussion and Conclusion

The present article has shown areas of application of SMA in public and private domains. SMA is used in many ways, especially for the interpretation of opinions and knowledge, as well as for the identification of moods and the development of forecasts. In sum, three major goals of SMA use can be identified: (1) disseminate information, (2) forecasting, and (3) monitoring.

Some hints for future research could be found. There is a growing lack of focus on ethical and data protection challenges in the application of SMA. None of the contributions found deals with this topic. Only one article examines the dissemination of content through spam and fake profiles [86]. Few of the articles deal with the representativeness and reliability of SM data although the society is constituted biased because young and old people are differently represented in SM [1]. Systems for the identification of misinformation are very important and thus need to be further investigated. In addition, the validity of data and SM platforms should be investigated [43], especially the differences between opinions in SM and public opinion. Other SM platforms need more attention like Instagram or Flickr. It became clear that most methods use semantic and syntactic characteristics. It would be interesting to include demographic characteristics such as age, gender and place of residence more strongly in the future. Besides text analysis, further dimensions could be added in the future (e.g. image and video), for example to analyze reaction on natural disasters. In addition, systems should be developed that combine SM data with conventional data (e.g. sensor data). Most of the SMA approaches found were developed for English text sources. They are unsuitable for many other languages. SMA approaches are also usually unable to detect irony or sarcasm in text contributions which can falsify results [52]. Sentiment analyses are often executed in short time intervals. The causality of the relationships should be investigated in long-term studies in the future. In sum, new questions arise from the listed results which could be solved in the future by interdisciplinary research in (business) computer science, statistics, networking, economics and social sciences. Other areas of application which need to be further investigated in research include terrorism, general event management, as well as supply chain management and logistics.

Even though, the review was carefully conducted, we have to admit some limitations. Due to the large number of reports, it was not possible to cover the entire literature. The list of areas of application is not an ultimate list. Moreover, due to a rising number of research, articles in other domains might already be in the making. Another aspect that could not be covered due to the novelty of the topic is the use of SMA to analyze data from internal enterprise SM platforms. In conclusion, SMA is an emerging interdisciplinary research discipline. Fields of research such as data protection, legal issues and ethics can be discussed and researched under various aspects in the future.

References

1. Kaplan, A.M., Haenlein, M.: Users of the world, unite! the challenges and opportunities of social media. Bus. Horiz. **53**, 59–68 (2010)
2. Stieglitz, S., Dang-Xuan, L., Bruns, A., Neuberger, C.: Social media analytics. Wirtschaftsinformatik **56**, 101–109 (2014)
3. Beier, M., Wagner, K.: Social media adoption: barriers to the strategic use of social media in SMEs. In: Proceedings of the European Conference of Information Systems, pp. 1–18. AIS, Istanbul (2016)
4. Treem, J.W., Leonardi, P.M.: Social media use in organizations: exploring the affordances of visibility, editability, persistence, and association. Ann. Int. Commun. Assoc. **36**, 143–189 (2013)
5. El-Haddadeh, R., Weerakkody, V., Peng, J.: Social networking services adoption in corporate communication: the case of China. J. Enterp. Inf. Manage. **25**, 559–575 (2012)
6. Baars, H., Kemper, H.-G.: Management support with structured and unstructured data - an integrated business intelligence framework. Inf. Syst. Manage. **25**, 132–148 (2008)
7. Zeng, D., Chen, H., Lusch, R., Li, S.-H.: Social media analytics and intelligence. IEEE Intell. Syst. **25**, 13–16 (2010)
8. Sinha, V., Subramanian, K.S., Bhattacharya, S., Chaudhary, K.: The contemporary framework on social media analytics as an emerging tool for behavior informatics, HR analytics and business process. Manage. J. Contemp. Manage. Issues **17**, 65–84 (2012)

9. Kurniawati, K., Shanks, G., Bekmamedova, N.: The business impact of social media analytics. In: Proceedings of the European Conference of Information Systems, pp. 48–61. AIS, Utrecht (2013)

10. Khan, G.F.: Seven Layers of Social Media Analytics: Mining Business Insights from Social Media; Text, Actions, Networks, Hyperlinks, Apps, Search Engine, and Location Data. CreateSpace Independent Publishing Platform (2015)

11. Kataria, D.: A review on social media analytics. Int. J. Adv. Res. Ideas Innov. Technol. **3**, 695–698 (2017)

12. Rathore, A.K., Kar, A.K., Ilavarasan, P.V.: Social media analytics: literature review and directions for future research. Decis. Anal. **14**, 229–249 (2017)

13. Kotov, A.: Social media analytics for healthcare. In: Reddy, C.K., Aggarwal, C.C. (eds.) Healthcare Data Analytics. Apple Academic Press Inc. (2015)

14. Wang, Z., Ye, X.: Social media analytics for natural disaster management. Int. J. Geograph. Inf. Sci. **32**, 49–72 (2018)

15. Sahatiya, P.: Big data analytics on social media data: a literature review. Int. Res. J. Eng. Technol. **5**, 189–192 (2018)

16. Stieglitz, S., Mirbabaie, M., Ross, B., Neuberger, C.: Social media analytics – challenges in topic discovery, data collection, and data preparation. Int. J. Inf. Manage. **39**, 156–168 (2018)

17. Sebei, H., Hadj Taieb, M.A., Ben Aouicha, M.: Review of social media analytics process and big data pipeline. Soc. Netw. Anal. Min. **8**, 30 (2018)

18. Webster, J., Watson, R.T.: Analyzing the past to prepare for the future: writing a literature. Rev. MIS Q. **26**, xiii–xxiii (2002)

19. Google Trends. https://trends.google.de/trends/explore?q=Social%20Media%20Analytics. Accessed 07 July 2018

20. Abbasi, A., et al.: Social media analytics for smart health. IEEE Intell. Syst. **29**, 60–80 (2014)

21. Achrekar, H., Gandhe, A., Lazarus, R., Yu, S.-H., Liu, B.: Predicting flu trends using Twitter data. In: Proceedings of the IEEE Conference on Computer Communications Workshops (INFOCOM), pp. 702–707. IEEE, Shanghai (2011)

22. Alimova, I., Tutubalina, E.: Automated detection of adverse drug reactions from social media posts with machine learning. In: van der Aalst, Wil M.P., et al. (eds.) AIST 2017. LNCS, vol. 10716, pp. 3–15. Springer, Cham (2018). https://doi.org/10.1007/978-3-319-73013-4_1

23. Akhtyamova, L., Alexandrov, M., Cardiff, J.: Review of trends in health social media analysis. In: 12th International Scientific and Technical Conference on Computer Sciences and Information Technologies. IEEE, Yerevan (2017)

24. Bello-Orgaz, G., Hernandez-Castro, J., Camacho, D.: Detecting discussion communities on vaccination in Twitter. Future Gener. Comput. Syst. **66**, 125–136 (2017)

25. Chee, B.W., Berlin, R., Schatz, B.: Predicting adverse drug events from personal health messages. In: AMIA Annual Symposium Proceedings Archive, pp. 217–226 (2011)

26. Harpaz, R., et al.: Text mining for adverse drug events: the promise, challenges, and state of the art. Drug Saf. **37**, 777–790 (2014)

27. Jin, J., Yan, X., Li, Y., Li, Y.: How users adopt healthcare information: an empirical study of an online Q&A community. Int. J. Med. Inf. **86**, 91–103 (2016)

28. Liu, X., Chen, H.: A research framework for pharmacovigilance in health social media: identification and evaluation of patient adverse drug event reports. J. Biomed. Inf. **58**, 268–279 (2015)

29. Patki, A., Sarker, A., Pimpalkhute, P., Nikfarjam, A., Ginn, R.: Mining adverse drug reaction signals from social media: going beyond extraction. In: Proceedings of BioLinkSig, Boston, MA, USA (2014)

30. Ritterman, J., Osborne, M., Klein, E.: Using prediction markets and Twitter to predict a Swine Flu pandemic. In: 1st International Workshop of Mining Social Media (2009)

31. Sarker, A., Gonzalez, G.: Portable automatic text classification for adverse drug reaction detection via multi-corpus training. J. Biomed. Inf. **53**, 196–207 (2015)

32. Waszak, P.M., Kasprzycka-Waszak, W., Kubanek, A.: The spread of medical fake news in social media – the pilot quantitative study. Health Policy Technol. **7**, 115–118 (2018)

33. Yang, M., Li, Y., Kiang, M.: Environmental scanning for customer complaint identification in social media. In: Proceedings of the International Conference on Information Systems. AIS, Shanghai (2011)

34. Yang, M., Wang, X., Kiang, M.: Identification of consumer adverse drug reaction messages on social media. In: Proceedings of the 17th Pacific Asia Conference on Information Systems. AIS, Jeju Island (2013)

35. Yang, C.C., Yang, H.: Exploiting social media with tensor decomposition for pharma-covigilance. In: Proceedings of the IEEE International Conference on Data Mining Workshop, pp. 188–195. IEEE Computer Society, Atlantic City (2015)

36. Cameron, M.A., Power, R., Robinson, B., Yin, J.: Emergency situation awareness from Twitter for crisis management. In: Proceedings of the Conference on World Wide Web, pp. 695–698. ACM, New York (2012)

37. Sakaki, T., Okazaki, M., Matsuo, Y.: Earthquake shakes Twitter users: real-time event detection by social sensors. In: Proceedings of the Conference on World Wide Web, pp. 851–860. ACM, New York (2010)

38. Li, L., Zhang, Q., Tian, J., Wang, H.: Characterizing information propagation patterns in emergencies: a case study with Yiliang earthquake. Int. J. Inf. Manage. **38**, 34–41 (2018)

39. Cheong, F., Cheong, C.: Social media data mining: a social network analysis of tweets during the 2010–2011 Australian floods. In: Proceedings of the Pacific Asia Conference on Information Systems. AIS, Brisbane (2011)

40. Dong, H., Halem, M., Zhou, S.: Social Media data analytics applied to Hurricane Sandy. In: Proceedings of the International Conference on Social Computing, pp. 963–966. IEEE Computer Society, Alexandria (2013)

41. Verma, S.: Natural language processing to the rescue? extracting "situational awareness" tweets during mass emergency. In: Proceedings of the International AAAI Conference on Weblogs and Social Media. AAAI, Barcelona (2011)

42. Ross, B., Potthoff, T., Majchrzak, T.A., Chakraborty, N.R., Lazreg, M.B., Stieglitz, S.: The diffusion of crisis-related communication on social media: an empirical analysis of Facebook reactions. In: Proceedings of the Hawaii International Conference on System Sciences. AIS, Waikoloa Village (2018)

43. Mendoza, M., Poblete, B., Castillo, C.: Twitter under crisis: can we trust what we RT? In: Workshop on Social Media Analytics, pp. 71–79. ACM, New York (2010)

44. Mukkamala, A., Beck, R.: Enhancing disaster management through social media analytics to develop situation awareness what can be learned from Twitter messages about Hurricane Sandy? In: Proceedings of the Pacific Asia Conference on Information Systems. AIS, Chiayi (2016)

45. Vieweg, S., Hughes, A.L., Starbird, K., Palen, L.: Microblogging during two natural hazards events: what Twitter may contribute to situational awareness. In: Proceedings of the SIGCHI Conference on Human Factors in Computing Systems, pp. 1079–1088. ACM, New York (2010)

46. Zin, T.T.: Knowledge based social network applications to disaster event analysis. In: Proceedings of the International MultiConference of Engineers and Computer Scientists, p. 6. IAENG, Hong Kong (2013)
47. Sen, A., Rudra, K., Ghosh, S.: Extracting situational awareness from microblogs during disaster events. In: Proceedings of the Communication Systems and Networks, pp. 1–6. IEEE Computer Society, Bangalore (2015)
48. Oh, C., Kumar, S.: How trump won: the role of social media sentiment in political elections. In: Proceedings of the Pacific Asia Conference on Information Systems, p. 48. AIS, Langkawi (2017)
49. Yaqub, U., Chun, S.A., Atluri, V., Vaidya, J.: Sentiment based analysis of tweets during the US presidential elections. In: Proceedings of the Annual International Conference on Digital Government Research, pp. 1–10. ACM, New York (2017)
50. You, Q., Cao, L., Cong, Y., Zhang, X., Luo, J.: A multifaceted approach to social multimedia-based prediction of elections. IEEE Trans. Multimedia 17, 2271–2280 (2015)
51. Tumasjan, A., Sprenger, T.O., Sandner, P.G., Welpe, I.M.: Predicting elections with Twitter: what 140 characters reveal about political sentiment. In: Proceedings of the International Conference on Weblogs and Social Media. AAAI, Washington (2010)
52. Stieglitz, S., Dang-Xuan, L.: Social media and political communication: a social media analytics framework. Soc. Netw. Anal. Min. 3, 1277–1291 (2013)
53. David, E., Zhitomirsky-Geffet, M., Koppel, M., Uzan, H.: Utilizing Facebook pages of the political parties to automatically predict the political orientation of Facebook users. Online Inf. Rev. 40, 610–623 (2016)
54. Agarwal, S., Sureka, A.: But I did not mean it! - intent classification of racist posts on Tumblr. In: European Intelligence and Security Informatics Conference, pp. 124–127. IEEE Computer Society, Uppsala (2016)
55. Stieglitz, S., Brockmann, T., Xuan, L.D.: Usage of social media for political communication. In: Proceedings of the Pacific Asia Conference on Information Systems. AIS, Ho Chi Minh City (2012)
56. Cheng, M., Edwards, D.: Social media in tourism: a visual analytic approach. Curr. Issues Tourism 18, 1080–1087 (2015)
57. Park, S., Ok, C., Chae, B.: Using Twitter data for cruise tourism marketing and research. J. Tourism Mark. 33, 885–898 (2016)
58. Marine-Roig, E., Anton Clavé, S.: Tourism analytics with massive user-generated content: a case study of Barcelona. J. Destination Mark. Manage. 4, 162–172 (2015)
59. Xiang, Z., Schwartz, Z., Gerdes, J.H., Uysal, M.: What can big data and text analytics tell us about hotel guest experience and satisfaction? Int. J. Hospitality Manage. 44, 120–130 (2015)
60. Chua, A., Servillo, L., Marcheggiani, E., Moere, A.V.: Mapping Cilento: using geotagged social media data to characterize tourist flows in Southern Italy. Tourism Manage. 57, 295–310 (2016)
61. Habib, M.B., Krol, N.C.: What does Twitter tell us about tourists' mobility behavior? a case study on tourists in The Netherlands and Belgium. In: Proceedings of the Pacific Asia Conference on Information Systems, p. 34. AIS, Langkawi (2017)
62. Brandt, T., Bendler, J., Neumann, D.: Social media analytics and value creation in Urban smart tourism ecosystems. Inf. Manage. 54, 703–713 (2017)
63. Leung, X.Y., Bai, B., Stahura, K.A.: The marketing effectiveness of social media in the hotel industry: a comparison of Facebook and Twitter. J. Hospitality Tourism Res. 39, 147–169 (2015)

64. Nann, S., Krauss, J., Schoder, D.: Predictive analytics on public data - the case of stock markets. In: Proceedings of the European Conference on Information Systems, p. 102. AIS, Utrecht (2013)
65. Bollen, J., Mao, H., Zeng, X.-J.: Twitter mood predicts the stock market. J. Comput. Sci. **2**, 1–8 (2011)
66. Zhang, X., Fuehres, H., Gloor, P.A.: Predicting stock market indicators through Twitter "I hope it is not as bad as I fear". Procedia Soc. Behav. Sci. **26**, 55–62 (2011)
67. Vu, T.T., Chang, S., Ha, Q.T., Collier, N.: An experiment in integrating sentiment features for tech stock prediction in Twitter. In: Proceedings of the Workshop on Information Extraction and Entity Analytics on Social Media Data, pp. 23–38. The COLING 2012 Organizing Committee, Mumbai (2012)
68. Adamopoulos, P., Todri, V.: Social media analytics: the effectiveness of promotional events on brand user base in social media. In: Proceedings of the International Conference on Information Systems. AIS, Auckland (2014)
69. Bekmamedova, N., Shanks, G.: Social media analytics and business value: a theoretical framework and case study. In: Proceedings of the Hawaii International Conference on System Sciences, pp. 3728–3737. IEEE Computer Society, Waikoloa (2014)
70. Ribarsky, W., Xiaoyu Wang, D., Dou, W.: Social media analytics for competitive advantage. Comput. Graph. **38**, 328–331 (2014)
71. Seebach, C., Beck, R., Denisova, O.: Sensing social media for corporate reputation management: a business agility perspective. In: Proceedings of the European Conference on Information Systems. AIS, Barcelona (2012)
72. Davcheva, P.: Identifying sports talents by social media mining as a marketing instrument. In: Annual SRII Global Conference, pp. 223–227. IEEE Computer Society, San Jose (2014)
73. He, W., Wang, F.-K., Zha, S.: Enhancing social media competitiveness of small businesses: insights from small pizzerias. New Rev. Hypermedia Multi. **20**, 225–250 (2014)
74. Rathore, A.K., Ilavarasan, P.V.: Social media analytics for new product development: case of a pizza. In: Proceedings of the International Conference on Advances in Mechanical, Industrial, Automation and Management Systems, pp. 213–219. IEEE, Allahabad (2017)
75. Misopoulos, F., Mitic, M., Kapoulas, A., Karapiperis, C.: Uncovering customer service experiences with Twitter: the case of airline industry. Manage. Decis. **52**, 705–723 (2014)
76. He, W., Zha, S., Li, L.: Social media competitive analysis and text mining: a case study in the pizza industry. Int. J. Inf. Manage. **33**, 464–472 (2013)
77. Ko, N., Jeong, B., Choi, S., Yoon, J.: Identifying product opportunities using social media mining: application of topic modeling and chance discovery theory. IEEE Access **6**, 1680–1693 (2018)
78. Su, C.J., Chen, Y.A.: Social media analytics based product improvement framework. In: International Symposium on Computer, Consumer and Control, pp. 393–396. IEEE Computer Society, Xi'an (2016)
79. Mirbabaie, M., Stieglitz, S., Eiro, M.R.: #IronyOff – understanding the usage of irony on Twitter during a corporate crisis. In: Proceedings of the Pacific Asia Conference on Information Systems, p. 66. AIS, Langkawi (2017)
80. Stieglitz, S., Mirbabaie, M., Potthoff, T.: Crisis communication on Twitter during a global crisis of volkswagen - the case of "Dieselgate." In: Proceedings of the Hawaii International Conference on System Sciences. AIS, Waikoloa (2018)
81. Melville, P., Sindhwani, V., Lawrence, R.D.: Social media analytics: channeling the power of the blogosphere for marketing insight. Proc. WIN **1**, 1–5 (2009)
82. Oh, C., Yergeau, S., Woo, Y., Wurtsmith, B., Vaughn, S.: Is Twitter psychic? social media analytics and television ratings. In: Proceedings of the International Conference on Computing Technology and Information Management, pp. 150–155 (2015)

83. Pensa, R.G., Sapino, M.L., Schifanella, C., Vignaroli, L.: Leveraging cross-domain social media analytics to understand TV topics popularity. IEEE Comput. Intell. Mag. **11**, 10–21 (2016)

84. Allcott, H., Gentzkow, M.: Social media and fake news in the 2016 election. J. Econ. Persp. **31**, 211–236 (2017)

85. Xiang, Z., Du, Q., Ma, Y., Fan, W.: A comparative analysis of major online review platforms: implications for social media analytics in hospitality and tourism. Tourism Manage. **58**, 51–65 (2017)

86. Aswani, R., Kar, A.K., Vigneswara Ilavarasan, P.: Detection of spammers in Twitter marketing: a hybrid approach using social media analytics and bio inspired computing. Inf. Syst. Frontiers **20**, 515–530 (2018)

Personalized Cloud Service Review Ranking Approach Based on Probabilistic Ontology

Emna Ben-Abdallah[(✉)], Khouloud Boukadi, and Mohamed Hammami

Mir@cl Laboratory, Sfax University, Sfax, Tunisia
emnabenabdallah@ymail.com

Abstract. Online cloud service reviews have recently gained an increasing attention since they can have a significant impact on cloud user' purchasing decision. A large number of cloud users consult these reviews before choosing cloud services. Therefore, identifying the most-helpful reviews is an important task for online retailers. The helpfulness of product/service reviews has been widely investigated in the marketing domain. However, these works do not pay attention to the following significant points: (1) the heterogeneity problem when extracting information from different Social Media Platforms (SMP), (2) the uncertainty judgment of review helpfulness and (3) the personalizing of review ranking by considering the context of the review. To tackle these three points we propose a new approach that relies on probabilistic ontology, called Context-aware Review Helpfulness Probabilistic Ontology (C-RHPO), to cope with the heterogeneity and uncertainty issues. In addition, the approach uses a personalized online review ranking method based on the end-user context. The herein reported experimental results proved the effectiveness and the performance of the approach.

Keywords: Online cloud service reviews · Helpfulness ·
Probabilistic ontology · Context

1 Introduction

With the advent of Internet in particular social media, consumers have been strongly influenced by electronic word-of-mouth (eWOM) from blogs, forums, and other social media websites. Business.com (2017) highlights that 77% of people read online reviews before buying as much as personal recommendations. However, the overwhelming amount of reviews can incur difficulty for consumers to filter relevant information. Hence, the listing of reviews plays an important role. Putting most relevant reviews on the top can minimize users' time because they can get their information from few relevant reviews. In general, online review platforms sorted consumer reviews based on newness or helpfulness. Helpful reviews are those that have obtained positive votes from other users [7]. Meanwhile, the most recent reviews are reviews that are sorted based

© Springer Nature Switzerland AG 2019
W. Abramowicz and R. Corchuelo (Eds.): BIS 2019, LNBIP 354, pp. 50–61, 2019.
https://doi.org/10.1007/978-3-030-20482-2_5

on the time at which the review was posted [8]. Helpful review identification has also attracted many researchers in the literature [7–9]. They have proposed several features, which are based on review content [7,11], review meta-data [9] and reviewer characteristics [8], considered as clues to recognize helpful reviews. In addition, they generally relied on classification [11] or regression [9] techniques to evaluate helpful reviews. However, previous studies did not pay attention to the following points: (1) the heterogeneity problem, for example the "profile" concept in Facebook is the same as the "account" concept in the review sites, where the "review" concept is similar to "feedback" concept in Facebook. Since the social media environment is open, distributed, and semantically active, there is no need not only to have helpful review prediction techniques but also to empower these techniques with semantics to facilitate the quality access and the retrieval of helpful reviews from any social media plateform; (2) the uncertainty judgment, helpfulness judgment is subjective and uncertain in nature. For example, if a review has some features showing that it is a helpful review, while others indicate that it is a not helpful one, which leads to a confusing situation; and (3) the personalized prediction, the relevance of a particular review is relative, and depends on the preferences, the needs and the context of a particular user. Especially in the cloud field where the user's context affects the service quality and the user requirements. For example, the availability of cloud service depends on the user location and the user requirements which differ according to their industries and use cases. In order to tackle the mentioned points, this paper presents a new approach that aims to rank online reviews based on their quality and the context of the end-user. To do this, we propose to rely on ontology to resolve the heterogeneity problem of social media platforms. However, traditional ontology does not support the uncertainty reasoning [6]. In fact, the probabilistic ontology has the merit of supporting uncertainty, which could be used to assess the helpfulness of reviews in SMPs. Besides, we rely on learning based method to generate the probability distribution of the review helpfulness. In order to personalize the review ranking, we promote reviews that evolve in similar contexts to that of the end-user.

The rest of the paper is organized as follows. Section 2 discusses the related works. Section 3 presents helpfulness features used in this paper as clues of helpfulness. The proposed C-RHPO based approach is depicted in Sect. 4. Section 5 presents experimental evaluations before drawing some conclusions and discussing future work in Sect. 6.

2   Related Work

Online cloud service reviews have become very popular because of their significant impact on cloud users' choice of cloud services. For this reason, several research studies have asserted the important role of cloud service reviews as they can provide cloud users and cloud providers with valuable information that can help the first one to make decisions for choosing the best cloud service [3,4] and to let the second one to be aware of consumers' perception in the market [10,15].

Due to the important role of online cloud service reviews, it is important to ensure their helpfulness for cloud users. However, most existing literature has focused on studying the helpfulness of online product reviews or other services such as hotels [9], restaurants [8]. Many researchers have investigated different factors that can affect the helpfulness of e-WOM in the business and marketing domain. Helpfulness of reviews is determined by different set of factors. Krishnamoorthy [11] examined the factors influencing the helpfulness of online reviews and built a predictive model. Their proposed predictive model extracts linguistic features such as adjectives, state verbs, action verb features and accumulates them to make linguistic feature (LF) value. They also used review meta-data (review extremity and review age.), subjectivity (positive and negative opinion words) and readability (Automated Readability Index, SMOG, Flesch–Kincaid Grade Level, Gunning Fog Index, and Coleman–Liau Index) related features in their model for helpfulness prediction. Authors in [7] have analyzed many characteristics of review text such as spelling errors, readability, subjectivity, etc. and examine its impact on sales. Linguistic correctness was found to be vital factor of effecting sales. The intuition that reviews of medium length with less spelling errors are more helpful to the naive buyers as compared to the reviews of very short or very large in length and having spelling errors. Considering these points, [7] has made three taxonomy for the characteristics of review text, which are: easiness to read a review, spelling mistakes in the review and subjectivity level. Hu and Chen in [9] have analyzed hotel review helpfulness based on four categories of input variables: review content, sentiment, author, and visibility. They have affirmed the interaction effect between hotel star rating and review rating on review helpfulness. They also proved the strong effect of review visibility features on review helpfulness. Three prediction techniques, including linear regression, model tree (M5), and support vector regression were selected to develop review helpfulness prediction models. The authors in [8] found an inconsistency between review-based helpfulness features and reviewer-based helpfulness features on how they influence on perceived review helpfulness. For that, they conducted a meta-analysis to reconcile the contradictory findings. The results proved the positive influence of review depth, review age, reviewer information disclosure, and reviewer expertise on the review helpfulness. While, it confirm the weak influence of review readability and review rating on the review helpfulness.

3 Helpfulness Features

To infer the degree of helpfulness of a review, this paper relies on helpfulness features (depicted in Table 1). We categorize helpfulness features on three categories according to their types:

- Review Readability: denotes the ease of reviews understanding and it is based on the review itself and extracted directly from review text.
- Review meta-data: This category is based on meta-data and not the review text itself.

- Review semantic richness: In which the reviewers give a very personal description of the item, and give information that typically does not appear in the official description of the item.
- Reviewer expertise: Besides the categories depicted above, we propose in this work a new category, named Reviewer expertise. Usually, the reviewer with high expertise is an experienced one that provides a concise, helpful information meanwhile the inexperienced one provides long but less useful content.

Table 1. Features for users and reviews in four defined categories (The calculated values are based on [7,12,14] and Figs. 1 and 4)

Feature category	Feature name	Feature description	Source
Review readability	Gunning's Fog Index (GFI)	The lower the measure, the more readable the text is	[7]
	Automated Readability Index (ARI)		
	Coleman-Liau Index		
Review meta-data	Review Age (RA)	The difference between review publication date and item release date	[12]
	Review Extremity (RE)	The difference between review valence (or rating) and mean item rating	[7]
Review semantic richness	Sentiment Score (SS)	# of subjective sentences/Number of sentences	[14]
	CHaracteristic Score (CHS)	# of sentences containing item characteristic/Number of sentences	This study
Reviewer expertise	R_CR	Reviewer Career level (Manager, Engineer, C-level executive, etc.)	This study
	R_JT	Reviewer JobType (Software Engineer, Web Developer, etc.)	
	R_Skl	Reviewer Skill (Java, HTML 5, etc.)	
	R_ETR	Reviewer Enterprise	
	R_IND	Reviewer Industry domain (Internet, Health-care, etc.)	

Indeed, we evaluate the expertise state by the reviewer career level, Job type, skills, enterprise and the reviewer industry domain to infer the degree of expertise of each reviewer.

4 Approach for Ranking Cloud Service Reviews

The main objective of the current research is to rank cloud service reviews according to their quality. Review quality is identified by evaluating the review helpfulness and computing the similarity between the reviewer context and that of the end-user. To do this, we define a probabilistic ontology, named C-RHPO. The helpfulness evaluation is performed using probabilistic reasoning, while semantic similarity is used to compute the context similarities. A detailed description of the C-RHPO and the ranking approach will be described in what follows.

4.1 C-RHPO Description

The C-RHPO is created using the Uncertainty Model for the Semantic Web (UMP-SW) [6] presented in our previous work [5]. The C-RHPO presents the conceptualization of context-aware review helpfulness evaluation domain. The proposed ontology encompasses three interrelated parts: (1) the review part which includes all necessary concepts and relations for modeling the review posted on social media platform (such as review, reviewer, Social_Media_Platform, etc.), (2) the helpfulness part, which depicts the main concepts and relations for describing the review helpfulness domain, and finally, the context part (see Fig. 2) which conceptualizes the main concepts and relations to design the reviewer's context (also called cloud user's context). In fact, the review is written by a cloud user which has a particular context. The context includes two categories of profiles: the Service Profile which describes how the cloud user used/will use the service (the functional and non-functional requirements adopted/desired and the use case) and the User profile which describes the professional characteristics of the cloud user, such as his career Level, his job position, the enterprise that belongs to and his geographic location. The review has a list of Helpfulness Features (already presented in Sect. 3). As illustrated in Fig. 1 using the red color, the probabilistic part consists of five probabilistic data properties: *helpfulness_Level, readability_Level, semanticRichnessLevel_Level, reviewMeta-Data_Level* and *expertise_Level*. The structure of these data properties is described by the MTheory presented in Fig. 3 whose objective is to answer the following question (query): What is the probability of a given review to be helpful? The answer is provided by the resident node helpfulnessLevel(r) which means: "may be the review has a high helpfulness level". This resident node was chosen to mention the uncertainty level associated with the relationship between the helpfulness level and high/low values.

An MTheory is a repeatable structure from which Situation-Specific Bayesian Networks (SSBNs) are generated according to the review and reviewer features

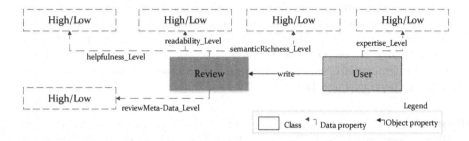

Fig. 1. The C-RHPO model with zoom on the probabilistic part

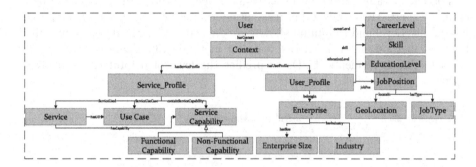

Fig. 2. Context model

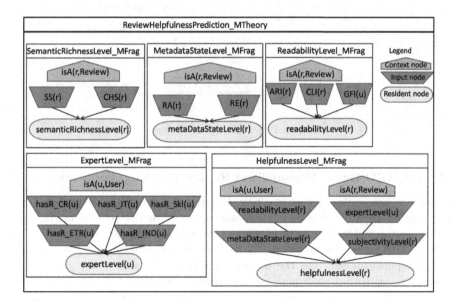

Fig. 3. MTheory for reasoning over uncertainty of review helpfulness.

of a given review. Thus, at run time, when there is a need to generate an SSBN to predict the reviews' helpfulness, i.e., when the end-user put his request, the network structure is generated dynamically according to MTheory. The proposed MTheory is composed of 15 fragments. As depicted in the Fig. 3, the main five fragments are SemanticRichnessLevel, MetadataStateLevel, ReadabilityLevel, ExpertLevel and HelpfulnessLevel MFrags.

Probability Distribution Learning: In order to generate the structure of the SSBNs, according to the MTheory, each resident node of an MFrag must also implement its own local probability distribution, whose values can be reported by a domain expert or can be generated through machine learning algorithms. In this paper, MetadataStateLevel's LPD and ExpertLevel's LPD are defined by a domain expert. Meanwhile, other resident nodes such as, SemanticRichnessLevel, ReadabilityLevel and HelpfulnessLevel, are defined automatically using machine learning method, named MEBN learning method [13]. Figure 4 presents an extract of each residents' LPDs to predict the level of helpfulness of a given review.

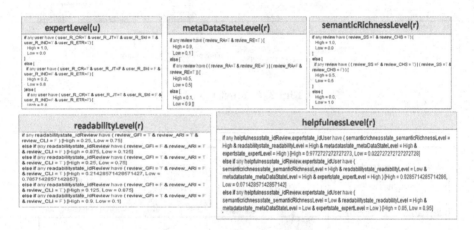

Fig. 4. LPDs for a given review

4.2 C-RHPO Based Personalized Review Ranking Approach

In order to personalize review ranking, this paper promotes reviews that have been written by users sharing context and interests similar to that of the end-user. Once the end-user introduces his context, including his profile as well as his interests about the cloud services, as the service capabilities (such as response time and performance) and the use case (such as web site, video games), we propose an algorithm that rank online reviews according to their helpfulness levels as well as their similarities with the end-user context. We present below the semantic-based context similarity as well as the personalized review raking algorithm.

1. Semantic-based context similarity: Two users are considered similar if they share a similar context. Context similarity is measured using the C-RHPO ontology where two contexts $context_u$ and $context_v$ are similar if they have maximum of equivalent relations' values. Assuming that a Context instance $context_u$ has a list of relations' values $R_u = \{r_{u1}, r_{u2}, ..., rum\}$ in the ontology such as "databases" instance in the "ServiceUseCase" relation and a Context instance $context_v$ has $R_v = \{r_{v1}, r_{v2}, ..., r_{vn}\}$, the similarity between the contexts $context_u$ and $context_v$ is computed as follows:

$$Sim(context_u, context_v) = \sum_{i=1}^{min(n,m)} exp\_w_i \times Eq(r_{ui}, r_{vi}) \qquad (1)$$

Where exp_w_i is the importance degree given by a cloud expert for each context element. In fact, we believe that the context elements have not the same impact on the quality of service. For example, the use case has the greatest importance compared to the others and the enterprise size has the least one according to the expert judgment. $Eq(r_{ui}, rvi)$ is equal to 1 if r_{ui} and r_{vi} have relation with the same instance of the class of the context element i.

2. Review ranking: The proposed Algorithm 1 provides a method to sort online reviews according to their helpfulness levels as well as their similarities with the end-user context. Its inputs are online reviews collected from different SMPs such as Facebook [2] and Trustradius [1], reviewers and the end-user related context. The algorithm firstly sorts online reviews based on their helpfulness level. Second, it computes the context_similarity_score between the end-user and each reviewer. Then, based on the obtained context_similarity_score, the algorithm re-sorts the reviews with high helpfulness level (*reviews_HL*) and reviews with low helpfulness level (*reviews_LL*) based on their context_similarity_scores. Indeed, online reviews are ranked as follows: *reviews_HL* with high context_similarity_score > *reviews_HL* with low context_similarity_score > *reviews_LL* with high context_similarity_score > *reviews_LL* with low context_similarity_score.

5   Experiments and Results

This section presents the experimental evaluation part of this study including the dataset description, the evaluation metrics and the experimental results. The main purpose of the experiments is to analyze the impact of helpfulness features as well as to examine the contribution of considering the context and personalizing the review ranking in the performance results.

Algorithm 1. Personalized-Review-Ranking

Input: List reviews, reviewers and the end-user
Output: List ranked_reviews

1: sort reviews by Pr(helpfulnessLevel=High)
2: **foreach** *reviwer rv ∈ reviwers* **do**
3:     $context\_similarity\_score_{rv} \longleftarrow Sim(context_{end-user}, context_{rv})$
4: **end foreach**
5: fill in List reviews_HL = reviews with Pr(helpfulnessLevel=High)>=0.5
6: sort reviews_HL by context_similarity_score
7: fill in List reviews_LL = reviews with Pr(helpfulnessLevel=High)< 0.5
8: sort reviews_LL by context_similarity_score
9: ranked_reviews ⟵ reviews_HL + reviews_LL
10: return ranked_reviews

Table 2. Review dataset

Dataset	Train	Test
#(Helpful)	4000	1000
#(Not helpful)	4000	1000
#Review with contextual information	5840	730
#Sentence	59297	14824
#(Avg. L/Sen)	22.66	22.68

5.1 Cloud Service Review Dataset

We collect cloud service reviews from different SMP categories, such as social media network (Facebook) and online review platforms like Trustradius and G2Crowd. More detail about the collected reviews is depicted in our previous work [5]. Table 2 includes a summary of the used dataset and its characteristics. It includes the reviewers' impressions and comments about the quality of cloud services. As shown in Table 2, the dataset contains 8000 reviews for training and 2000 reviews for testing, considered as balanced, i.e., half reviews are helpful and the others are not helpful. Due to the lack of annotated cloud service reviews, the annotation of the dataset is conducted in conjunction with cloud instructors from the IT department of the University of Sfax (considered as experts). The goal was to annotate the collected cloud service reviews with helpful reviews or not. To this end, the instructors organized themselves into four groups, where each group examined around 1000 reviews to decide if each review is helpful or not according to their expertise on the field of cloud computing. Afterwards, they conducted a cross-validation process among the different groups. In order to evaluate the impact of considering the context, we apply the proposed approach using 50 different contexts where the experts execute each time a query containing each context and then judge the result and decide if they are satisfied or not.

The prediction results performance was evaluated using four evaluation measures: recall, precision, accuracy and F-measure. Recall measures the percentage of helpful online reviews that have been correctly identified, and precision

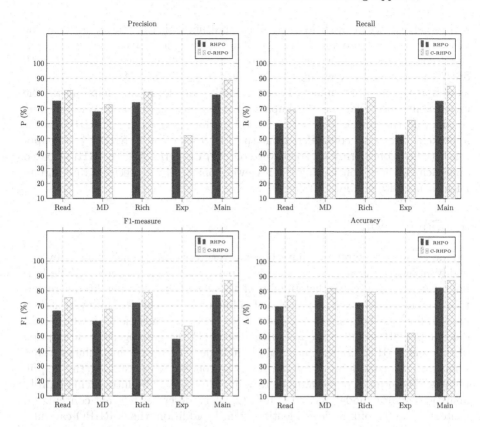

Fig. 5. Helpful review detection performance with and without considering the context. **Read:** Review Readability; **MD:** Review Meta-Data; **Rich:** Review Richness; **Exp:** User Expertise; **Main:** All helpfulness features.

measures the degree to which the predicted helpful reviews are indeed helpful. Accuracy is defined as the percentage of online cloud service reviews that were correctly classified as helpful or unhelpful reviews. F-measure is defined as the mean between precision and recall.

5.2 Experimental Results

This section demonstrates the C-RHPO based approach effectiveness and performance. We observe the positive impact of the review semantic richness (around 80% of accuracy). This can be explained by the fact that the sentiments and the service characteristics description have a high influence on the customer satisfaction. In our study, the service characteristics are presented by the functional and non-functional cloud service properties and the sentiment polarity describes the consumption emotion of customers and the emotions expressed through their textual reviews which highly influence customer satisfaction. We also observe that the review age feature (around 82.1% of accuracy) has an impact on online review helpfulness in the field of cloud computing. This can be explained by the fact that the cloud providers regularly lunch new cloud service releases. Indeed, cloud consumers consider the old reviews as not helpful ones. Consistent with other previous studies, we observe the negative influence of review extremity as well as the positive impact of the readability on the review helpfulness. The experiments also examine the impact of personalizing the review ranking by considering the reviewer's context in the performance of the C-RHPO based approach. It can be concluded from the Fig. 5 that the C-RHPO based approach (when considering the reviewer context) outperformed the RHPO based approach (without considering the reviewer context) by a precision rate and an F1-measure rate of about 10%.

6   Conclusion

Online cloud service reviews have become very popular recently since they can highly influence cloud user's choice of a particular cloud provider. While it is important to ensure the helpfulness of these reviews, existing literature has focused on analyzing the helpfulness of online product reviews. However, little is known about the helpfulness of cloud service reviews. In addition, previous works did not pay attention to the following issues: (1) the heterogeneity description of different SMPs, (2) the uncertainty judgment of review helpfulness evaluation, (3) the consideration of the reviewer context when ranking the helpful reviews. The present paper aims at tackling these points by introducing a probabilistic semantic model, named C-RHPO, which describes relevant concepts for context-aware helpful reviews identification. In addition, the C-RHPO can infer the degree of helpfulness given incomplete information about helpfulness features thanks to the machine learning based probabilistic reasoning. In order to personalize the review ranking, the C-RHPO based approach promotes helpful reviews existing in a similar context as the end-user. The experiments demonstrated the performance and the effectiveness of the C-RHPO based approach for the helpful review prediction from real cloud service reviews extracted from different SMPs. This study can be further extended in several ways. First, we could examine the validity of our model on larger dataset. Second, we could integrate an importance degree to each helpfulness feature to improve the accuracy of helpfulness evaluation.

References

1. Amazon web services reviews & ratings — trustradius. trustradius.com/products/amazon-web-services/reviews, Accessed 10 Mar 2019
2. Digitalocean   -   avis.   facebook.com/pg/DigitalOceanCloudHosting/reviews, Accessed 10 Mar 2019
3. Alkalbani, A.M., Ghamry, A.M., Hussain, F.K., Hussain, O.K.: Predicting the sentiment of SaaS online reviews using supervised machine learning techniques. In: 2016 International Joint Conference on Neural Networks (IJCNN), pp. 1547–1553, July 2016. https://doi.org/10.1109/IJCNN.2016.7727382

4. Ben-Abdallah, E., Boukadi, K., Hammami, M.: SMI-based opinion analysis of cloud services from online reviews. In: Abraham, A., Muhuri, P.K., Muda, A.K., Gandhi, N. (eds.) ISDA 2017. AISC, vol. 736, pp. 683–692. Springer, Cham (2018). https://doi.org/10.1007/978-3-319-76348-4_66

5. Ben-Abdallah, E., Boukadi, K., Hammami, M.: Spam detection approach for cloud service reviews based on probabilistic ontology. In: Panetto, H., Debruyne, C., Proper, H., Ardagna, C., Roman, D., Meersman, R. (eds.) On the Move to Meaningful Internet Systems, OTM Confederated International Conferences., LNCS, pp. 534–551. Springer, Cham (2018). https://doi.org/10.1007/978-3-030-02610-3_30

6. Carvalho, R.: Probabilistic ontology: representation and modeling methodology (2011)

7. Ghose, A., Ipeirotis, P.G.: Estimating the helpfulness and economic impact of product reviews: mining text and reviewer characteristics. IEEE Trans. Knowl. Data Eng. **23**(10), 1498–1512 (2011). https://doi.org/10.1109/TKDE.2010.188

8. Hong, H., Xu, D., Wang, G.A., Fan, W.: Understanding the determinants of online review helpfulness: a meta-analytic investigation. Decis. Support Syst. **102**, 1–11 (2017). https://doi.org/10.1016/j.dss.2017.06.007. http://www.sciencedirect.com/science/article/pii/S0167923617301197

9. Hu, Y.H., Chen, K.: Predicting hotel review helpfulness: the impact of review visibility, and interaction between hotel stars and review ratings. Int. J. Inf. Manage. **36**(6, Part A), 929–944 (2016). https://doi.org/10.1016/j.ijinfomgt.2016.06.003. http://www.sciencedirect.com/science/article/pii/S0268401215301845

10. Jayaratna, M.S.H., Bouguettaya, A., Dong, H., Qin, K., Erradi, A.: Subjective evaluation of market-driven cloud services. In: 2017 IEEE International Conference on Web Services (ICWS), pp. 516–523, June 2017. https://doi.org/10.1109/ICWS.2017.60

11. Krishnamoorthy, S.: Linguistic features for review helpfulness prediction. Expert Syst. Appl. **42**(7), 3751–3759 (2015). https://doi.org/10.1016/j.eswa.2014.12.044. http://www.sciencedirect.com/science/article/pii/S0957417414008239

12. Ngo-Ye, T.L., Sinha, A.P.: The influence of reviewer engagement characteristics on online review helpfulness: a text regression model. Decis. Support Syst. **61**, 47–58 (2014). https://doi.org/10.1016/j.dss.2014.01.011

13. Park, C.Y., Laskey, K., Costa, P., Matsumoto, S.: Multi-entity Bayesian networks learning in predictive situation awareness (2013)

14. Riloff, E., Wiebe, J.: Learning extraction patterns for subjective expressions. In: Proceedings of the 2003 Conference on Empirical Methods in Natural Language Processing, EMNLP 2003, pp. 105–112. Association for Computational Linguistics, Stroudsburg (2003). https://doi.org/10.3115/1119355.1119369

15. Taghavi, M., Bentahar, J., Otrok, H., Wahab, O.A., Mourad, A.: On the effects of user ratings on the profitability of cloud services. In: 2017 IEEE International Conference on Web Services (ICWS), pp. 1–8, June 2017. https://doi.org/10.1109/ICWS.2017.8

Influential Nodes Detection in Dynamic Social Networks

Nesrine Hafiene[1,2(✉)], Wafa Karoui[1,3], and Lotfi Ben Romdhane[1,2]

[1] Laboratoire MARS LR17ES05, ISITCom, Université de Sousse,
4011 Sousse, Tunisie
hafiene.nesrine@gmail.com, karoui.wafa@gmail.com,
lotfi.ben.romdhane@gmail.com
[2] ISITCom, Université de Sousse, 4011 Sousse, Tunisie
[3] ISI, Université de Tunis El Manar, 2080 Tunis, Tunisie

Abstract. The influence maximization problem aims to identify influential nodes allowing to reach the viral marketing objectives on social networks. Previous researches are mainly concerned with the static social network analysis and the development of algorithms in this context. However, when network changes, those algorithms must be updated. In this paper, we offer a new interesting approach to study the influential nodes detection problem in changing social networks. This approach can be considered to be an extension of a previous static algorithm SND (Semantic and structural influential Nodes Detection). Experimental results prove the effectiveness of SNDUpdate to detect influential nodes in dynamic social networks.

Keywords: Dynamic social networks · Influence maximization ·
Influence propagation · Influential nodes

1 Introduction

In recent years, with the increasing popularity of social networks like Facebook and Twitter, more and more scientists who study the influence maximization problem pay their attention to this field. Indeed this problem has drawn much attention and many researches have proposed various algorithms to detect influential nodes, most of these methods are based on static social networks. However, real social networks keep changing during time. New connections between users are created and some other users lose contact. Since many influential nodes can be detected by carefully studying the relationships among the links, these ones play an important role in dynamic social network analysis.

A social network looks like a graph structure consisting of nodes and edges, where nodes represent users and edges represent the interactions between the connected users. Users develop their connections to each other, while interactions between them vary over time. A changing social network consists of social

---

MARS—Modeling of Automated Reasoning Systems.

© Springer Nature Switzerland AG 2019
W. Abramowicz and R. Corchuelo (Eds.): BIS 2019, LNBIP 354, pp. 62–73, 2019.
https://doi.org/10.1007/978-3-030-20482-2_6

networks observations at different time stamps $\{G1, G2, ..., Gn\}$ and contains not only a set of relationships between nodes, but also information on how these relationships change in each time stamp.

To resolve the above drawbacks, we propose an extension of SND to tackle the problem of important nodes detection under changing social networks. The remainder of this paper is organized as follows. The related works are reviewed in Sect. 2. In Sects. 3 and 4, the proposed extension SNDUpdate for maximizing the influence propagation is described. The details of the experiments are presented in Sect. 5. Finally, the paper is concluded in Sect. 6.

2   Related Work

This section discusses the review of different researches done on dynamic social networks. In this paper, we concentrate on the problem of influential nodes detection in edges changing social networks.

2.1   Methods Based on a Non-linear Model

Aggarwal et al. proposed [7] the first paper which deals with the problem of information flow authority determination in dynamic networks based on temporary interactions. Moreover, this paper studies both the problem of influence propagation, and that of tracking back from a given pattern of spread the influential nodes. They considered how to discover a set of nodes having the highest influence within a time. They modelled the influence spread as a non-linear system which is very different from triggering models like the Linear Threshold model or the Independent Cascade model. The algorithm in [7] is heuristic and the produced results have not any provable quality guarantee. There are only a few papers focusing on Influence Maximization under changing networks. Aggarwal et al. [7] focuses on finding a seed set at time t, that maximizes the influence spread at some $[t + \Delta]$ given the dynamics of the evolution of networks over interval time $[t, t + \Delta]$. In this paper we consider to maximize the influence under a serie of snapshots taken from a social network. Zhuang et al. [18] study the influence maximization problem under dynamic networks where the changes can be only detected by occasionally probing some nodes. The main idea of their paper is how to choose a subset of nodes so that the actual influence propagation is improved.

2.2   Methods Based on Metrics

During the years, some centrality metrics have been proposed to estimate node importance. It is well-known that degree centrality, betweenness centrality and closeness centrality are three basic centrality measures to identify influential nodes. However, all of these centrality measures are designed for static networks and they looked over the feature that networks are usually dynamic. Motivated by this deficiency, to apply those metrics in changing networks we added information about the time of edges evolution.

Kitsak et al. [1] proposed *k-shell* decomposition and they discover that the influential nodes are those positioned in the core of the network. The *k-shell* decomposition assigns many nodes in the same K-shell. Basaras et al. [6] proposed an alternative measure, the μ-power community index, that is an combination of coreness and betweenness centrality, μ-PCI is calculated in a totally localized manner and thus is appropriated for any type of networks ignoring its size or dynamicity. Wei et al. [15] proposed two classes of dynamic metrics to estimate temporal evolution of agents with regard to persistence and emergence. Here, the network activities are measured per time stamp using static network metrics such as degree centrality, authority centrality or clustering coefficients. Ren et al. [14] proposed a new approach to identify influential nodes in a complex network called Evidential K-shell centrality based on edge weight. The authors study only undirected networks, while many surveys could be done on identifying influential nodes in directed networks. As reported by Wang et al. [5], nodes in the network can be evaluated by centrality metrics. Therefore, they defined influential nodes in a dynamic community as nodes which have the most important centrality scores higher than the other nodes and have comparatively long life in this community.

2.3   Methods Based on a Diffusion Model

Chen et al. [8], included time in the influence diffusion model to track influential nodes in changing social networks. The main idea is to choose a budgeted subset to maximize the influence propagation φ time stamps later. However, they assume that the dynamic network is completely observed. This is impossible in many real situations. Ohsaka et al. [16] studied a related problem, maintaining some RR sets over a stream of networks updates under the IC model such that approximation influence maximization can be completed with a fixed probability. Tong et al. [13] proposed the Dynamic Independent Cascade (DIC) model by extending the classic IC model, DIC is able to better take control of the dynamic aspects of real social networks. Liu et al. [4] proposed an incremental approach, IncInf, which can efficiently locate the top-k influential nodes in changing social networks based on previous information. Song et al. [3] proposed Upper Bound Interchange Greedy (UBI) algorithm for the Influential Nodes Tracking (INT) problem, in which they found the seed set that augments the influence under G^{t+1} based on the seed set S^t that they have effectively found in G^t. Then they proposed UBI+ algorithm that enhances the computation of the upper bound and achieves better influence propagation. The authors in [9] modelled a changing network as a stream of edge weight updates. In [10], the authors have systematically tackle two essential tasks of tracking $top - k$ influential nodes. Their goal is to control the error incurred in their algorithm [9], so that even without prior knowledge about the data, they can still obtain meaningful results by setting a relative error threshold.

3 Preliminaries and Problem Statement

In this section, we first introduce the previous diffusion model for static networks. We then present our extension algorithm as a generalization of the influential nodes detection problem to dynamic social networks.

3.1 Problem Statement

The previous algorithm SND aims to detect influential nodes for only static social networks. However, real social networks are dynamic so both the structure and also the influence propagation associated with the edges are constantly changing. As a result, according to the evolution of the network structure and the influence propagation, the leader nodes that maximizes the influence propagation should be changed. In this paper, we model the dynamic social network as a group of snapshot graphs where nodes remain the same while the edges in each snapshot graph change over different time intervals. We denote the snapshot graph as $G^t = (V, E^t)$, where $V = \{v_1, ..., v_i, ..., v_n\}$ represents nodes and E represents edges appearing during time intervals. Notation G^t, $t = 0, ..., T$ defines the snapshot graph over time. Our main idea is to identify a set of leading nodes, denoted as NL^t; $t = 1, ..., T$, that maximizes the influence propagation in each of the snapshot graph G^t. Each user is represented by an attributes vector $X_i = (x_{i1}, ..., x_{ij})$, where x_{ij} is the value taken by the attribute j of the vertex v_i. In SND this value is binary, either 1 (if the user likes a center of interest) otherwise 0. This approach exploits, on the one hand, the relations between the vertices of the network and, on the other hand, the attributes that characterize them. Table 1 lists the notations to be used extensively in the sequel of this paper.

Table 1. Notation explanation

Notations	Descriptions
$G^t = (V, E^t)$	The snapshot graph
V	The nodes in G^t
E^t	The edges in G^t
Δt	The distance between two consecutive snapshots
b	The number of E^t in G^t
C_i	The community in G^t
M	Number of communities
NL^t	The leaders nodes in G^t
NA^t	The active nodes in G^t
N^t_{inf}	The influential nodes in G^t
A_v	The active nodes

4   Proposed Algorithm

The main objective of SNDUpdate is to detect the most influential nodes in a dynamic social network. It exploits the structural and semantic aspects of the network. For this reason, the main idea is to propose an approach that contains two phases. Indeed, the first phase of SNDUpdate explores the structural aspect of the network and the second phase concentrates on the semantic aspect. In the sequel of this paper, each user is described by a set of interests that are represented as an attributes vector. In the previous work, the influential nodes are detected in a static social network and thus the weight of the link between two nodes remains unchangeable. In this paper, the network is dynamic so the structure of the network changes and thus the link between two nodes belonging to a snapshot graph G^t can be removed in the snapshot graph G^{t+1} which generates a modification in the value of the semantic similarity between two nodes as well as a modification in the set of leader nodes.

4.1   Phase 1: Community Detection

In this phase, we propose to use Combo [11]. This algorithm handles the community detection problem which is able to deal with various objective functions. The majority of search strategies take one of the next steps to improve the quality of partition: merging two communities, splitting a community into two and moving nodes between two distinct communities. Combo covers all these possibilities.

4.2   Phase 2: Influential Nodes Detection

The first part of the phase 2 enables to generate a set of leader nodes. A node is a leader if its degree centrality is greater or equal then its neighbors. The degree centrality enables to measure the total of the node connections with its neighbors. In this paper, we model the dynamic social network as a group of snapshot graphs while the edges in each snapshot graph change over different time intervals and thus the leader nodes change in each snapshot graph. This equation is used to calculate the degree centrality of a node v of a given snapshot graph $G^t = (V, E^t)$:

$$dc(v) = \frac{deg(v)}{|V| - 1} \tag{1}$$

In the second part of this phase, we use the diffusion model in [2] to define the inactive nodes that can be activated. In G^t, the link between two nodes is associated with a weight which is defined by the semantic similarity of their information. The semantic similarity related to each snapshot graph G^t is given by following equation:

$$sim^t_{u,v} = \frac{Common(u, v)}{long(u) + long(v)} \tag{2}$$

Every node v in a partition $\mathcal{P} = \{C_1, ..., C_r\}$ is related to a degree of centrality. Taking into consideration the degree of centrality of each node in G^t and a set of active nodes, a node v becomes active if the total of the similarity or the total weight of its active neighbors overrides the threshold value $dc(v)$ associated with a node v. Our diffusion model is defined as follows:

$$\sum_{u \in \mathbb{A}_v} w_{v,u}^t > dc(v) \tag{3}$$

When we apply our diffusion model in each snapshot G^t, we can get a set of active nodes NA^t which allows us to identify the set of influential nodes. Our main objective is to identify from the active nodes those that maximize the influence propagation. In each community, we determine its influence degree which is given by the Eq. 4 and is equal to the number of active nodes in each community (V_{active}) divided by the sum of nodes in the graph $G(V)$.

$$R(C_r) = V_{active}/V \tag{4}$$

Firstly, the proposed approach determines influential nodes from the community that contains the value of the highest influence degree. Secondly, our objective is to calculate for each active node its closeness centrality given by the Eq. 5. It represents the total of the length of the shortest paths between the node and all other nodes in the graph. The closeness centrality is defined by the following equation:

$$CCenter(v_i) = \frac{1}{\sum_{v_j \epsilon V} |ShortPath(v_i, v_j)|} \tag{5}$$

Once we have calculated the closeness centrality of every leader node, we classify these nodes according to an increasing order. The influential node is that admits the highest closeness centrality. The algorithm of SNDUpdate is described in Algorithm 1. Firstly, it detects communities using Combo algorithm. Secondly, on each snapshot graph G^t, it generates a set of nodes playing the role of a leading nodes. Once we have generated the set of the leader nodes that are the initiators, we apply our diffusion model. Finally, we apply our diffusion model to determine the set of active nodes and then identify the influential nodes. In the proposed algorithm we find the set of the active nodes that maximizes the influence within each snapshot graph G^t with updating b (Update b) the function which allow us to update the value of edges over each time stamp t. About the SNDUpdate time complexity, because it is difficult to analyse the complexity of dynamic edges update operations [16], we will only analyze each step in a static snapshot. SNDUpdate consists of T snapshots consisting each one of two phases. The first phase admits a complexity of $O(V^2 \log M)$ as we used the Combo algorithm. The second phase is composed of two steps. The first generates leader nodes and admits a complexity of $O(MV^2)$. The second calculates the temporary shortest path of active nodes and corresponds to a complexity of $O(E^t(NA^t) \log(NA^t))$.

Algorithm 1. SNDUpdate

Data: An initial snapshot graph $G^t = (V, E^t)$, T, b, Δt: the distance between two consecutive snapshots;

Result: A set N_{inf}^t of influential nodes at $t = 1, ..., T$;

1   $N_{inf}^t \leftarrow \emptyset$;
2   $NL^t \leftarrow \emptyset$;
3   $NA^t \leftarrow \emptyset$;
4   **Begin**
5   **for** *t=0 to T* **do**
6   **for** *b edges in G^t* **do**
7   Apply **Combo** to detect communities;
8   **foreach** *community C in G^t* **do**
9   Determine the degree centrality of each node in G applying the equation 1 ;
10   **foreach** *node $v \in V$ of a community C* **do**
11   **if** *isLeader(v)* **then**
12            $NL^t \leftarrow NL^t \cup \{v\}$;
13   **end**
14   **end**
15   **end**
16   **foreach** *Node u in C* **do**
17   Allocate a weight for each node that represents the degree of centrality;
18   **end**
19   **foreach** *Edge (u, v) in C* **do**
20   Determine the similarity sim (u, v) using equation 2;
21         $w_{u,v}^t \leftarrow sim_{u,v}^t$;
22   **if** $\left(\sum_{u \in A_v} w_{v,u}^t > dc(v)\right)$ **then**
23           $NA^t \leftarrow NA^t \cup \{u\}$;
24   **end**
25   **end**
26   **end**
27      $M = |C|$;
28   **while** $M \neq \emptyset$ **do**
29   Determine the degree of influence of each community applying the equation 4;
30       $A_{C_{max}} \leftarrow \{$active nodes in C_{max} $\}$;
31   **while** $(C_{max} \neq \emptyset) et (A_{C_{max}} \neq \emptyset)$ **do**
32   Calculate the closeness centrality of each node in $A_{C_{max}}$ using equation 5;
33   Classify these nodes in an ascending order of closeness centrality;
34         $v_{max} \leftarrow$ the node having the highest closeness centrality;
35         $N_{inf}^t \leftarrow N_{inf}^t \cup \{v_{max}\}$;
36   **end**
37   **end**
38   **return** N_{inf}^t;
39   **end**
40     $t \leftarrow t + \Delta t$;
41   Update b;
42 **End**

---

5 Experiments

In this section, we estimate the effectiveness and efficiency of our proposed SNDUpdate to identify the influential nodes in dynamic social networks on three real networks. The experimental results demonstrate that our algorithm is both efficient and effective.

5.1 Experiment Settings

In our experiments, we use three real social networks as shown in Table 2. For each data set, we used 3 different social networks, updated edges, and thus executed the experiments 3 times. For the 3 examples, however the original networks and updated edges are different, the combination of the 3 groups snapshots is similar to the data set itself. Let b denote the number of edges in each snapshot graph, using different parameters, b and Δt, we can create a family of snapshots with many properties for our next experiments. In Table 2, to simulate dynamic networks, we partitioned all edges into 3 time stamps: 25% of edges in time stamp $t = 5$, 50% of edges in time stamp $t = 10$ and 85% of edges in time stamp $t = 15$. The evolution of edges can reflect that the real-world social networks change rapidly during the considered time periods.

Table 2. Datasets and edge information

Networks	Nodes	Edges	Edge information		
Flixster	99,825	978,265	t = 5	t = 10	t = 15
			1,222,831	1,467,398	1,809,790
NetHept	15,634	62,836	78,545	94,254	116,247
CA-HEP-PH	12,008	237,010	296,263	355,515	438,469

Table 2 shows the number of edges in each snapshot graph created from networks. We make a group of snapshot graphs from three networks by changing the number of edges with a constant time difference $\Delta t = 5\,\text{min}$. The metrics applied in this paper to evaluate the performance of our algorithm are: influence propagation and running time. Influence propagation is the total number of influenced nodes activated by leader nodes and the running time is the time to identify the most influential nodes.

5.2 Experiment Results

In this paper, we compare our algorithm with two static influence maximization algorithm IMM, Degree and one dynamic algorithm DIM [19]. As shown in Figs. 1 and 2, the influence propagation of SNDUpdate algorithm outperforms that of IMM and DIM algorithms on Flixster and NetHept datasets. Varying the network size, Degree algorithm can not detect influential nodes anymore, while SNDUpdate finds better values of the influence propagation than those of both IMM and DIM. According to the results of this evaluation, while increasing the size of the network and updating the number of edges at each time stamp, SNDUpdate is still able to have a better influence propagation. Thus, it covers a large number of influential nodes in any type of networks. As conclusion, on the Flixster and NetHept networks, SNDUpdate is more efficient since the diffusion strategy based on edges change can narrow the search space of influential nodes.

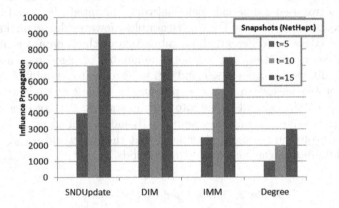

Fig. 1. Detecting influential nodes under NetHept network

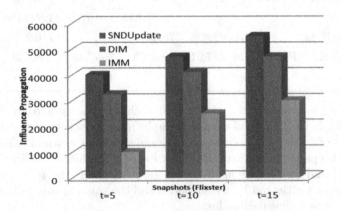

Fig. 2. Detecting influential nodes under Flixster network

5.3   Comparison with Methods Based on Metrics

We compared our algorithm SNDUpdate with three algorithms based on metrics which are: K-shell, MDD [17] and CN [12]. Our study is based on metrics to detect influential nodes in dynamic social networks. In the K-shell method, some nodes with large degree are not under consideration. When examining the exhausted degree in the network decomposition, the ranking of these nodes are ameliorated by the MDD algorithm. That is the reason why MDD method outperforms the K-shell method. In spite of the fact that the MDD method can improve the K-shell method in running time, it is still not the best way to deal with the problem of identifying influential nodes. According to this evaluation, when updating the size of the network SNDUpdate still has a better running time than the three other algorithms K-shell, CN and MDD. We explain this result by the fact that our method covers a considerable number of influential nodes in any type of networks. As indicated in Table 3, SNDUpdate is the best one regarding to the running time.

Table 3. Evaluation of the running time in different algorithms

Networks	K-shell	CN	MDD	SNDUpdate
CA-HEP-PH	2037	3003	8772	100
NetHept	4048	5929	16,195	120
Flixster	8110	10,269	32,442	1000

5.4   Comparison with Methods Based on a Diffusion Model

To detect influential nodes, all the algorithms are running under a diffusion model: Independent Cascade model (IC), Linear Threshold model (LT) and the Weighted Cascade (WC). Our diffusion model is based on the semantic similarity between nodes. We evaluate the running time of SNDUpdate, DIM and IMM algorithms on NetHept and Flixster networks in three snapshots. As illustrated in Figs. 3 and 4, the running time of SNDUpdate is better than the other two algorithms on the two networks. Among the three compared algorithms, we can easily find that IMM is very slow, while SNDUpdate performs well in terms of running time on the two networks. The reason is that IMM is a static algorithm so, when the network changes, DIM and SNDUpdate which are two dynamic algorithms run faster than the static algorithm IMM on the two networks. On Flixster network, the SNDUpdate running time increases, but is still better than the running time of DIM algorithm on large dynamic network. On NetHept and Flixster networks, SNDUpdate is faster than DIM algorithm. As conclusion, we can see that SNDUpdate has the best running times. The reason is that we use a diffusion model that improves the influence propagation as well as the running time of SNDUpdate. Note that IMM and DIM cannot detect influential nodes behind any diffusion model because they do not know the influence propagation of each node.

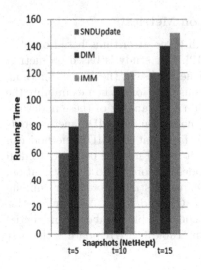
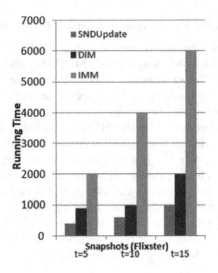

Fig. 3. Running time on NetHEPT   **Fig. 4.** Running time on Flixster

6   Conclusion

In this paper, we study the problem of identifying influential nodes in a changing social networks. We propose an approach called SNDUpdate algorithm and describe how to integrate it into a classical SND algorithm. The goal of this method is to detect influential nodes in dynamic social networks where edges are evolving over time. To prove the performance of the proposed approach, we compare it with approaches based on metrics and diffusion model in three real networks. Experimental studies carried out on the selected networks shown how it is possible to obtain good results in the detection of nodes that maximize the influence propagation. Empirical studies show that our algorithm has great improvement on the influence propagation compared with IMM, Degree and DIM algorithms. In the future, we plan to predict the change of influential nodes where both nodes and edges evolve in dynamic social networks.

References

1. Kitsak, M., Gallos, L., Havlin, S.: Identification of influential spreaders in complex networks. Nature Phys. **6**, 888–893 (2010)
2. Hafiene, N., Karoui, W.: A new structural and semantic approach for identifying influential nodes in social networks. In: IEEE/ACS International Conference of Computer Systems and Applications AICCSA, pp. 1338–1345 (2017)
3. Chen, X., Song, G., He, X., Xie, K.: On influential nodes tracking in dynamic social networks. IEEE Trans. Knowl. Data Eng. **29**, 359–372 (2015)
4. Liu, X., et al.: On the shoulders of giants: incremental influence maximization in evolving social networks. Complexity 1–14 (2017)

5. Wang, T., Dai, W., Jiao, P., Wang, W.: Identifying influential nodes in dynamic social networks based on degree-corrected stochastic block model. Int. J. Mod. Phys. B **30**(16), 1–18 (2016)
6. Basaras, P., Katsaros, D., Tassiulas, L.: Detecting influential spreaders in complex, dynamic networks. Computer **46**, 24–29 (2013)
7. Aggarwal, C.C., Lin, S., Yu, P.S.: On influential node discovery in dynamic social networks. In: International Conference on Data Mining, pp. 636–647 (2012)
8. Chen, W., Lu, W., Zhang, N.: Time-critical influence maximization in social networks with time-delayed diffusion process. In: International Conference on Data Mining, pp. 636–647 (2012)
9. Yang, Y., Wang, Z., Pei, J., Chen, E.: Tracking influential nodes in dynamic networks. IEEE Trans. Knowl. Data Eng. **29**, 2615–2628 (2017)
10. Yang, Y., Wang, Z., Jin, T., Pei, J., Chen, E.: Tracking top-k influential vertices in dynamic networks. IEEE Trans. Knowl. Data Eng. **29**, 1–14 (2018)
11. Sobolevsky, S., Ratti, C., Campari, R.: General optimization technique for high-quality community detection in complex networks. Phys. Rev. **90**, 1–19 (2014)
12. Feng, S., Wang, L., Sun, S., Xia, C.: Synchronization properties of interconnected network based on the vital node. Non Linear Dyn. **93**(2), 335–347 (2018)
13. Tong, G., Weili, W., Tang, S., Du, D.-Z.: Adaptive influence maximization in dynamic social networks. IEEE/ACM Trans. Netw. **25**(1), 112–125 (2017)
14. Ren, J., Wang, C., Liu, Q., Wang, G., Dong, J.: Identify influential spreaders in complex networks based on potential edge weights. Int. J. Innov. Comput. Inf. Control **12**(2), 581–590 (2016)
15. Wei, W., Carley, K.: Measuring temporal patterns in dynamic social networks. J. ACM Trans. Knowl. Discov. Data **10**(1), 1–27 (2015)
16. Ohsaka, N., Akiba, T., Yoshida, Y., Kawarabayashi, K.: Dynamic influence analysis in evolving networks. J. Proc. VLDB Endow. VLDB **9**(12), 1077–1088 (2016)
17. Zeng, A., Zhang, C.-J.: Ranking spreaders by decomposing complex networks. Phys. Lett. **377**, 1031–1035 (2013)
18. Zhuang, H., Sun, Y., Tang, J., Zhang, J., Sun, X.: Influence maximization in dynamic social networks. In: International Conference on Data Mining, pp. 636–647 (2013)
19. Wang, Y., Zhu, J., Ming, Q.: Incremental influence maximization for dynamic social networks. In: Zou, B., Han, Q., Sun, G., Jing, W., Peng, X., Lu, Z. (eds.) ICPCSEE 2017. CCIS, vol. 728, pp. 13–27. Springer, Singapore (2017). https://doi.org/10.1007/978-981-10-6388-6_2

A Fuzzy Modeling Approach for Group Decision Making in Social Networks

Gulsum Akkuzu$^{(\boxtimes)}$, Benjamin Aziz, and Mo Adda

School of Computing, University of Portsmouth, Portsmouth, UK
{gulsum.akkuzu,benjamin.aziz,mo.adda}@port.ac.uk

Abstract. Social networks have been commonly used, people use social networks with various purposes, such as, enjoying time, making business, and contacting their friends. All these activities are mainly based on sharing data. In social networks, making decision on data sharing process has become one of the main challenge because it involves people who have different opinions on the same problem. Diversified opinions cause uncertainties in decision making process. Fuzzy logic is used to overcome uncertainties' situations. In this work, we provide a fuzzy logic based decision making framework for SNs. The proposed fuzzy logic based framework uses data sensitivity value and trust value (confidence value) to make the group decision. Users express their opinions on data security features to obtain aggregated decision. Facebook data sharing process is chosen as a case study.

Keywords: Aggregated group decision making · Social network · Fuzzy systems

1 Introduction

Social networks (SNs) enable users to communicate with each other via data sharing [1]. The common issue for SNs is to make decision on data which is related to more than one user [2–4], the reason is to reach aggregated decision on the data sharing process.

To obtain aggregated decision, group decision making (GDM) is proposed. GDM is a process that involves a group of people who state their opinions on different options in order to chose the best option [5–7]. In the traditional group decision making, decisions are made by administrators or experts even if the case related to different people. This case is still seen in many organisations where important decisions are made by restricted board of people. Urena et al. [9] compares the traditional group decision making, in which the experts have just right to express their opinions on alternative situations, with social network. Based on their comparison, SNs bring the global group decision making which means all people can give their opinions on a case which is related to them.

Supported by organization x.

W. Abramowicz and R. Corchuelo (Eds.): BIS 2019, LNBIP 354, pp. 74–85, 2019.
https://doi.org/10.1007/978-3-030-20482-2_7

Even though group decision making is possible in SNs, it is still a problem in many SNs especially in online social network (OSN) while data is shared. People either use the traditional decision making process even if the data is owned by different users. These confusions are because of the data is owned more than one user (data is called co-owned data), and having different decision criteria to share the data in OSNs. They also cause vagueness on decision making process. To overcome uncertainty situations fuzzy logic was introduced by Zadeh [8], the fuzzy logic resolves the uncertainties particularly in decision making process.

We introduce an aggregated group decision making system for SNs data sharing process, and introduce a fuzzy logic approach to deal with uncertainty situations while the decision is made. Two factors are important when the sharing decision is made on data. The first factor is the data sensitivity value and second one is the trust in the group of people who will have access the shared data. The proposed system provides alternatives to co-owners on the data security features which have effects on the data sensitivity value. Based on co-owners' choices the proposed fuzzy logic based system makes the final decision.

The rest of the paper is organised as follows. Section 2 gives similar research papers. Section 3 presents the proposed work's framework and its mathematical expressions. We introduce our fuzzy system in Sect. 4 with the experimental results. We finalise the work in Sect. 5.

2 Related Work

Group decision making is an important and challenging process, because it includes decision makers' doubts, problems, and uncertainties [10]. Therefore, finding appropriate ways to help decision makers is one of the key and critical point. The consensus-reaching process, which is an approach to get aggregated decision on final decision in group decision making problems, has been provided by researcher to help decision makers in social networks [5,11,12]. Wu and Chiclana [11] propose a trust based consensus approach to tackle group decision making problem. In work [12] consistency is used as an approach to control consensus-reaching process. Liang et al. [10] introduces an approach in which social connections of decision makers effect to get final aggregated decision in social networks. This work supports Liang et al. [10] on the point that shows users' relations have effects to make decision in SNs. Beside the users' relations, we also introduce the data sensitivity value has effect on decision making in SNs.

Fuzzy logic is an approach to tackle within ability of binary logic which is underlying on modern computer. Fuzzy logic is used to describe fuzziness, therefore, it can easily be applied to decision making. Fuzzy logic approach has been commonly used to tackle decision making problems with different alternatives in different areas such as education [13], health [14], Internet of Things [15], social networks [5,9,16,17]. Due to fuzzy logic the effectiveness in decision making, it has also been applied to solve group decision making issues. Thirumalai and Senthilkumar [18] propose a fuzzy model to resolve the group decision making problems in business area, the proposed approach uses membership and non-membership attributes to make the group decision. Similarly, Kahraman et al. [19] used the fuzzy logic to overcome group decision problems in facility location selection. This paper uses the fuzzy logic approach to remove uncertainties in group decision making process for SNs.

3   Background

This section introduces the models and the framework of this work.

3.1   CIAPP Security Model

To ensure the protection of data security; Confidentiality, Integrity, and Availability (CIA) model was developed, which is the model to guide policies to ensure the information security [20]. In CIA model, confidentiality is a boundary to limit access to information, integrity is a guarantee of limited access to the information, and availability is assured that the information is only accessed by authorised people [20, 21]. The information security is also needed in SNs in order to protect users' sensitive data [22]. The data sensitivity is a measurement that is calculated with the number of authorised people, however, Akkuzu et al. proposed a model in which Privacy and Possession features are added to extend CIA model [4]. The proposed model is CIAPP model in which Privacy and Possession are added to CIA model. Hence Privacy and Possession features are used to control information and network security. In SNs, users are asked directly to set the data sensitivity value [23] to define the level of data privacy. However, users may not be enough knowledgeable to set the data sensitivity value. It might be easier to ask their choices on the data security features, with their choices on the data security features the data sensitivity value can be calculated. To do so, we provide a model the CIAPP data security features are used to calculate the data sensitivity value. Table 1 indicates the related features to data sensitivity in OSNs. Table 1's features are deduced from [21], the information security subjects are divided into five circles based on the goals and disciplines. Deduced five features are combined to measure the data sensitivity in OSNs.

Table 1. Related information security features to SNs

Subject of protection	Discipline
Confidentiality, Integrity, Availability, Privacy	Information
Possession	Information and Network

3.2   Framework

Figure 1 introduces simply the structure of the framework for group decision. First, the data is uploaded by owner (the person who starts sharing process), he decides the targeted group for the data, and lastly the owner notifies the co-owners to get their opinions on the data sharing process. Then, the process which is given in Fig. 1 starts, co-owners are notified with the data and the targeted group which the data will be available for them. Once co-owners know which data is intended to be shared with whom, they select individually data security features (CIAPP) [21] that are seen as a threat for their privacy if the data is shared.

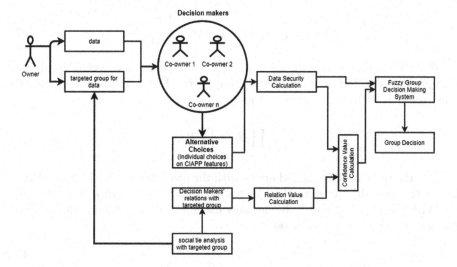

Fig. 1. The framework for group decision making with fuzzy system

3.3   Collective Measures for Decision Making

Collective measures are the models that are the main requirements for making decision, this is because they are used input values for fuzzy system.

$$S_d = \frac{\sum_{i=1}^{m}(P_i * (w_i))}{f} \tag{1}$$

S_d represents the data sensitivity, it ranges between [0, 1]. The numerator gives the summation of the data Confidentiality, Integrity, Availability, Privacy, and Possession (CIAPP) [21] probabilities, in which P_i indicates the probability of CIAPP concerns that is chosen by co-owners and w_i is the weight of the properties. The denominator f indicates the total number of the features. Model 1 clearly indicates that the more worried users on the data sharing the higher sensitivity value. Also, more worrying data security features cause the higher data sensitivity value.

We model confidence value with the owner trust relation in the targeted group, co-owner trust relation in the targeted group, and the sensitivity value. We first show the calculation of trust relation;

$$R_o : f(r_{o1}, r_{o2}, ..., r_{osi}) = \frac{\sum_{j=1}^{s_i}(r_{oj})}{s_i} \tag{2}$$

R_o represents the owner's trust in each member of the targeted group and s_i represents the size of the targeted group. $f(r_{o1}, r_{o2}, r_{o3}, ..., r_{osi})$ represents the relation value between the data owner and each member in the targeted group.

$$R_{ci} : f(r_{c1}, r_{c2}, ..., r_{csi}) = \frac{\sum_{j=1}^{s_i}(r_{cj})}{s_i} \tag{3}$$

R_{ci} represents the co-owner's trust in each member of targeted group and s_i represents the size of the targeted group. From Eqs. 2 and 3, we finalise the trust relation with the following formula;

$$R = \prod_{l=1}^{c} R_{li} * \prod_{k=1}^{c} R_{ki} \tag{4}$$

R is the trust in the targeted group with the owner's trust in the group i R_{oi}, also with the each co-owner's trust in group i R_{ci}. R_{oi}, R_{ci} and R range $\in [0, 1]$.

With the Eqs. 1, 2 and 3 we can now calculate the Confidence value ($C_f \in [0, 1]$) in targeted group as follows;

$$C_f = 1 - S_d * (1 - R) \tag{5}$$

3.4 Proposed System's Social Network Analysis for GDM

A social network is a platform in which users communicate with each other via data. It is represented with a graph $G(V,E)$, with nodes V representing users $V = V_1, V_2, ..., V_n$ and $E = E_1, E_2, ..., E_n$ are edges indicating the relations between users [24]. Social networks are classified into two classes, namely directed social network and undirected social network [25]. While the direction of edges is important in directed social network, the edges do not have direction in undirected network. This work includes an undirected social network dataset. We use the Stanford University Facebook large network dataset [26], which has 4039 nodes, 88234 undirected edges, and average clustering coefficient 0.6. The representation of dataset's nodes and edges is shown in Fig. 2.

In the dataset, nodes represent the users and edges represent the relation between nodes. Let us assume that User 0 wants to share the data ($data_{id} = d_1$) with his friends (346 people, in this case the network depth is 1), the data is related to User 1 and User 2. User 0 notifies the User 1 and User 2 by giving them the *data id* and the *targeted group*.

Fig. 2. The SNAP Facebook dataset network representation

User 1 and User 2 now need to choose which data security features are worrying them if the d_1 is shared with User 0' friends. Their choices are used to get the data sensitivity value (see Eq. 1) which is one of the input variable for our fuzzy system to make group decision. Table 2 represents users' choices on CIAPP features of d_1.

Table 2. User 1 and User 2 relation values

User id	Confidentiality	Integrity	Availability	Privacy	Possession
User 1	✓	X	X	X	✓
User 2	✓	✓	✓	✓	✓

With CIAPP security features selections (the weights of features are set 1) on Table 2 and Eq. 1, the d_1's sensitivity value becomes 0, 7.

The relation values calculation is computed with 3. Table 3 indicates the relation values for each user.

Table 3. User's choices on CIAPP features for d_1

User id	Relation value with targeted group
User 1	0,04
User 2	0,02

S_i is 347 since User 0 has 347 friends, therefore, the targeted group size is equal to the number of User 0's friends. User 1 has connection with 16 people from User 0' friends. Similarly, User 2 has connections with 9 people from User 0' friend group. Table 4 represents the numbers of known people for each user. The dataset's (Facebook dataset) relations between nodes and targeted group representations are given in Figs. 3, 4 and 5.

Table 4. User's choices on CIAPP features for d_1

User id U_i	The number of known people by U_i
User 0	346
User 1	16
User 2	9

Fig. 3. User 0's relations: Targeted group for data

Fig. 4. Exist relations User 1's in targeted group

User 0's relation with the targeted group is equal to 1. We now can use Eq. 4 to find the relation value that is used to find out the confidence value (Eq. 5). The relation value becomes $R = 1 * 0,06 = 0,06$. The last calculation is confidence value, which is the second input variable for our fuzzy system. It is computed with the data sensitivity value and the relation value (see Eq. 5). This confidence value is calculated as $Cf = 1 - S_d(1 - R) = 0,65$. The input values for the fuzzy system to make decision are $0,7$ and $0,65$. The decision out of these two input values are given in Sect. 4.1 (see Fig. 7).

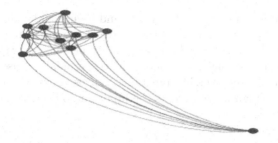

Fig. 5. Exist relations User 2's in targeted group

4   Our Fuzzy-Based Group Decision Making Model

We start with defining the key components for determining the data sensitivity value and the trust (we use confidence in this paper) in targeted group. For our problem, there are five data security features that have effects to calculate the data sensitivity value. We use five data security features from Cherdanseva et al.' s work [21], these are namely, confidentiality, integrity, availability, privacy, and possession (CIAPP) (see Eq. 1). For example, a user can be worried about his data's confidentiality if the data is viewed by people who may cause a threat for him. The second key component is confidence value in targeted group, we calculate the confidence value by using relations between user and targeted group (see Eq. 5).

As we mentioned earlier, our fuzzy system has two inputs and one output, data sensitivity and confidence in targeted group are inputs and decision is output variables. In the fuzzy set, there is no predefined boundary between objects, therefore, each element of the set is associated with a value which indicates to what degree the element is a member of the set. Fuzzy decision is based on the fuzzy logic in which the decision values range [0, 1] rather than binary values (0 or 1). Table 5 lists the input and output variables and their ranges.

Table 5. Membership database

Linguistic variables	Type	Membership functions (Linguistic)	Membership values (Python values)
Sensitivity value & Confidence value	Inputs	Low	Range [0, .2, .3, .4]
Sensitivity value & Confidence value	Inputs	Medium	Range [.4, .5, .6, .7]
Sensitivity value & Confidence value	Inputs	High	Range [.6, .8, .9, 1]
Decision	Output	No	Ranges [0, 0, .2, .4]
Decision	Output	Maybe	Ranges [.2, .4, .5, .7]
Decision	Output	Yes	Ranges [.6, .8, 1, 1]

The next step is to define the fuzzy sets and their membership function values, the membership function returns the degree of membership for a given value within a fuzzy set. Fuzzy sets can have different shapes such as trapezoidal, triangular, gaussian, and rectangle. We choose the trapezoidal, we use the clustering method to define the membership functions' ranges (Fuzzy c-means clustering technique is used). Figure 6 represents the input and output variables' membership function values.

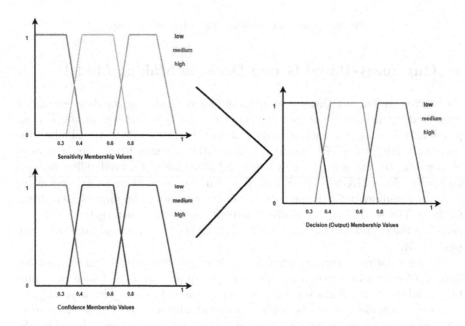

Fig. 6. Fuzzy input-output membership functions

We can now the define the our system's fuzzy rules, we use the expert knowledge to define the fuzzy rules. In our system, there are two input variables and each input variable have three different membership value, therefore we have maximum nine rules (3 * 3). Table 6 indicates the rules.

4.1   Using the Proposed Fuzzy System to Make Group Decision

The fuzzy system, which is represented in Sect. 4, has fuzzification, rule evaluation, aggregation of the rules, output, and defuzzification steps.

- Fuzzification: Obtains membership values from crisp values.
- Rule evaluation: Obtains the consequence of each rule, then combines output of each rule into a single fuzzy set with fuzzy aggregation operator.
- Aggregation: is the process to unify the outputs of all rules.
- Defuzzification: Converts fuzzy quantities into crisp numbers as the output.

Table 6. Fuzzy system decision making rules

Rule number	Rules
1	If x_1 is low AND x_2 is low then decision=maybe
2	If x_1 is low AND x_2 is medium then decision=maybe
3	If x_1 is low AND x_2 is full then decision=yes
4	If x_1 is medium AND x_2 is low then decision=maybe
5	If x_1 is medium AND x_2 is medium then decision=maybe
6	If x_1 is medium AND x_2 is full then decision=yes
7	If x_1 is high AND x_2 is low then decision=no
8	If x_1 is high AND x_2 is medium then decision=maybe
9	If x_1 is high AND x_2 is full then decision=yes

We give a sample output of our fuzzy system in Fig. 7. Given decision output value is obtained with the *sensitivity variable value = 0.7 and the confidence variable value = 0.65*. The output value is *Maybe with its degree = 0.45*.

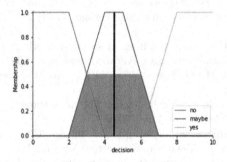

Fig. 7. Decision value

5 Conclusion

Making decision on a co-owned data in SNs has been a problem, SNs' users (data owners) either ignore other users' (co-owners, also known decision makers) opinions on co-owned data or have difficulties to decide which co-owners' decisions are more important than others. Therefore, it is necessary to develop a system which can give co-owners' aggregated opinions on co-owned data to help data owners to make decision. To do so, in this contribution we develop a framework in which co-owners' express their opinions on co-owned data security features, co-owners' relations with the targeted group are calculated. We represent the aggregation of the co-owners' choices on CIAPP features. With co-owners' choices and their relation values, the developed fuzzy system gives the final decision.

In the future work, we aim to extend the work with adding the trust values between users to show whose decision is more important than the others on decision making process. Then, see the effects of trust values on final group decision. And, also use the consensus reaching techniques to extend the work.

Acknowledgements. The authors would like to acknowledge the anonymous reviewers for providing their precious comments and suggestions. Also acknowledge is given to Turkish Education Embassy for their financial supports.

References

1. Scott, J.: Social Network Analysis: A Handbook, p. 210. Sage Publications, London (1991)
2. Liu, Y., Fan, Z.P., Zhang, X.: A method for large group decision-making based on evaluation information provided by participators from multiple groups. Inf. Fusion **29**, 132–141 (2016)
3. Herrera-Viedma, E., Cabrerizo, F.J., Chiclana, F., Wu, J., Cobo, M.J., Konstantin, S.: Consensus in group decision making and social networks (2017)
4. Akkuzu, G., Aziz, B., Adda, M.: Fuzzy logic decision based collaborative privacy management framework for online social networks. In: 3rd International Workshop on FORmal Methods for Security Engineering: ForSE 2019. SciTePress, January 2019
5. Dong, Y., et al.: Consensus reaching in social network group decision making: research paradigms and challenges. Knowl. Based Syst. **162**, 3–13 (2018)
6. Cook, W.D., Kress, M.: Ordinal ranking with intensity of preference. Manag. Sci. **31**(1), 26–32 (1985)
7. Hochbaum, D.S., Levin, A.: Methodologies and algorithms for group-rankings decision. Manag. Sci. **52**(9), 1394–1408 (2006)
8. Zadeh, L.A.: Is there a need for fuzzy logic? Inf. Sci. **178**(13), 2751–2779 (2008)
9. Urena, R., Chiclana, F., Melancon, G., Herrera-Viedma, E.: A social network based approach for consensus achievement in multiperson decision making. Inf. Fusion **47**, 72–87 (2019)
10. Liang, Q., Liao, X., Liu, J.: A social ties-based approach for group decision-making problems with incomplete additive preference relations. Knowl. Based Syst. **119**, 68–86 (2017)
11. Wu, J., Chiclana, F.: A social network analysis trust-consensus based approach to group decision-making problems with interval-valued fuzzy reciprocal preference relations. Knowl. Based Syst. **59**, 97–107 (2014)
12. Herrera-Viedma, E., Alonso, S., Chiclana, F., Herrera, F.: A consensus model for group decision making with incomplete fuzzy preference relations. IEEE Trans. Fuzzy Syst. **15**(5), 863–877 (2007)
13. Al-Samarraie, H., Teng, B.K., Alzahrani, A.I., Alalwan, N.: E-learning continuance satisfaction in higher education: a unified perspective from instructors and students. Stud. High. Educ. **43**(11), 2003–2019 (2018)
14. Ekin, T., Kocadagli, O., Bastian, N.D., Fulton, L.V., Griffin, P.M.: Fuzzy decision making in health systems: a resource allocation model. EURO J. Decis. Process. **4**(3–4), 245–267 (2016)

15. Thota, C., Sundarasekar, R., Manogaran, G., Varatharajan, R., Priyan, M.K.: Centralized fog computing security platform for IoT and cloud in healthcare system. In: Exploring the Convergence of Big Data and the Internet of Things, pp. 141–154. IGI Global (2018)

16. Capuano, N., Chiclana, F., Fujita, H., Herrera-Viedma, E., Loia, V.: Fuzzy group decision making with incomplete information guided by social influence. IEEE Trans. Fuzzy Syst. **26**(3), 1704–1718 (2018)

17. Martinez-Cruz, C., Porcel, C., Bernabé-Moreno, J., Herrera-Viedma, E.: A model to represent users trust in recommender systems using ontologies and fuzzy linguistic modeling. Inf. Sci. **311**, 102–118 (2015)

18. Thirumalai, C., Senthilkumar, M.: An assessment framework of intuitionistic fuzzy network for C2B decision making. In: 2017 4th International Conference on Electronics and Communication Systems (ICECS), pp. 164–167. IEEE, February 2017

19. Kahraman, C., Ruan, D., Dogan, I.: Fuzzy group decision-making for facility location selection. Inf. Sci. **157**, 135–153 (2003)

20. Samonas, S., Coss, D.: The CIA strikes back: redefining confidentiality, integrity and availability in security. J. Inf. Syst. Secur. **10**(3), 21–45 (2014)

21. Cherdantseva, Y., Hilton, J.: The Evolution of Information Security Goals from the 1960s to today, February 2012, Unpublished

22. Hu, H., Ahn, G.J., Jorgensen, J.: Detecting and resolving privacy conflicts for collaborative data sharing in online social networks. In: Proceedings of the 27th Annual Computer Security Applications Conference, pp. 103–112. ACM, December 2011

23. Petkos, G., Papadopoulos, S., Kompatsiaris, Y.: PScore: a framework for enhancing privacy awareness in online social networks. In: 2015 10th International Conference on Availability, Reliability and Security, pp. 592–600. IEEE, August 2015

24. Boyd, D.M., Ellison, N.B.: Social network sites: definition, history, and scholarship. J. Comput. Mediat. Commun. **13**(1), 210–230 (2007)

25. Scott, J.: Social Network Analysis, pp. 12–25. Sage, London (2017)

26. McAuley, J., Leskovec, J.: Learning to discover social circles in ego networks. In: NIPS (2012)

System Modeling by Representing Information Systems as Hypergraphs

Bence Sarkadi-Nagy$^{(\boxtimes)}$ and Bálint Molnár⬤

Eötvös Loránd University, Budapest, Hungary
`bence.sarkadi@outlook.com`

Abstract. Hypergraph as a formal model offers a sound foundation for representing information systems. There are several issues that are worth observing during analysis, design, and operation of information systems such as consistency, integrity, soundness of control and security mechanisms. The improvement and advancement of machine learning and data science algorithms provide the opportunity to spot patterns, to predict and to prescript some activities within complex environments that can depict huge sets of data. Our proposal is that the available algorithms can be applied on hypergraphs through profound customization whereby the capability of algorithms can be exploited for Business Information Systems.

Keywords: Information system · Set systems · Hypergraph theory · Hypergraph algorithms · Graph representation

1 Introduction

The proposal that we want to present is to use hypergraphs in information systems modeling to support various viewpoints and perspectives of Information Systems Architecture. We think that hypergraphs yield a unified and uniform framework to handle the complexity and variety of models employed during the analysis, design, and operation of information systems. From formal viewpoints, hypergraphs can give better insight into the architecture of information systems, and their structuring principles. In practice, this approach may provide clues for principles of designing and operating information systems; furthermore, it takes care of controlling and security mechanisms of information systems.

2 Mathematical Background

There are several conceptual formalizations that are mentioned in other papers [1,8] which can be described by a set of relationships from individual models (like UML-based class-diagram, work-flows, etc.).

© Springer Nature Switzerland AG 2019
W. Abramowicz and R. Corchuelo (Eds.): BIS 2019, LNBIP 354, pp. 86–95, 2019.
https://doi.org/10.1007/978-3-030-20482-2_8

2.1 Hypergraphs in Information Systems

The before-mentioned models are representing different facets of perception of IS, and they represent a complex system through a set of complex, heterogeneous relationships. This set of relationships can be described by directed hypergraphs; the directed hypergraph applies the same basic notions as the generalized hypergraphs with the extension of direction. In this set we can separate the elements in two subsets:

1. hierarchical
2. network-like relationships.

Hypergraphs can be used because of their versatility to represent complex data models, views and also their relationships [4,5]. A detailed description of the definitions and opportunities about the generalized hypergraphs and their usage as Architecture Describing Hypergraphs can be found in [3,8–10].

2.2 Verification and Validation of the Model for Information Systems

We have analyzed in previous papers the document-centric modeling of information systems on the basis of hypergraphs [2,10,12]. We follow the hypergraph definition by Bretto [3], and we use the homomorphism and morphism definition by Bretto and Voloshin [14].

The main goal of the research dedicated to modeling of information systems and architecture and to look for adequate algorithms is to aid in the model checking, verification and validation of information systems model in all "seasons" of the analysis and development [11]. One of the major issues is the consistency checking and consistency enforcement among the elements of the model and architecture. Coercing consistency can happen through compelling constraints on elements of representation. The question is how the requirement for constraint satisfaction can be formulated mathematically in hypergraphs. The advantage of hypergraphs can be grasped in their capabilities to describe heterogeneous finite structures, in our case various models and relationships by exploiting the flexibility of generalized hypergraphs. Thereby keeping up the consistency of the whole model means maintaining a sound set of relationships among models with heterogeneous structures. We can formulate this phenomenon mathematically that isomorphism between sub-hypergraphs is not sufficient for representing and enforcing constraints on various model elements and their relationships. However, there is a mathematical concept for mapping between mathematical structures that save the significant properties of the structures, and besides simplifies the investigation; this mathematical mapping is called homomorphism. Our following task is to define the notion of homomorphism in our Architecture Describing Hypergraph environment for model verification and validation [2].

As the Architecture Describing Hypergraph contains by definition several well-defined, and disparate sub-hypergraphs, we define the *homomorphism* between hypergraphs as it follows:

Definition 1. *Homomorphism between two simple hypergraphs is a map f from V_1 to V_2 for $\forall e_1 \in E_1 \implies f(e_1) = \{f(x)|x \in e_1\} \subseteq e_2 \in E_2$ where $H_1 = (V_1; E_1), H_2 = (V_2; E_2)$ two simple hypergraphs, V_j is the finite set of vertices, E_j are the set of hyperedges.*

$$E_j = \{e_i \subseteq \wp(V_j)|i \in I_j\} \tag{1}$$

where $\wp(V_j)$ is the power set of V_j.

Definition 2. *Architecture Describing Hypergraph is a **generalized** hypergraph that can be extended by some functions and operations:*

> *$label_{node} : V \rightarrow L_{node}$, where L is a set of labels, it is a vertex labeling function;*
> *$label_{edge} : E \rightarrow L_{edge}$, where L is a set of labels, it is a edge labeling function;*
> *$source_E : E \rightarrow V$, $target_E : E \rightarrow V$; these functions return the source and target vertices of an edge E in case of a directed hypergraph;*
> *$attr_V : Attr \rightarrow V$ attribute assignment function, $Attr = \{T_1, \ldots, T_n\}$ that consist of the attribute types;*
> *The finite set of domains is $DOMSET = \{D_1, \ldots, D_k\}$ that contains the domain of each single type, T_i, i.e. $dom(T_i) = D_i$ [13];*
> *$source_{attr} : Attr \rightarrow V$; The vertex that owns the attribute returned;*
> *$target_{attr} : Attr \rightarrow D$; The data values of attributes that are yielded, where $D \in DOMSET$ represents the set of data;*

D can be grasped again as vertices (efficiency of the representation is left out of the investigation) within the hypergraph and it can be interpreted as variables.

Over D as the data value of a set of variables, set of operations (OP) can be defined that can be used to describe constraints and rules within formulas.

Definition 3. *A directed hypergraph is an ordered pair*

$$\overrightarrow{H} = (V; \overrightarrow{E} = \{\overrightarrow{e_i}|i \in I\}) \tag{2}$$

Where V is a finite set of vertices and $\overrightarrow{E}$ is a set of hyperarcs with finite index set I. Every hyperarc $\overrightarrow{e_i}$ can be perceived as an ordered pair

$$\overrightarrow{e_i} = (\overrightarrow{e_i^+} = (e_i^+; i); \overrightarrow{e_i^-} = (e_i^-; i)) \tag{3}$$

Where $e_i^+ \subseteq V$ is the set of vertices of $\overrightarrow{e_i^+}$ and $e_i^- \subseteq V$ is the set of vertices $\overrightarrow{e_i^-}$. The elements of $\overrightarrow{e_i^+}$ (hyperarcs and/or vertices) are called tail of $\overrightarrow{e_i}$, while elements of $\overrightarrow{e_i^-}$ are called head [3]. We may use as shorthand notation for ordered pairs, e.g. a vertex and a directed hyperarc as ordered pair $\langle v_i, e_j \rangle$.

Definition 4. *We can formulate the definition of directed hypergraph the following way:*
$\overrightarrow{H} = (V; \overrightarrow{E} = \{\overrightarrow{e_i} | i \in I\}; tail, head)$ *is a four-tuple where* $tail, head : \overrightarrow{E} \to V$ *are two maps that give back the set of vertices in the* $e_i^+ = (e_i^+; i)$, *i.e. in the tail and* $\overrightarrow{e_i} = (e_i^-; i)$, *i.e. in the head, respectively.*

Definition 5. *Total hypergraph homomorphism is between two directed hypergraphs (for short dirhypergraphs):*
$\overrightarrow{H_1} = (V_1; \overrightarrow{E_1} = \{\overrightarrow{e_i})| i \in I_1\}; tail_1, head_1)$ *and* $\overrightarrow{H_2} = (V_2; \overrightarrow{E_2} = \{\overrightarrow{e_i})| i \in I_2\}; tail_2, head_2)$.
$f : \overrightarrow{H_1} \to \overrightarrow{H_2}$ *is a total hypergraph homomorphism,* $f|_V = f_V : V_1 \to V_2$ *and* $f|_E = f_E : E_1 \to E_2$; *then* $tail_2 \circ f_E = f_V \circ tail_1 : E_1 \to V_2$ *and* $head_2 \circ f_E = f_V \circ head_1 : E_1 \to V_2$, *i.e. the direction of hyperarcs and target set of vertices are saved.*

Definition 6. *Partial hypergraph homomorphism, f is between two dirhypergraphs:* $f : \overrightarrow{H_1} \to \overrightarrow{H_2}$, *if* $f|_{\overrightarrow{H_{1f}}} = f' : \overrightarrow{H_{1f}} \to \overrightarrow{H_2}$ *is a total hypergraph homomorphism, where* $\overrightarrow{H_{1f}} \subset \overrightarrow{H_1}$.

There are many alternative ways to formalize paths in directed hypergraphs [5]. Here we only use one of them, the so called *simple path*.

Definition 7. *A simple hyperpath, is a* $s = v_0, \overrightarrow{e_0}, v_1, \overrightarrow{e_1}, \ldots, \overrightarrow{e_k} v_{k+1} = t$ *sequence, where* $s \in tail(\overrightarrow{e_0})$, $t \in head(\overrightarrow{e_k})$ *and* $v_i \in head(\overrightarrow{e_{i-1}}) \cap tail(\overrightarrow{e_i}), j = 2, \ldots, k$.

3   Representing Information Systems by Hypergraphs

Information Systems modeling is a challenging task, thereby several modeling frameworks were proposed to address it. Enterprise Architecture frameworks offer deep insight into the inherently complex nature of it, such as the Zachman and TOGAF methods [4,7]. Blokdijk's *Four Models in Three Views* of information systems depicting approaches for analyses and design of information systems [6]. Modeling process should consider some reasonable guidelines so that the complexity of the whole activity can be handled: (1) graphical modeling languages provide the most interpretable description of applications in practical use; (2) the model instances should be transformed into refined model instances. The documents in the form of XML should be joined to some model instances that represent the significant business processes that use the documents as input and output for transformation. A unification of Zachman Enterprise Architecture, the axiomatic design, document object model (DOM) and UML as visual language give a solution that can be used for modeling IS [18–20] (Table 1).

Document-centric approaches give a natural way to transform information systems into hypergraphs. Both individual and complex elements of information systems can be represented by a vertex. This leads to the fact that those vertices

Table 1. Mapping the concepts of information systems onto the notion of hypergraph.

Information system (IS)	We create an abstraction of IS in the form of a generalized hypergraph that consists of vertices, hyperedges, hyperarcs
Vertex in a hypergraph	Each vertex corresponds to an element within an IS, e.g. documents, elements of documents (constituting a tree structure), data items, data collections (i.e. collection of data items), business processes, processes in scenarios and scenes, workflows, layers of workflows, web services, networks of web services, etc.
Arc in a hypergraph	Arc is a specific hyperarc with cardinality equal to two. Arc denotes binary relationships between two nodes, as e.g. free documents (with unbounded variables) is consumed by a certain business process (e.g. function/Web service), a generic document is the ancestor of an intensional documents, a free-document resulted in a ground-document after binding, after valuating of variables etc.
Hyperarc	A Hyperarcs depict some kind of containment relation let that be a Web services belonging to a specific workflow, business process containing workflows, nodes of a document tree structure composing a document etc.
System graph	A hypergraph that includes a disjoint node dedicated to the *modeling environment* of the system, plus all the nodes, hyperarcs along with the generalized hyperarcs of the IS
Sub-system	A subset of nodes and their incident hyperarcs, generalized hyperarcs. A vertex is an incident to a hyperarc if the hyperarc contains the vertex. A sub-system may be composed of documents, Web services, and related entities out of data model etc. These components are represented by vertices and their grouping into hyperarcs describes the relationships among them. Subsystem - that may be created by operations -, that conforms with graph theoretical concepts as: (1) Induced subhypergraph; (2) Subhypergraph; (3) Partial hypergraph
Interconnecting sub-systems	It is composed of hyperarcs of the generalized hypergraph. A graph consisting of all the vertices in sub-systems and all hyperarcs connecting together subsystems along with the *disjoint, environment node*

describe items on different levels of abstraction, but they each represent easily comprehensible concepts. Hyperarcs can then be applied to the model so that the model describes relations between sets of vertices.

The hypergraph representation provides some benefits such as:

1. using different levels of abstraction in the same model, and similarly handling whole sets of concepts during modeling, can aid the overall design process, the consistency check and human apprehension of the model

2. Hypergraph algorithms can be directly applied in the model. Specifically, homomorphism operators now consider the whole set relations
3. Several existing hyperpath definitions can be used for addressing different kinds of requirements (Here we only investigate the properties of simple hyperpaths).

4   Handling Hypergraphs Computationally

Storing and handling set systems in an efficient and meaningful manner imposes a lot of difficulties. One way to produce an easy-to-use, redundancy-free representation is to transform the hypergraph into a directed graph. This is achieved by the so-called Bipartite Incidence Structure (BPIS). This particular model uses the $G(A, B)$ bipartite graph to store the set relations of $H = (V_H, \overrightarrow{E_H})$, where $\forall v \in A : v \in V_H$, $|V| = |V_H|$, $\forall v \in B : v \in E_H$, $|E| = |E_H|$, and there is a directed arc $(u, v), u \in A, v \in B$ in G if and only if the corresponding $u \in V_H$ vertex is in the tail set of the corresponding $v \in E_H$ hyperarc, while the direction is in the other way around if it is in the head set.

The BPIS not only provides a container type for hypergraphs but infers useful generations of graph algorithms to the notion of simple hyperpaths. Although - to the best of our knowledge - the underlying graph data structures have not been used to create a consistently built up algorithmic framework. To achieve this, we used the arguably most prominent graph algorithms (BFS, DFS, Tarjan's connected components, and maximum flow algorithms) as a starting point and tried to generalize those for the hypergraph setting. Of course, it is paramount that the generalized version of an algorithm yields the same result on the original problem too (i.e. when the hypergraph is a graph).

Traversing a hypergraph is the same as running BFS or DFS on the BPIS and emitting every "hyperarc" vertex. Since these vertices are all in the B set of $G(A, B)$, we can run the traversal and discard every other item (when started from $v \in A$). The algorithms have the same properties as their graph counterparts because BPIS only changes every direct connection between hypergraph vertices to a two-long path (where the intermediate node is representing the hyperarc). Simply omitting this middle vertex, not from the result but the graph, and changing the path going through it to an arc yields a graph with the same traversal.

Tarjan's algorithm returns the strongly connected components in a graph. One can think of a strongly connected component of a graph as a subgraph in which there is a path from any node to any other. This definition can be used without any modifications on a hypergraph too. Running Tarjan's algorithm on the BPIS gives us exactly those components in which there is a directed simple hyperpath between any two vertices which is just what we wanted. The resulting components contain hyperarc vertices too, so we have to filter those out at the end. Since Tarjan's algorithm uses the DFS as a subroutine to label vertices, we can use those assigned the DFS labels and our observations about the distribution of hyperarc vertices during the DFS algorithm for the filtering.

The generalization is also correct, as changing an arc to a two long path does not change the reachability of any vertices.

For maximum flow algorithms, we started off with two possible generalizations. The only difference between them was that in one version the capacity of a hyperarc was bounding the flow between the set of tail and head, while in the other, it was only bounding the flow between any two vertices of the hyperarc. Since the second version can be - easily and without much overhead - modeled using the first, we decided to adopt that one. (Consider the hypergraph in which instead of a given directed hyperarc from vertex set A' to vertex set B' with a capacity of c there is an auxiliary hyperarc from each individual item of A' to each individual item of B' with the same capacity.) In the BPIS representation, it only requires to introduce another type of pseudo vertex a' which we insert between each hyperarc vertex a and its outgoing neighbors, i.e. for a given a and a': $E_{G(A,B)} = E_{G(A,B)} \cup (a, a') \cup \{(a', v) | (a, v) \in G(A, b)\} \setminus \{(a, v) | (a, v) \in G(A, b)\}$. Assigning the capacity of hyperarc a to the arc (a, a') while infinity to the incoming arcs of a and outgoing arcs of a'. In fact, since the cost of the (a, a') arc is an upper boundary (or a possible min cut for the subgraph), we can use the same capacity on any incoming/outgoing arcs too. One can see right away that this way we capped the flow on every hyperarc by its capacity which is exactly what we wanted to do. The same line of thoughts can be extended to the minimal circulation problem too.

One important remark is that for the maximum flow problem we generalized the whole problem setting, instead of an existing graph algorithm. Some of these algorithms require special prerequisites (such as acyclicity or the use of specific data structures) which we did not deal with. In spite of that, there are algorithms that do not require any further considerations, such as Edmonds-Karp or preflow-push [16], same for the minimal circulation [17].

There is one exception among the algorithms we wanted to generalize for directed hypergraphs. Up till now, we have looked at algorithms that have a corresponding graph algorithm too, in contrast, GYO reduction [15] is a process for undirected hypergraphs.

Input: undirectedHypergraph $H(V,E)$
Output: boolean $isAcyclic$
Result: Decides whether an undirected hypergraph is acyclic or not
for *vertex v exists that is only in one hyperarc* **do**
> Remove any such v from H (in one batch);
> Remove any hyperarc that is contianed in another;

end
if $|E_H| > 0$ **then**
> set $isAcyclic$ to *false*;

else
> set $isAcyclic$ to *true*;

end

Algorithm 1. GYO reduction for undirected hypergraphs

We can similarly detach sink and source vertices in directed hypergraphs too.

Input: directedHypergraph $H(V,A)$
Output: boolean *isAcyclic*
Result: Decides whether a directed hypergraph is acyclic or not
for *vertex v exists that only occurs in tail sets* **do**
   | Remove any such v from H (in one batch);
   | Remove any hyperarc that has an empty tail set;
end
if $|A_H| > 0$ **then**
   | set *isAcyclic* to *false*;
else
   | set *isAcyclic* to *true*;
end

Algorithm 2. GYO reduction for directed hypergraphs

One important side note is that while the original GYO provides a cycle in the end (if one exists), the directed GYO yields a cycle and all branches that are starting from that cycle. To fix this problem - when it is needed - after removing the tail sets we can also run the directed GYO for removing the head sets. Also - in most implementations - removing the vertices that are not in any hyperarcs (or tail/head sets) can speed up the computation.

It is clear that the given algorithm tries to create a topological ordering of all vertices which is a possible generalization for acyclicity and it is in accordance with the undirected case.

Based on this observation we conclude that the generalized GYO does not have any advantage over the topological ordering of the BPIS, at least not for the simple path definition.

5 Optimizing for Resource Restrictions on Hypergraphs

The listed algorithms are generalized in a manner that makes it possible to use them as powerful modeling and optimization tools for the used IS model. Since the model does not depend on n-ary relations, the proposed - relatively simple - algorithms can meaningfully operate in the conceptually easy-to-handle hypergraph setting, once the necessary problem formulations are in order.

Cycles in the hypergraphs (equivalently in the BPIS) are circular dependencies in the document model. This can either be a modeling error or an episodic process. For example let us think about an ID renewal process, during which we use our old ID card or alternatively about Scrum sprint. These circular dependencies turn up as strongly connected components (i.e. cycles) in the BPIS. It is important to note that episodic processes have a specific starting point which does not explicitly appear in the model. Topological ordering answers whether there are any such cycles in the hypergraph, while Tarjan's can identify the individual episodes too, with little overhead. When the overhead is negligible or we want to run Tarjan's way we can check the number of strongly connected components as a means to check full acyclicity. If they are equal to the number of vertices (and we did not allow loops) then the hypergraph is acyclic. During modeling the strongly connected components have to be handled with special care to incorporate any episodic starting point in the model too. Using the episodic starting points we can unroll any such cycle in sufficient depth, creating a fully acyclic hypergraph model.

Because hyperarcs express dependencies, acyclic hypergraphs can be used for planning with PERT models. Any subhypergraph i.e. subsystem can have associated resources such as time, money or other implications. Using the PERT method we can find the cumulative need for a specific resource and find the critical paths on which it can be reduced. To use PERT we need a topological ordering on BPIS and the acyclicity of the graph as a prerequisite, both which have been ensured earlier.

Network flow and circulation algorithms can be used for solving a great many problems, here we only venture to describe one such possibility. Circulations can be used to explore both the sufficiency and the excess of specific regenerative resources (i.e. resources that do not deplete when used but are finite or infinite in quantity). Taking a subsystem and using the necessary resource consumption of specific processes both as their upper and lower boundary we can identify where the excess of resources occur. If the circulation is infeasible, then we do not have sufficient resources, whilst the difference of available resources to the value of the circulation is the excess we can freely reallocate.

6   Conclusion

While the hypergraph representation for information systems make it easier for the modeling process, the BPIS representation of hypergraphs provides a natural way to generalize graph algorithms to (directed) hypergraphs when we use simple path definitions. Using these algorithms on existing information system models utilizing hypergraphs, several useful structural information can be extracted. These pieces of information then can be used for both transforming the underlying model to adhere to stricter structural constraints and to support resource allocations of enterprises.

Acknowledgements. The project has been supported by the European Union, co-financed by the European Social Fund (EFOP-3.6.3-VEKOP-16-2017-00002).

References

1. Molnár, B.: Applications of hypergraphs in informatics: a survey and opportunities for research. Ann. Univ. Sci. Budapest. Sect. Comput. **42**, 261–282 (2014)
2. Molnár, B., Benczúr, A.: Facet of modeling web information systems from a document-centric view. Int. J. Web Portals (IJWP) **5**(4), 57–70 (2013). https://doi.org/10.4018/ijwp.2013100105

3. Bretto, A.: Hypergraph Theory: An Introduction. Springer, Heidelberg (2013). https://doi.org/10.1007/978-3-319-00080-0
4. Zachman, J.A.: A framework for information systems architecture. IBM Syst. J. **26**(3), 276–292 (1987)
5. Gallo, G., Longo, G., Pallottino, S., Nguyen, S.: Directed hypergraphs and applications. Discrete Appl. Math. **42**(2), 177–201 (1993)
6. Blokdijk, A., Blokdijk, P.: Planning and Design of Information Systems. Academic Press, London (1987)
7. Open Group: TOGAF. The Open Group Architecture Framework, TOGAF® Version 9 (2010). http://www.opengroup.org/togaf/
8. Ausiello, G., Franciosa, P.G., Frigioni, D.: Directed hypergraphs: problems, algorithmic results, and a novel decremental approach. In: Theoretical Computer Science. LNCS, vol. 2202, pp. 312–328. Springer, Heidelberg (2001). https://doi.org/10.1007/3-540-45446-2_20
9. Iordanov, B.: HyperGraphDB: a generalized graph database. In: Shen, H.T., et al. (eds.) WAIM 2010. LNCS, vol. 6185, pp. 25–36. Springer, Heidelberg (2010). https://doi.org/10.1007/978-3-642-16720-1_3
10. Molnár, B., Benczúr, A., Béleczki, A.: Formal approach to modelling of modern information systems. Int. J. Inf. Syst. Proj. Manag. (2016). http://www.sciencesphere.org/ijispm/archive/ijispm-040404.pdf
11. Bell, M.: Service-Oriented Modeling (SOA): Service Analysis, Design, and Architecture. Wiley, Hoboken (2008)
12. Molnár, B., Benczúr, A.: Modeling information systems from the viewpoint of active documents. Vietnam J. Comput. Sci. **2**(4), 229–241 (2015)
13. Mitchell, J.C.: Type systems for programming languages. In: Formal Models and Semantics, pp. 365–458 (1990)
14. Voloshin, V.I.: Introduction to Graph and Hypergraph Theory. Nova Science Publ., New York (2009)
15. Goodman, N., Shmueli, O., Tay, Y.C.: GYO reductions, canonical connections, tree and cyclic schemas and tree projections. In: Proceedings of the 2nd ACM SIGACT-SIGMOD Symposium on Principles of Database Systems (PODS 1983), pp. 267–278. ACM, New York (1983). https://doi.org/10.1145/588058.588089
16. Goldberg, A.V., Tarjan, R.E.: A new approach to the maximum-flow problem. J. ACM **35**(4), 921–940 (1988). https://doi.org/10.1145/48014.61051
17. Tardos, É.: A strongly polynomial minimum cost circulation algorithm. Combinatorica **5**, 247 (1985). https://doi.org/10.1007/BF02579369
18. Marini, J.: Document Object Model. McGraw-Hill Inc., New York (2002)
19. Molnár, B., Benczúr, A., Tarcsi, Á.: Formal approach to a web information system based on story algebra. Singidunum J. Appl. Sci. **9**, 63–73 (2012)
20. Molnár, B., Tarcsi, Á.: Design and architectural issues of contemporary web-based information systems. Mediterranean J. Comput. Netw. **9**, 20–28 (2013)

Development of a Social Media Maturity Model for Logistics Service Providers

Axel Jacob[1(✉)] and Frank Teuteberg[2]

[1] Osnabrück University of Applied Sciences,
Caprivistr. 30a, 49076 Osnabrück, Germany
a.jacob@hs-osnabrueck.de
[2] Osnabrück University, Katharinenstr. 1, 49074 Osnabrück, Germany
frank.teuteberg@uni-osnabrueck.de

Abstract. Logistics service providers (LSPs) conduct their business in an environment of steadily changing stakeholders and business models. Social media (SM) has become an important communication tool and source for new business models for LSPs. Nevertheless, a lot of LSPs struggle with the utilization of SM. In this paper, we develop an SM maturity model (MM) for LSPs. By doing so, our research sheds light on the use of SM at LSPs and reveals impediments. Thus, the developed MM will help researchers better understand the utilization of SM at LSPs and practitioners to improve their business processes.

Keywords: Social media · Web 2.0 ·
Logistics service providers · Organizational adoption · Maturity model

1 Introduction

The digitization of society and the economy is progressing steadily [1]. The logistics service industry is one of the most affected sectors. Although it has always been influenced by new technological developments and a competition-induced need for cost-effectiveness [2, 3], new internet-based business concepts such as crowd logistics [4], blockchain technology [5] or Uber-inspired platforms for decentralized logistics [6] seem to affect the industry disruptively. In order to face these developments, LSPs seek to enforce existing cooperations [2, 3] and include crowd concepts themselves. SM seems to be a promising technology to achieve these goals. However, there is a lack of understanding with regards to how LSPs apply SM in a way that it strategically benefits the organization.

Therefore, this paper aims to address this lack of research pertinent to academics and practitioners, by developing an SM MM for LSPs that helps to improve the current use of SM at LSPs. Furthermore, the MM may be used as a typology for classifying organizations based on the potential of their use of SM.

By doing so, we also respond to calls from Aral et al. [7] to research on (a) how companies should organize, govern, fund, and evolve their SM capabilities; (b) which skill and culture changes are needed to best adapt to a social world; (c) which skills, talent, or human resources companies should develop; and (d) how companies should create incentives to guide SM activities.

W. Abramowicz and R. Corchuelo (Eds.): BIS 2019, LNBIP 354, pp. 96–108, 2019.
https://doi.org/10.1007/978-3-030-20482-2_9

The remainder of the paper is structured as follows. With the help of a literature review, Sect. 2 lays the theoretical foundation for assessing SM maturity and understanding the importance of SM, especially for LSPs. Subsequently, in Sect. 3, the research methodology and the data sample are explained. The findings of our study are presented in Sect. 4. These are discussed and summarized in propositions in Sect. 5. In the final section, we summarize the contributions of our study and address implications for further research.

2 Theoretical Background

2.1 Social Media in Logistics Services

Following Kaplan and Haenlein [8], "social media is a group of Internet-based applications" such as collaborative projects, blogs, content communities, social networking sites, virtual game worlds, and virtual social worlds that allow the creation and exchange of user-generated content and that are built on the foundations of web 2.0, which is described as a participatory and collaborative use of the internet and a continuously modification of content and applications by all users. SM applications are used in different fields, such as the tourism industry, medicine, and public relations and can be classified by their purposes: marketing, managing customer relationship, knowledge sharing, collaborative activities, organizational communication, education and learning [9]. In this paper, we follow the aforementioned definition of SM for explaining how LSPs make use of SM.

LSPs perform all or part of the logistics functions of a client, such as managing and operating transportation and warehousing, inventory management, information-related services (e.g. tracking and tracing), value-added services (e.g. pre-assembly), and order processing [10]. SM can provide advantages for LSPs in two different ways. First, as a source for big data analytics and second, as a communication tool [11]. Furthermore, it can be applied in the B2C context as well as in the B2B context [12]. Gunasekaran et al. [13] mention for example SM-based big data analytics as a future possibility to determine critical success factors for greening supply chains. Daugherty et al. [11] as well as Bhattacharjya et al. [14] describe SM as an important communication tool for LSPs that helps to create a positive consumer experience in online retail. Likewise, Neaga et al. [15] provide a framework for the analysis of SM data in order to align logistics services. Daugherty et al. [11] also explain that SM can enforce the communication between LSPs and shippers and by doing so, reach a deeper understanding of each other's processes, which may lead to an improvement of the services. Ding et al. [16] specify this idea and provide a framework for an SM-based system for the coordination of an integrated production and transportation. Another significant influencing factor of SM on the business models of LSPs derives from the linkage of SM and crowd logistics. Crowd logistics is considered to be of high importance for LSPs [4, 17]. SM capabilities are an important antecedent for the integration of people into crowd logistics concepts [18]. Thus, the ability to create, improve, and develop SM activities might be a key capability for LSPs in the near future.

However, the aforementioned examples are more conceptual than empirical and the practical use of SM at LSPs is in general rather low [12] and sometimes less successful [14]. Therefore, it is helpful for researchers and practitioners to understand how LSPs successfully apply SM. The development of an SM MM is reasonable in this case, since no comparable studies on the problems described have been conducted so far. An SM MM can help academics and practitioners in different ways. First, it can be used to classify companies regarding its SM capabilities. This is relevant for academics who seek to understand how companies deal with SM, but it is also important for practitioners in order to rank their own companies as well as competitors regarding their SM capabilities. Second, in the case that existing SM capabilities are perceived to be insufficient, especially practitioners can use SM MM to evaluate, improve, and benchmark existing SM capabilities. Third, an SM MM can help to derive courses of action for the development of the capabilities. Thus, referring to the aforementioned shortcomings with regards to the current use of SM at LSPs and the upcoming challenges, especially in the case of crowd logistics, an SM MM for LSPs will be very beneficial for researchers and practitioners.

2.2   Social Media Maturity

MMs gained a lot of attention in information systems research and practice [19]. With the help of MMs, the development of organizational capabilities are described as a typical sequence of stages that depict a predictable, desired, or logical development path from an initial to a desired level of maturity [20]. Thus, MMs can be used for the following purposes: (a) the classification of the current capabilities, (b) the evaluation, continuous improvement, and benchmark of the capabilities, and (c) as a roadmap for the development of the capabilities [21]. Wang et al. [22] provide the most recent overview of the literature on SM capability to our knowledge. Following them, organizational SM capabilities can generally be categorized in four dynamical levels (technological, operational, managed and strategic).

To our knowledge, there is no MM in the field of logistics and LSPs that focuses on SM. However, a specific MM is necessary because of the versatile character of the LSPs industry. LSPs offer widely varying services to a high number of very different and changing customers and compete with a high number of changing competitors [3]. The integration into existing (information) systems and the alignment with other parties in the supply chain are inevitable, despite the organizational boundaries between the different parties of such a supply chain. Fast and direct communication is a key competence in such an environment and SM can act as a boundary spanner in such inter-organizational networks [23]. Other existing SM MMs (cf. Sect. 4.1) do not capture this specific nature of the LSPs industry. Thus, the development of a specific SM MM for LSPs is reasonable. In view of the aforementioned importance of SM in the logistics service industry, such an SM MM overcomes a research deficit and provides a comprehensive description of the development stages of LSPs with regard to their SM capabilities.

3   Research Process

3.1   Methodological Approach

In order to be able to understand the SM capabilities of LSPs, we developed an MM as proposed by de Bruin et al. [24] and Becker et al. [25]. For the development of our MM, we first carried out a structured literature review [26] and then conducted a qualitative-explorative study in which the fields of application of SM technologies at German LSPs were analyzed by means of expert interviews. A qualitative-explorative approach is appropriate in this case, since no comparable studies on the problem described have been available so far and qualitative-explorative studies are therefore particularly suitable [27].

3.2   Sample Description

The interviewees are experts who hold managerial positions at German LSPs and have expertise in operational logistics activities. The respondents were named by the companies themselves after they had been randomly selected from a ranking of German LSPs [28]. Of a total of 17 companies contacted, seven agreed to participate. The interviewees were informed about the intentions of the study before the start of the interviews and the interviews were conducted on the basis of a guideline which was made available to the interviewees one week before the interview. The interviews took place in July 2017. They were recorded on tape and later transcribed. Within the scope of a qualitative content analysis, the statements of the interviewees were manually paraphrased and coded [29]. Table 1 gives an overview of the interviewed experts and the companies.

Table 1. Descriptive details on the seven semi-structured expert interviews.

No.	Role of the interviewee	Duration of employment in years	Employees
1	Factory Manager	7	7,500
2	Commercial Director	2	680
3	General Manager	2	11,000
4	Head of Key Account Management	2	11,000
5	Key Account Manager	2.5	12,500
6	Head of Marketing	2.5	800
7	Head of Online Marketing	4	16,000

3.3   Development of the Maturity Model

The analysis of the related MMs substantiates the need for a specific MM [25]. Following de Bruin et al. [24] the nature of MMs can be (a) descriptive, (b) prescriptive or (c) comparative. Descriptive MMs capture the current state of the art of a domain, prescriptive MMs also correlate the domain to business performance and indicate possible improvements, and comparative MMs enable benchmarking across industries

[24]. In this paper, we develop a prescriptive MM that explains the current state of SM maturity at LSPs and enables the development of a roadmap for improvement.

Furthermore, de Bruin et al. [24] propose six phases for the development of an MM: (a) scope, (b) design, (c) populate, (d) test, (e) deploy, and (f) maintain. Although the order of the phases is important, the progression through some phases may be iterative [25]. The scope of our MM focuses on the SM capabilities of LSPs, based on reviewed literature and the experience of practitioners. Within the design phase, the audience and the main architecture of the MM are determined. The audience of our MM are academics and practitioners. A deeper understanding of the development stages of SM capabilities will help them to advance SM-based business processes and business models as well as to improve business performance. In order to keep our MM robust, comprehensive, and explanatory, we interviewed specialized managers at seven German LSPs and defined five development stages in a top-down procedure [24]. In the populate phase we then identified attributes for each development stage. Afterwards, we tested the drafted MM for relevance and rigor [24] and refined it iteratively [25]. The phases of deployment and maintenance are not completed with this paper.

4 Findings

4.1 Levels of Social Media Maturity for Logistics Service Providers

Following Webster and Watson [26], we first identified the relevant literature in leading journals and recognized conferences in the categories business administration; logistics; service management; technology, innovation & entrepreneurship and business information systems according to VHB-JOURQUAL3 ranking [30]. In the second step, we conducted a search in the reference section of studies identified, in order to find further relevant studies. Finally, we conducted a forward search based on the studies identified in the previous two steps.

As there are numerous MMs for SM maturity in general but none in the field of logistics services, we confined to the most relevant and most cited MMs and identified four studies that focus on SM maturity in business literature. Jussila et al. [31] were the first who provided a preliminary MM for SM adoption. This model basically consists of two defined maturity levels. Besides, three intermediate levels, which are not described in detail, indicate that maturity does not emerge in a dichotomy. Duane and O'Reilly [32] offer a conceptual stages-of-growth model for managing an organization's SM business profile, which was empirically validated by Chung et al. [33]. The MM of Lehmkuhl [34] consists of five maturity stages that are defined by the dimensions strategy, processes, IT systems, culture, and governance. Geyer and Krumay [35] propose that SM maturity is in a wider sense determined by the demographics of an organization, the organizational readiness and the SM maturity as such, which consists of six specific dimensions. Table 2 gives an overview of related MMs.

All described MMs consist of five or six levels, which define the maturity stage. The bottom levels stand (a) for no SM activities or (b) for an initial state that can be characterized by an organization having little SM capabilities. In contrast, the highest levels represent a conception of full maturity. The levels between the two extremes

Table 2. Overview of related maturity models.

Authors	Year	Levels/dimensions
Jussila et al. [31]	2011	1. Initial (non-existing/ad-hoc)
		2.-4. [not defined]
		5. Fully mature
Duane and O'Reilly [32]	2012	1. Experimentation and learning
		2. Rapid growth
		3. Formalization
		4. Consolidation and integration
		5. Institutional absorption
Lehmkuhl et al. [34]	2013	0. No degree of maturity
		1. Low degree of maturity
		2. Rather low degree of maturity
		3. Medium degree of maturity
		4. Rather high degree of maturity
		5. High degree of maturity
Geyer and Krumay [35]	2015	1. Demographics
		2. Organizational Readiness
		3. Maturity
		3.1. Operational social media maturity
		3.2. Human resource management
		3.3. Social listening and monitoring
		3.4. Social media integration
		3.5. Social media strategy
		3.6. Guidelines for responsible behavior

involve a continuous progression regarding an organization's SM capabilities [25]. We used these theoretical insights and combined them with an initial desk research on the companies of the interviewees in the sense of a goal-oriented procedure [36]. Consequently, we decided to follow this pattern and defined five stages of maturity: 1. Beginner; 2. Explorer; 3. Utilizer; 4. Enabler; 5. Forerunner.

4.2 Dimensions of a Social Media Maturity Model for Logistics Service Providers

The different levels represent evolutionary development stages. Each level consists of a set of attributes of different dimensions that are characteristic but not mandatory for a company, which achieved a certain level. LSPs can use this set of attributes to (a) define their current maturity level; (b) evaluate, continuously improve, and benchmark their capabilities; and (c) identify the attributes that must be met to reach a new maturity level. In the following, we present a brief description of the five maturity-influencing dimensions and their attributes, which we evaluated descriptively in our MM.

Corporate Culture. In general, the corporate culture plays an important role in the digital maturity of a company [37]. With regards to the use of SM, the main aspects of corporate culture are (a) the attitude towards change; (b) the acceptance by employees and decision-makers; and (c) a responsible use of the tools. A positive attitude towards change is fundamental for the development and thus for the sustainable growth of any organization [38]. Likewise, the acceptance by employees and decision-makers is crucial for the successful adoption of any new technology [39]. The characteristic "responsible use of the tools" describes the assumption to what extent SM distract the employees and interfere the work performance.

Know-How. The ability to choose the right SM and to apply it in a fruitful way requires a distinct know-how [8]. Likewise, a procedural know-how is necessary [33]. In order to evaluate the ability of a company with regards to its SM know-how, we distinguish between (a) fragmentary; (b) incomplete; (c) purposeful; (d) broad; and (e) comprehensive know-how.

Resource Allocation. This dimension refers to the degree of the assignation of personnel and capital. The allocation of resources is substantive for the adoption and further use of any technology [40]. To our mind, companies may (a) assign no resources ("none"); (b) assign randomly occurring slack resources; (c) extend existing resources; (d) specialize existing resources; and (e) provide new resources.

Integration into Systems and Processes. SM can influence the business processes of a company in different ways [33]. In order to achieve a positive influence, SM has to be integrated into the system of value creation and the related processes [37]. We use the attributes (a) ad hoc; (b) additional; (c) aligned; (d) partially integrated; and (e) fully integrated to describe the intensity of the integration. "Ad hoc" means that SM is not integrated and only used in very few instances. If SM is not integrated but used repeatedly in addition to the normal business processes, we call it "additional". The attribute "aligned" represents a frequent use of SM in order to support business processes. If specific tasks of a business process are designed to be fulfilled by the use of SM, we use the term "partially integrated". Consequently, business processes that are based on SM refer to the attribute "fully integrated".

Scope. This dimension describes the use of SM by the participants involved. The participants may be internal or external stakeholders of a company [33]. However, this differentiation does not influence our MM, as the structure of the communication partners is more significant, whether they are internal or external. During our research, we found evidence that the use of SM may (a) be restricted to certain individuals ("isolated"); (b) take place in a group with changing members; (c) be an accepted communication tool in a formal unit (e.g. department or company); (d) be used as an accepted communication tool in a company and its environment; and (e) be used as an accepted communication tool in supply chains and its environments.

Figure 1 shows our final MM with its levels and dimensions.

		Levels				
		Beginner	Explorer	Utilizer	Enabler	Forerunner
Dimensions	Corporate culture	Reluctant, reactive	Hesitant, cautious	Active, open, integrative	Cooperative, promoting	Challenging, proactive, innovative
	Know-how	Fragmentary	Incomplete	Purposeful	Broad	Comprehensive
	Resource allocation	None	Use slack resources	Extend resources	Specialize resources	Create resoures
	Integration into systems and processes	Ad hoc	Additional	Aligned	Partially integrated	Fully integrated
	Scope	Isolated	Group	Department, company	Company and its environment	Supply chain and its environment

Fig. 1. Social media maturity model for LSPs.

5 Discussion

5.1 Social Media Maturity of Logistics Service Providers

Our MM gives an overview of the current state of art regarding the use of SM at LSPs. In the following, we will first discuss the MM itself, and then address the stages of maturity. For the development of our prescriptive MM, we have applied the process proposed by de Bruin et al. [24] and Becker et al. [25]. During the design and iterative populate phases we relied on a systematic literature review [26] and qualitative expert interviews. Thus, we consider our MM to be robust, comprehensive, and explanatory. Nevertheless, it could be even more robust and rigorous, if we had applied more elaborated tools for the determination of the maturity levels and the dimensions with its attributes, e.g. a set theoretical approach [41]. However, we still consider our MM to be meaningful as it is to our knowledge the first study that sheds light on the SM maturity at LSPs, which cannot be described with already existing MMs.

The first maturity level ("beginner") describes a LSP in its typical initial state. The corporate culture of these LSPs in the context of digitization can be described as rather reluctant. The handling of social web tools and the associated changes is characterized by mistrust. Behavior of the LSP is reactive, i.e. changes are triggered by external requirements. Accordingly, the know-how in the area of SM is fragmentary and no resources are allocated for the development of SM capabilities. If SM is already in use, it will be ad hoc between certain users in addition to the normal business activities and therefore not embedded in systems or processes.

The second maturity level describes explorers. This means that LSPs discover the benefits of SM. At this stage of maturity, the culture of the company begins with a gradual opening towards SM and companies are rather in an experimental phase. This means that LSPs are approaching the topic, but are still hesitant in their behavior. The interest in SM is growing and consequently the LSPs begin to use slack resources, e.g. by participating in workshops for the acquisition of know-how, which is still incomplete. The use of SM takes place in varying groups parallel to existing processes.

The corporate culture of utilizers is much more open-minded and thus enables them to act actively, generate new ideas and integrate them into the operational environment. The interest in SM is pronounced and results in an increasing use of resources for the acquisition of the required know-how and the purposeful application of SM. The LSPs begin to use SM to support processes efficiently. The processes may concern different departments in the company. Resulting feedback is used and processed as part of an improvement process.

At the enabler phase, the implementation of SM in processes has already taken place and the focus is on its continuous further development. The corporate culture of the LSP is very open and motivates the employees to generate new application possibilities of SM. The know-how of the LSP with regard to SM is thus continuously increasing. Through the partial integration of SM into processes that may also concern customers and partners, they gain in efficiency and can generate cost advantages.

As a forerunner, the company has a proactive, innovative corporate culture. The company is seeking for new uses of SM. This challenging attitude can be found at all levels of the company and depicts the curiosity and interest of employees in the continuous development of professional SM usage in the company. The LSP has comprehensive know-how about SM and continue to invest in the development of expertise. SM is fully integrated into the processes and in some cases are even the core element of the business processes. The use of SM refers to the company, but also to the interactions of the LSP within supply chains and their environments.

5.2   Fields of Application of Social Media at Logistics Service Providers

During our research, we identified a gap between the fields of application described in the literature and the practical use of SM. In practice, the internal applications include knowledge management or e-learning and the handling of administrative processes such as personnel deployment planning using web applications. This kind of usage goes along with the literature [9]. However, we could not find cases where SM is used for formal internal communication, as it is proposed by the research of others [12]. Nevertheless, one expert reported that some informal groups exist in SM.

In the area of external applications, SM is frequently used for public relations, networking and recruiting, as described in the literature [35]. Data exchange with customers and partners takes place in practice primarily via standardized interfaces (usually via EDI). Only in a few cases, process-relevant information is exchanged via SM, as described by Daugherty et al. [11], for example. Likewise, the external application of SM includes the operation of collaborative wikis with customers in several cases. These insights lead us to the following propositions:

Proposition 1: The internal and external use of SM in support activities is common among LSPs.

Proposition 2: In Practice, only very few LSPs make use of SM in primary activities, as proposed in theory.

In contrast to the halting development regarding the application of SM, LSPs attribute a high beneficial effect to SM. Reasons for this contradiction can be found in different impediments. First, necessary investments and resources as well as the acquisition of knowledge represent obstacles. Following Culnan et al. [42], we argue that these inhibiting factors can be overcome with moderate efforts. Thus, our third proposition is:

Proposition 3: The resource requirements in respect of knowledgeable labor and finance induced by SM is a minor inhibitor for LSPs.

Second, a perceived lack of data security and privacy hinders the application in primary processes. Interestingly, some interviewees also stated that they fear a too transparent company, which reveals inefficiencies or problems and thus has a negative impact on customers and partners. Following Ngai et al. [9], the attitude towards data security and privacy are part of the corporate culture. In most LSPs the corporate culture seems to be hesitant or even reluctant. We base this assumption on statements of most of the interviewees who made the point that the rather high average age of employees in the companies leads to a low acceptance and a slower adoption of new technologies. Furthermore, they fear a non-professional use of SM, which reduces productivity and thus leads to inefficiencies. Referring to Ngai et al. [9] and Venkatesh [39], we draw our fourth proposition:

Proposition 4: A hesitant corporate culture and leadership are major impediments for beneficiary use of SM at LSPs.

Finally, with regards to the aforementioned tremendous future influence of SM on LSPs [11, 12] we come to the last proposition:

Proposition 5: Most LSPs are inappropriately prepared for the arising challenges due to internet-based business concepts.

6   Conclusion

In our study, we examined the contemporary phenomenon of SM maturity in the real-life context of German LSPs. We developed an MM with five levels and five dimensions, which characterize the different levels. By doing so, we shed light on how companies organize, fund, and evolve their SM capabilities, which culture changes are needed to make use of SM [7]. Furthermore, we revealed a need for a higher engagement of LSPs in SM.

However, our study as presented in this paper has certain limitations. Our sample is rather small and confined to German LSPs. As the use of SM varies in different cultures [9], additional research is necessary to prove our findings in other cultural settings. With regards to the proposed development process of de Bruin et al. [24], we were only able to conduct the phases "scope", "design", "populate", and "test". The phases "deployment" and "maintenance" were not part of this study and have to be addressed in future. Furthermore, we did not evaluate the levels and the dimensions of our MM empirically with the help of specific measurements, which would improve the value of our MM [25, 41]. This matter should also be addressed in future. Further research could examine the potential of SM in depth for new business models of LSPs, e.g. in the context of crowd logistics. Future research could also investigate which types of SM are most beneficial for LSPs and if the use of SM is related to the size of an LSP.

References

1. Legner, C., et al.: Digitalization: opportunity and challenge for the business and information systems engineering community. Bus. Inf. Syst. Eng. **59**, 301–308 (2017)
2. Wang, X., Persson, G., Huemer, L.: Logistics service providers and value creation through collaboration: a case study. Long Range Plan. **49**, 117–128 (2016)
3. König, C., Caldwell, N.D., Ghadge, A.: Service provider boundaries in competitive markets: the case of the logistics industry. Int. J. Prod. Res. **11**, 1–16 (2018)
4. Carbone, V., Rouquet, A., Roussat, C.: The rise of crowd logistics: a new way to co-create logistics value. J. Bus. Logist. **22**, 3 (2017)
5. Korpela, K., Hallikas, J., Dahlberg, T.: Digital supply chain transformation toward blockchain integration. In: Hawaii International Conference on System Sciences (2017)
6. Gallay, O., Korpela, K., Tapio, N., Nurminen, J.K.: A peer-to-peer platform for decentralized logistics. epubli (2017)
7. Aral, S., Dellarocas, C., Godes, D.: Introduction to the special issue—social media and business transformation: a framework for research. Inf. Syst. Res. **24**, 3–13 (2013)
8. Kaplan, A.M., Haenlein, M.: Users of the world, unite! The challenges and opportunities of Social Media. Bus. Horiz. **53**, 59–68 (2010)
9. Ngai, E.W.T., Tao, S.S.C., Moon, K.K.L.: Social media research: theories, constructs, and conceptual frameworks. Int. J. Inf. Manag. **35**, 33–44 (2015)
10. Lai, K.-h.: Service capability and performance of logistics service providers. Transp. Res. Part E Logist. Transp. Rev. **40**, 385–399 (2004)
11. Daugherty, P.J., Bolumole, Y., Grawe, S.J.: The new age of customer impatience. Int. J. Phys. Dist. Log Manage. **37**, 44 (2018)
12. Kucukaltan, B., Irani, Z., Aktas, E.: A decision support model for identification and prioritization of key performance indicators in the logistics industry. Comput. Hum. Behav. **65**, 346–358 (2016)
13. Gunasekaran, A., Subramanian, N., Rahman, S.: Green supply chain collaboration and incentives: current trends and future directions. Transp. Res. Part E Logist. Transp. Rev. **74**, 1–10 (2015)
14. Bhattacharjya, J., Ellison, A., Tripathi, S.: An exploration of logistics-related customer service provision on Twitter. Int. J. Phys. Dist. Log Manage. **46**, 659–680 (2016)
15. Neaga, I., Liu, S., Xu, L., Chen, H., Hao, Y.: Cloud enabled big data business platform for logistics services: a research and development agenda. In: Delibašić, B., et al. (eds.) ICDSST 2015. LNBIP, vol. 216, pp. 22–33. Springer, Cham (2015). https://doi.org/10.1007/978-3-319-18533-0_3
16. Ding, K., Jiang, P., Su, S.: RFID-enabled social manufacturing system for inter-enterprise monitoring and dispatching of integrated production and transportation tasks. Robot. Comput. Integr. Manuf. **49**, 120–133 (2018)
17. Akeb, H., Moncef, B., Durand, B.: Building a collaborative solution in dense urban city settings to enhance parcel delivery: an effective crowd model in Paris. Transp. Res. Part E Logist. Transp. Rev. **119**, 223–233 (2018)

18. Le, T.V., Ukkusuri, S.V.: Modeling the willingness to work as crowd-shippers and travel time tolerance in emerging logistics services. Travel. Behav. Soc. **15**, 123–132 (2019)
19. Wendler, R.: The maturity of maturity model research: a systematic mapping study. Inf. Softw. Technol. **54**, 1317–1339 (2012)
20. Fraser, P., Moultrie, J., Gregory, M. (eds.) The use of maturity models/grids as a tool in assessing product development capability. In: IEEE International Engineering Management Conference (2002)
21. Patas, J.: Towards maturity models as methods to manage IT for business value - a resource-based view foundation Washington, USA, August 9–11, 2012. In: 18th Americas Conference on Information Systems, AMCIS 2012, Seattle, Washington, USA, 9–11 August 2012. Association for Information Systems (2012)
22. Wang, Y., Rod, M., Ji, S., Deng, Q.: Social media capability in B2B marketing: toward a definition and a research model. J. Bus. Indus. Market. **32**, 1125–1135 (2017)
23. Galaskiewicz, J.: Studying supply chains from a social network perspective. J. Supply Chain. Manag. **47**, 4–8 (2011)
24. de Bruin, T., Freeze, R., Kulkarni, U., Rosemann, M.: Understanding the main phases of developing a maturity assessment model. In: Australasian Conference on Information Systems (2005)
25. Becker, J., Knackstedt, R., Pöppelbuß, J.: Developing maturity models for IT management. Bus. Inf. Syst. Eng. **1**, 213–222 (2009)
26. Webster, J., Watson, R.T.: Analyzing the past to prepare for the future: writing a literature review. MIS Q. **26**, xiii–xxiii (2002)
27. Schultze, U., Avital, M.: Designing interviews to generate rich data for information systems research. Inf. Organ. **21**, 1–16 (2011)
28. Kille, C., Schwemmer, M.: Top 100 in European transport and logistics services. Market sizes, market segments and market leaders in the European logistics industry; [a study by Fraunhofer Center for Applied Research on Supply Chain Services SCS]. DVV Media Group, Hamburg (2015)
29. Saldaña, J.: The Coding Manual for Qualitative Researchers. Sage, London (2009)
30. VHB: VHB-JOURQUAL3.   https://vhbonline.org/vhb4you/jourqual/vhb-jourqual-3/gesamtliste/
31. Jussila, J.J., Kärkkäinen, H., Lyytikkä, J.: Towards maturity modeling approach for social media adoption in innovation. In: The 4th ISPIM Innovation Symposium, Wellington, New Zealand, 29 November–2 December 2011. International Society for Professional Innovation Management, Manchester (2011)
32. Duane, A.M., O'Reilly, P.: A conceptual stages of growth model for managing an organization's social media business profile (SMBP). In: Proceedings of the International Conference on Information Systems, ICIS 2012, Orlando, Florida, USA, 16–19 December 2012 (2012)
33. Chung, A.Q.H., Andreev, P., Benyoucef, M., Duane, A.M., O'Reilly, P.: Managing an organisation's social media presence: an empirical stages of growth model. Int. J. Inf. Manag. **37**, 1405–1417 (2017)
34. Lehmkuhl, T., Baumol, U., Jung, R.: Towards a maturity model for the adoption of social media as a means of organizational innovation. In: 2013 46th Hawaii International Conference on System Sciences, pp. 3067–3076. IEEE (2013)
35. Geyer, S., Krumay, B.: Development of a social media maturity model – a grounded theory approach. In: 2015 48th Hawaii International Conference on System Sciences (HICSS), pp. 1859–1868 (2015)
36. Solli-Sæther, H., Gottschalk, P.: The modeling process for stage models. J. Organ. Comput. Electron. Commer. **20**, 279–293 (2010)

37. Rossmann, A.: Digital maturity: conceptualization and measurement model. In: Proceedings of the International Conference on Information Systems - Bridging the Internet of People, Data, and Things, ICIS 2018, San Francisco, CA, USA, 13–16 December 2018 (2018)
38. Carvalho, J.V., Rocha, Á., van de Wetering, R., Abreu, A.: A Maturity model for hospital information systems. J. Bus. Res. **94**, 388–399 (2019)
39. Venkatesh, V., Morris, M.G., Davis, G.B.: User acceptance of information technology: toward a unified view. MIS Q. **27**, 425 (2003)
40. Rogers, E.M.: Diffusion of Innovations. Free Press, New York (1995)
41. Lasrado, L.A., Vatrapu, R., Andersen, K.N.: A set theoretical approach to maturity models: guidelines and demonstration. In: Proceedings of the International Conference on Information Systems - Digital Innovation at the Crossroads, ICIS 2016, Dublin, Ireland, 11–14 December 2016 (2016)
42. Culnan, M., McHugh, P., Zubillaga, J.I.: How large U.S. companies can use Twitter and other social media to gain business value. MIS Q. Exec. **9** (2010)

Potential Benefits of New Online Marketing Approaches

Ralf-Christian Härting[1]([⊠]), Christopher Reichstein[2],
and Andreas Müller[1]

[1] Business Administration,
Aalen University of Applied Sciences, Aalen, Germany
{ralf.haerting,andreas.mueller}@kmu-aalen.de
[2] Business Informatics, Baden-Wuerttemberg Cooperative State University,
Heidenheim, Germany
christopher.reichstein@dhbw-heidenheim.de

Abstract. This study examines the potential benefits of new approaches in online or digital marketing. In the course of this study, the research design and the new approaches in online marketing are considered. In a specially prepared quantitative study, experts were questioned about the individual approaches by means of a questionnaire. The questionnaire is based on derived hypotheses from the literature. The study focuses on the analysis of the survey results using the SmartPLS software. After data analysis using structural equation modeling, the results show that Mobile and Data-driven Marketing as well as Programmatic Advertising do have a significant influence on the potential of the new approaches in online marketing. The results are used to recommend actions for enterprises.

Keywords: Marketing 4.0 · Online marketing · Quantitative study · Expert survey

1 Introduction

Online or digital marketing can be defined as an umbrella term for all marketing activities done through the World Wide Web. In addition, the internet is the differentiator to traditional marketing activities, which are considered offline [1]. The objective of online marketing is getting prospects to perform the desired action such as purchasing a product or service through online channels. Twenty years ago, online marketing experts handled everything by themselves. However, this rapidly changed through advances in the technologies of the broadband internet and other technological advances. Due to those technological advances, the internet became accessible to a broader mass with ease [2]. As a result, online marketing grew and became more sophisticated through its ability to reach a wider audience. Driven by growth, several new areas developed such as SEO, SEA, affiliate marketing and more. Each one of these newly developed areas requires significant and detailed expertise [3]. A further considerable impact came through the development and introduction of the smartphone in 2010 [31].

© Springer Nature Switzerland AG 2019
W. Abramowicz and R. Corchuelo (Eds.): BIS 2019, LNBIP 354, pp. 109–117, 2019.
https://doi.org/10.1007/978-3-030-20482-2_10

By the year 2015, the individual sections are barely visible. The topics in each area blurred, and marketing experts are speaking of marketing automation and cross-channel strategies [3]. With today's fast-growing digitization, new approaches have emerged in online marketing. This study identifies six determinants and examines them. Enterprises can select from a wide range of potential marketing approaches. It takes time to determine all of them and their potential benefits. The study attempts to identify critical procedures and their potential benefits. The intent is to generate recommendations for actions for enterprises. Furthermore, the study aims to enhance the exciting research on the new online marketing approaches.

The following chapter specifies the Study Design, Research Method and Data Collection. It is followed by the Results of the survey, and ends with the Conclusion of the paper.

2   Research Design and Method

In the first step relevant hypotheses will be derived from a literary review. The proposed hypotheses are introduced and different items which lead to the hypotheses. At last, the authors introduces the research method and the data collection process.

2.1   Study Design

This study is based on an extensive literary review that identified several approaches of online marketing and their potential benefits. This was sought from international journals, papers and empirical studies. The selected approaches are reflected in six determinants (Table 1).

Table 1. Overview of literature review and determinants with selected items

Derterminant [References]
Artificial Intelligence [3–5]
Data-driven Marketing [6–9]
Dynamic Pricing [10–14]
Mobile Marketing [15]
– enables improved pre-purchase Information
– strengthens viral marketing effects in social media
– enables time-independent usability
– enables a spatially independent usability
New Google Services [16–19]
– improve market research
– such as Google My Business increases the online visibility of companies
– enable an increase in customer lifetime value
Programmatic Advertising [20–23]

The first hypothesis is Artificial Intelligence (AI). Artificial Intelligence is (separated from data driven marketing) connected to methods of Data Mining (also Text and Image Mining), such as decision trees or clustering and machine learning. The methods allow the replication of human cognitions by certain algorithms. The literature review showed that AI might help to provide more value to customers through discovering of meaningful patterns among the data available to enterprises [5]. As a result, the advertising could become more meaningful to the customers. Furthermore, a study indicates that AI will outperform statistical-based supporting tools. The systems are applied to various issues that companies face and improve the decision-making process [4]. Moreover, AI could even develop a true relationship with customers [5]. Voice assistance such as Amazon Echo, Siri use AI. The more the customers communicate with such devices the better it gets in identifying their abilities, beliefs, and intentions [5]. The key advantage here is that the machine does not forget the past conversations and becomes more individualized to the customers' need that might lead in a higher usage of the product. This leads to the first hypothesis:

H1: Artificial Intelligence has a positive influence on the potential benefits of new online marketing approaches.

The second hypothesis is grounded on Data-driven Marketing. Data-driven Marketing deals with the collection, analysis and interpretation of data in order to use it strategically for marketing purposes. For analyzing mainly methods of Web-Analytics are used. Marketing used to be a field, which was based on intuition and the gut feeling of marketers [6]. New approaches of Data-driven Marketing enable companies to track and measure campaign performance. Therefore marketing activities will change [7]. As a result, Data-driven Marketing strengthens the position of marketers [8]. In addition, sales and revenues can be increased through the knowledge that channels perform well and which do not. This helps companies to find the most lucrative way to advertise their products and services [9]. The information available to companies combined with Data-driven Marketing can increase the personalization of the advertising campaigns [9]. All of the aspects lead to the second hypothesis:

H2: Data-driven Marketing has a positive influence on the potential benefits of new online marketing approaches.

The third proposed hypothesis reflects the following findings of the literature review. First, a study revealed that Dynamic Pricing helps to discover the demand of a product [10]. Another study discovered that Dynamic Pricing strategies lead to an increase in sales compared to static pricing strategies [11, 12]. This finding is confirmed by a different study that indicates that an advanced pricing strategy combined with performance measurements increases financial results [13]. Nevertheless, enterprises have to take the different fair price zones in consideration. If they ignore it and customers recognize the strategy, it can damage the satisfaction of the customers [14]. Hence, the following hypothesis is developed:

H3: Dynamic Pricing has a positive influence on the potential benefits of new Online marketing approaches.

Another promising approach is Mobile Marketing which could have potentially a positive influence. Pre-purchase information is important in the purchasing process. Mobile marketing can increase the amount of available information for customers [15]. Furthermore, Mobile Marketing opens up extended services. For instance, enterprises can provide the customers apps and provide more services such as the product information, reservation handling, ticket purchasing, and follow up mails [15]. Researchers found out that the consumer loyalty increases if Mobile Marketing and mobile storefront applications are applied. Especially, when it is perceived as convenient and does not require a lot of search or cognition [15]. Further it simplifies the post purchase interactions with the customers. Viral effects can be reinforced through Mobile Marketing [15]. The consumers are communicating over the phones that makes them receptive for marketing messages [15] leading to hypothesis four:

H4: Mobile marketing has a positive influence on the potential benefits of new online marketing approaches.

The fifth selected proposed hypothesis is based on the approach of selected New Google Services (e.g. Google Now, Google MyBusiness or Trends), which are focusing on the business transactions. An earlier study about Google Trends concluded that search for equipment features are a reliable indicator of future car sales [30]. Enterprises can utilize this knowledge to emphasize these features in their campaign [16, 17]. It can be implied that these features are important factors to the customers and should be considered in the product development and future marketing campaigns. The researches from the study "Forecasting German Car Sales Using Google Data and Multivariate Models" used Google Trend data and the car sales to check and compared it to other multivariate models. They found out that Google was the most accurate model to forecast car sales [18]. Enterprises can improve their visibility through Google My Business. For instance, they can announce new promotions, increase the visibility of upcoming events and so on [19]. Further, the study found out that the customer lifetime value is higher for customers that are acquired through Google search advertising compared to conventional methods [19]. This leads to the hypothesis:

H5: New Google Services (e.g. Google MyBusiness) has a positive influence on the potential benefits of new online marketing approaches.

The last hypothesis examines if programmatic advertising has a positive impact. The hypothesis consists of several different aspects. A conducted study reveals that programmatic advertising is more effective, when no ad-blocker is used [20, 21]. Ad-blockers give the users the power to harm the performance of ad campaigns [21]. Past studies indicate that the effectiveness of advertising campaigns increase by programmatic advertising. The authors derived from the finding that the competitiveness increases for enterprises that apply programmatic advertising [22]. Moreover, the importance of the individuals increases. The data-growth helps companies to understand their customer on an individual level [21]. Through programmatic advertising, a more individual approach is possible. In programmatic advertising look-a-like, targeting can be applied to identify similar customers and exclude already existing ones. As a result, programmatic advertising increases the amount of people that can be targeted, which is beneficial for the business [23]. Programmatic advertising saves time by delivering automated online advertising through reduction of the obsolete buying process [24]. Therefore, the following hypothesis is formulated:

H6: Programmatic Advertising has a positive influence on the potential benefits of new online marketing approaches.

2.2 Research Method and Data Collection

A survey based on the introduced hypotheses builds the starting point of data collection process. To carry out the survey the open source Lime Survey was used, and experts were targeted. The authors predetermined the experts as people that have experience in the field of digital marketing for at least one year or have a leadership position in marketing. The experts were convened through different channels such as online marketing groups on Facebook, XING, LinkedIn or via direct mail. A pretest with almost 100 experts ensured the quality of the survey and its comprehensibility.

A total number of 102 experts participated and fully completed the survey (n = 102). All the other ones were removed in the data cleaning process. To identify the impact of the selected determinants a Likert scale of five was applied starting from 1 totally agree, 2 somewhat agree, 3 undecided, 4 somewhat disagree to 5 totally disagree.

The authors choose structural equation modeling (SEM) with Partial Least Squares (PLS) out of the available options to analyze the survey. These analyzation methods visualize the relationships among the variables [25]. Through this method, the effects of the clusters or hypotheses on the dependent variable can be shown and analyzed. The path significance is measured by Bootstrapping. This type of analyzation is especially useful for small data sets and it does not require making any assumptions upfront [26, 27]. The causal model includes single items as well as multi items. Therefore, the typical quality criteria Cronbach´s Alpha (CA), Average Variance Extracted (AVE), and Composite Reliability (CR) are applied [28].

3 Results

This chapter will give an overview over the results discovered through the survey and the analyzation method structural equation modeling. The following figure (Fig. 1) shows the model and the impact of each cluster on the dependent variable.

The value of R square for this model is 0,382, which indicates a sufficient model according to the requirements of Chin [26].

The result show that *Artificial Intelligence* negatively affects (−0,122) the potential benefits of new online marketing approaches. However, this result is not significant which means that it has no proven relationship between both variables. Thus, the hypothesis is rejected.

Data-driven Marketing clearly shows a positive (+0,526) and highly significant (p = 0,000 < 0,01) influence on the potential benefits of new online marketing approaches. It indicates the increasing importance of data in the decision-making

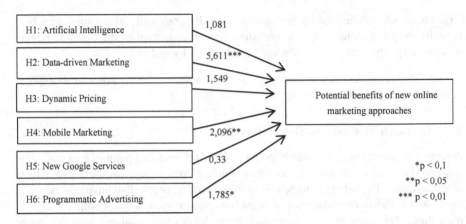

Fig. 1. Structural equation model with coefficients

process of marketing actions. This promising approach has the potential to deliver better results for businesses and marketing campaigns. The second hypothesis is confirmed.

Dynamic Pricing on the other hand shows that it has a positive influence (+0,144) but is not significant (p = 0,125 > 0,1). Therefore, the approach must be rejected to have a positive relationship. It might come from the risks that come with using this approach. Customers could discover the strategy and withdraw from buying from a business.

The construct *Mobile Marketing* shows a positive (+0,230) and significant (p = 0,039 < 0,05) influence. This confirms the increasing importance of Mobile Marketing and that enterprises can have influence on the pre-purchase information available to customers. Enterprises can strengthen viral effects and make use of the time and spatially independent usability of mobile devices.

The authors cannot confirm hypothesis 5 regarding the construct *New Google Services*. The analyzation revealed a positive (0,035) with no significance (p = 0,742 > 0,1). Google tools do not seem to play a big role in online marketing as might expected. Through the results of the analyzation, the hypothesis is rejected.

The last determinant *Programmatic Advertising* indicates a negative relationship to the potential benefits of new online marketing approaches with a value of −0,147. Moreover, the relationship is weak significant with a p-value of 0,077. However, the authors expected initially a positive relationship so that the hypothesis has to be rejected as well.

The results of the structural equation model are detailed in Table 2. The potential benefits of new online marketing approaches have been abbreviated to PB. The results show that all SEM paths with multi-item measurements exceed the thresholds defined by literature of .70 for CA, .60 for CR and .50 for AVE [29].

Table 2. Results of the structural equation model

	SEM-Path	Path coefficient	T statistics	P values	CA	AVE	CR
H1	Artificial Intelligence → PB	−0,122	1,081	0,282	1,0	1,0	1,0
H2	Data-driven Marketing → PB	0,526	5,611	0,000	1,0	1,0	1,0
H3	Dynamic Pricing → PB	0,144	1,549	0,125	1,0	1,0	1,0
H4	Mobile Marketing → PB	0,230	2,096	0,039	0,816	0,641	0,877
H5	New Google Services → PB	0,035	0,330	0,742	0,701	0,625	0,833
H6	Programmatic Advertising → PB	−0,147	1,785	0,077	1,0	1,0	1,0

4 Conclusion

As this research and previous works show, online marketing becomes an increasingly important topic to enterprises. New approaches have a wide range of potential benefits for businesses. Given the growing dynamics of the market over the past decade and the increasing complexity of new digital marketing strategies, many companies face the challenge of how to increase their profitability by using these new marketing tools. Therefore, enterprises will have to find appropriate approaches to their activities. This paper gives an overview of the different new available methods and their potential benefits.

The focus of the enterprises should be on the two approaches of Data-driven Marketing and Mobile Marketing. Both of these approaches are promising and have a positive and significant influence on the benefits of new online marketing approaches. Data-driven marketing might strengthen the position of the marketers in the business because the marketing campaigns are based on data and its impact be analyzed. Moreover, this approach increases sales and personalization of the campaigns. It is recommended for enterprises to get an overview over the topic of Data-driven Marketing and to start with new applications step by step. It will lead to enhanced marketing, and enterprises can make use of all potential benefits that the approach provides. The second approach Mobile Marketing can increase the available information, services extend the offered services, simplifies the past purchase communication and so on. It has many benefits and enterprises should start with a goal in mind what they want to achieve.

The study has its limitation. First, it was only carried out in German-speaking countries. Further research could be carried out internationally and compare its outcome with the results on hand. Furthermore, the survey period and the sample size could be increased to ensure the correctness of the results. Alternatively, another literature review could find more items that describe the different approaches and used to come up with new questions to measure the influence of the method on the potential benefits of new online marketing approaches. Moreover, further studies could extend the period to check if this changes the outcome. If researchers research the potential benefits of new online marketing approaches in other countries, they can use the data to examine variations among them. New items and questions could enlarge the existing research. These items and questions could possibly lead to a greater understanding of why the experts prefer specific approaches to other ones.

Online marketing will enhance even more in the future. New procedures will emerge and change the way of digital marketing entirely. Other approaches available today will improve and become more sophisticated. It is necessary for enterprises to stay up to date on these developments. If they do, they might increase the personalization of their marketing campaigns, improve their product developments, and base their decisions mostly on data.

References

1. Härting, R.-C., Mohl, M., Steinhauser, P., Möhring, M.: Search engine visibility indices versus visitor traffic on websites. In: Abramowicz, W., Alt, R., Franczyk, B. (eds.) BIS 2016. LNBIP, vol. 255, pp. 91–101. Springer, Cham (2016). https://doi.org/10.1007/978-3-319-39426-8_8
2. Härting, R., Mohl, M., Bader, S.: Digitalisierung als Treiber für Marketing 4.0. Industrie 4.0 und Digitalisierung – Innovative Geschäftsmodelle wagen! Ralf-Chrisitan Härting, Aalen (2016)
3. Lammenett, E.: Praxiswissen Online-Marketing – Affiliate- und E-Mail-Marketing, Suchmaschinenmarketing, Online-Werbung, Social Media, Facebook-Werbung, 6. Auflage, Springer Gabler, Wiesbaden (2016)
4. Wang, R.J., Malthouse, E.C., Krishnamurthi, L.: On the go: how mobile shopping affects customer purchase behavior. J. Retail. **91**, 217–234 (2015)
5. van Doorn, J., Mende, M., Noble, S.M., Hulland, J., Ostrom, A.L., Grewal, D.: Domo Arigato Mr. Roboto: emergence of automated social presence in organizational frontlines and customers' service experiences. J. Serv. Res. **20**, 43–58 (2017)
6. Doyle, S.: The role of social networks in marketing. J. Database Mark. Cust. Strat. Manag. **15**, 60–64 (2007)
7. Brynjolfsson, E., Hitt, L.M., Kim, H.H.: Strength in numbers: how does data-driven decision making affect firm performance? (2011)
8. Järvinen, J., Karajaluoto, H.: The use of web analytics for digital marketing performance measurement. Ind. Mark. Manage. **50**, 117–127 (2015)
9. Akter, S., Wamba, S.F.: Big data analytics in E-commerce: a systematic review and agenda for future research. Electron. Mark. **26**, 173–194 (2016)
10. Lamas, A., Chevalier, P.: Joint dynamic pricing and lot-sizing under competition. Eur. J. Oper. Res. **266**, 864–876 (2017)
11. Sato, K., Sawaki, K.: A continuous-time dynamic pricing model knowing the competitor's pricing strategy. Eur. J. Oper. Res. **229**, 223–229 (2013)
12. den Boer, A.V.: Dynamic pricing and learning: historical origins, current research, and new directions. Eur. J. Oper. Res. **20**, 1–18 (2015)
13. Davidson, A.: Pricing Strategy and execution: an overlooked way to increase revenues and profits. Strat. Leadersh. **33**, 25–33 (2005)

14. Xia, L., Monroe, K.B., Cox, J.L.: The price is unfair! A conceptual framework of price fairness perceptions. J. Mark. **68**, 1–15 (2004)
15. Wang, F., Xu, B.: Who needs to be more visible online? The value implications of web visibility and firm's heterogeneity. J. Inf. Manag. **54**, 506–515 (2017)
16. Binder, J., Weber, F.: Data Experience-Marktforschung in den Zeiten von Big Data, Marketing Review St. Gallen, vol. 32, pp. 30–39 (2015)
17. Du, R.Y., Hu, Y., Damangir, S.: Leveraging trends in online searches for product features in market response modeling. J. Mark. **79**, 29–43 (2015)
18. Fantazzini, D., Toktamysova, Z.: Forecasting German car sales using Google data and multivariate models. Int. J. Prod. Econ. **170**, 97–135 (2015)
19. Chan, T.Y., Xi, Y., Wu, C.: Measuring the lifetime value of customer acquired from Google search advertising. J. Mark. Sci. **79**, 757–944 (2011)
20. Withmer, C.: 5 Programmatic Mistakes That You're Probably Making, American Marketing Association (2016). https://www.ama.org/publications/eNewsletters/B2BMarketing/Pages/5-Pro-grammatic-Mistakes-You're-Probably-Making.aspx. Accessed 12 Dec 2018
21. Seitz, J., Zorn, S.: Perspectives of programmatic advertising. In: Busch, O. (ed.) Programmatic Advertising: Management for Professionals, pp. 37–51. Springer, Cham (2016). https://doi.org/10.1007/978-3-319-25023-6_4
22. Bishop, T.: As programmatic advertising becomes the new normal, how can advertisers create greater consumer engagement and publishers ensure greater return? J. Digit. Soc. Media Mark. **5**, 6–17 (2017)
23. Stevens, A., Rau, A., McIntyre, M.: Integrated campaign planning in a programmatic world. In: Busch, O. (ed.) Programmatic Advertising: Management for Professionals, pp. 193–210. Springer, Cham (2016). https://doi.org/10.1007/978-3-319-25023-6_16
24. Valle, M.: The Secret is Out, American Marketing Association (2014). https://www.ama.org/publications/MarketingInsights/Pages/secret-is-out.aspx. Accessed 10 Aug 2018
25. Alhassany, H., Faisal, F.: Factors influencing the internet banking adoption decision in North Cyprus: an evidence from the partial least square approach of the structural equation modelling. Financ. Innov. **4**, 29 (2018)
26. Chin, W.W.: The partial least squares approach to structural equation modeling. In: Marcoulides, G.A. (ed.) Modern Methods for Business Research, pp. 295–336. Lawrence Erlbaum Associates, Mahwah (1998)
27. Ringle, C.M., Wende, S., Becker, J.M.: SmartPLS. SmartPLS GmbH, Boenningstedt (2015)
28. Ringle, C.M., Sarstedt, M., Straub, D.: A critical look at the use of PLS-SEM in MIS quarterly. MIS Q. (MISQ) **36**(1), iii–xiv (2012). SSRN: https://ssrn.com/abstract=2176426
29. Hair Jr, J.F., Hult, G.T.M., Ringle, C., Sarstedt, M.: A Primer on Partial Least Squares Structural Equation Modeling (PLS-SEM). Sage Publications, Los Angeles (2016)
30. Bitterich, B., Möhring, M., Härting, R.: Google Trends im Geomarketing – als Hebel für ein analytisches CRM. In: ERP-Management, February 2014, pp. 20–22. Sage Publications (2014)
31. Härting, R.: Digitalisierung und Smart Service World – Potenziale und internetbasierte Dienste am Beispiele Marketing. In: Borgmeier, A., Grohmann, A., Gross, S. (eds.) Smart Services und Internet der Dinge: Geschäftsmodelle, Umsetzung und Best Practices, München 2017, Carl Hanser Verlag GmbH und Co. KG, September 2017. ISBN 978-3-446-45184-1

Applications, Evaluations and Experiences

Using Blockchain Technology
for Cross-Organizational Process
Mining – Concept and Case Study

Stefan Tönnissen[✉] and Frank Teuteberg

Universität Osnabrück, Osnabrück, Germany
{stoennissen, frank.teuteberg}@uni-osnabrueck.de

Abstract. Business processes in companies lead to an enormous number of event logs in their IT systems. Evaluating these event logs using data mining can provide companies with valuable process analysis information which can uncover process improvement potentials. However, media breaks frequently occur in these processes, so that there is a risk of optimizing isolated sub-processes only. Blockchain technology may avoid these media breaks and thus create the basis for complete event log analysis. The focus of our paper is to investigate existing requirements and to identify a blockchain based solution scenario evaluated by experts.

Keywords: Blockchain · Process mining · Data science · Process analytics

1 Introduction

Multinational companies (MNEs) have increased their global trade significantly in recent years [1]. A multinational (MNE) is characterized by distributed value creation in factories outside the home country [2]. Due to highly-developed division of labour, value creation in a multinational corporation takes place in various decentralized units. There are numerous intercompany supply relationships within a multinational company in the production of goods, semi-finished goods and intermediates. The transport of goods around the globe today requires a great deal of time and money and is therefore the subject of a continual search for opportunities to reduce both time and costs. Data that can help with this can be found in the intercompany business processes of the various ERP and IT systems of the group. Process mining is where this data is collected, analysed, weaknesses identified and optimization potentials determined. The challenge today is that heterogeneity in ERP and IT systems, common in multinational corporations, makes consistent process analysis by a central authority much more difficult [3]. In such cases, weak points in processes can be determined by means of process mining carried out on results from data science coupled with the real-time data of the blockchain [3]. The blockchain is the gateway to connect the processes and deliver the relevant data for the process mining. It has against traditional databases the advantage, that in additional to internal units, external parties like customers or suppliers can be integrated, without the development of a further technology. All the process participants are equal partners and can therefore trust the neutral blockchain

© Springer Nature Switzerland AG 2019
W. Abramowicz and R. Corchuelo (Eds.): BIS 2019, LNBIP 354, pp. 121–131, 2019.
https://doi.org/10.1007/978-3-030-20482-2_11

technology. At the end, after the integration of all participants of the process, the end-to-end process analysis with process mining is possible and brings all the participants a competitive advantage.

The purpose of this paper is to examine whether blockchain technology is a solution to the challenge of transparency of multinational companies' intercompany business processes. Our research question is therefore:

How can blockchain technology be used for Cross-Organizational Process Mining in a multinational corporation to meet the challenge of transparency of intercompany processes?

In order to answer our research question, we design and develop an artifact, following the design science research paradigm of Gregor et al. [4]. We therefore adapt existing knowledge about the blockchain to new operational problems. We develop the artifact in a case study based on a real-world problem at a multinational company in the commercial vehicle industry and evaluate it by interviewing experts.

2   Theoretical Background

2.1   Process Mining

The analysis of processes based on their event data is a process mining technique used for checking compliance, identifying and analysing bottlenecks, comparing process variants within a benchmark and identifying potentials for improvement [3]. Process mining closes the gap between data mining and process analysis and describes activities related to searching large amounts of data for relevant or significant information [5]. The idea behind data mining is that companies create huge amounts of automatically generated homogeneous data every day, which can generate decision support issues for decision makers [6]. Process analysis, on the other hand, deals with the course of a business process [7] and consists of a series of functions in a specific order, ultimately providing value for an internal or external customer [8]. Cross-functional end-to-end processes within a multinational company should be considered as a whole to avoid improving only isolated sub-processes. Process analysis is based on event logs that are generated during process execution and is thus adapted to the real-word situation [8]. Currently, process analysis is mostly based on data available within organizations [9]. For cross-organizational process mining, for example, in the analysis of supply chains, data can even be spread across multiple organizations [3]. Today's information systems log enormous amounts of events, but such information is usually unstructured, for example, event data in SAP R/3 is spread across many tables or must be retrieved from subsystems that exchange messages. In such cases, event files are present, but some effort is required to extract the data. Data extraction is an integral part of all process mining efforts and is not possible without corresponding event logs [3] which allow process-level process analysis to be performed and the setting up and calculation of indicators based on process execution traces [8]. However, dependency on the event logs limits the process mining techniques to identifying activities that are not included in the event logs, such as manual activities performed in the process [10]. The reality is that event data is typically distributed across different data sources, and often some effort is required to gather the relevant data [3]. The blockchain keeps records of executed processes and can provide valuable information to assess case load, duration, frequency of paths, parties involved, and the correlation between unencrypted data elements. This information can be used to detect processes, detect deviations, and conduct root cause analysis [3], ranging from small business groups to an entire industry.

Process Mining, with its combination of event data and process models, enables both a data-driven and process-oriented view. It can be used to answer numerous compliance and performance questions. Control of conformity is achieved by comparing observed behaviour with modelled behaviour [3]. In this way, compliance violations can be identified as well as detected for other inconsistencies in the processes [11]. The investigation of weaknesses and the detection of bottlenecks belong to performance questions about processes in a company [3].

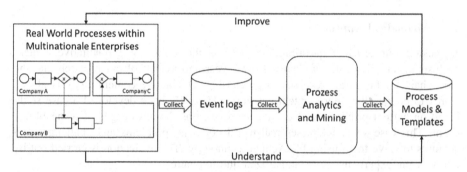

Fig. 1. The basic process of Process Mining, based on Hof (2018) [27].

Basic Process Mining begins with the transfer and extraction of event data from the various IT systems into an event log database. Then this data is adapted to the company needs and to take into account both syntactic and semantic requirements. Finally, the data is loaded into a data warehouse system [3] and is available for analysis [12]. The knowledge gained can be used to adapt and improve processes.

2.2 Blockchain Technology

Blockchain technology became popular with the introduction of the crypto-currency Bitcoin 10 years ago and is now used in numerous use cases. A blockchain as distributed ledger is a concatenation of data based on transactions between subscribers that are aggregated into individual blocks and stored on all users' computers in a peer-to-peer network. The data is concatenated using cryptographic methods, creating a chronological chain of immutable data. The inclusion of a new record in the blockchain requires the passage of a so-called consensus mechanism that runs across the network of all subscribers and is used to reach an agreement between all members of the blockchain network about the correct status of the data. This ensures that the data is the same on all nodes in the network [13]. The best-known consensus mechanism is a proof-of-work method that requires the computer to perform a complicated mathematical algorithm at great expense. Only after successful execution, a new data block can be generated in the blockchain, which must be checked by the other computers in the peer-to-peer network before being included in the blockchain [14]. In addition to the data, each block contains a timestamp as well as the hash value of the previous block. The blocks are protected by cryptographic methods against subsequent changes, so that a coherent chain of linked data blocks forms over time [14]. The data exchange between a blockchain and an ERP system could, for example, be done via the Unibright Connector. The Unibright Connector (UBC) is based on a Microsoft .NET class library and establishes a connection between the blockchain and external systems [15]. The Unibright Framework cross-blockchain and cross-system connections provide a blockchain-based business integration process [16].

2.3 Methodical Approach

To answer our research question, we first use the results of a qualitative content analysis of interviews that we carried out a year ago as part of our research into the blockchain for business processes in purchasing. Qualitative content analysis allows words to be classified into content categories [17]. We have developed a case study based on current problems and challenges as well as the advantages of using blockchain. This case study addresses real-world company problems and challenges and attempts to solve them using blockchain technology. The evaluation is carried out by interviewing experts using a standardized questionnaire.

2.4 Related Works

We were able to find related studies based on a previous review of the literature. Mendling et al. (2018) suggest that blockchain technology has the potential to drastically change inter-organizational processes. The need for drastic change arises, among other things, from the lack of a global view of processes today. The fragmentation of processes across countries and their systems leads to misunderstandings and blame if there is a conflict. The reason is that companies often use systems for the implementation and execution of processes only for intra-organizational processes [7]. Rbigui and Cho (2018) perform performance analysis on a process mining process and come to the conclusion that process mining yields good results only where processes are complete [18]. In our work we develop a concept to integrate the blockchain technology with existing IT-systems in firms as the basis for a inter-organizational process mining. The concept sees the blockchain as the bridge to close the gap between different steps in a process. Therefore, the process mining is able to analyse the whole process.

3 Analysis of Interviews

It was important for our concept to record the existing challenges in logistics processes of multinational enterprises (current situation) as well as recording the possible results of improvements by the integration of a blockchain into the logistics process (target situation). For the assessment of the current situation, we entered the search string "Blockchain" and "Interview" and "Supply chain" or "Logistics *" in Google for the period 01.01.2017 to 31.01.2018 and received 35,400 results. Based on our assumption that the results of the first pages reflect the relevance of Google's search algorithms [19], we used the titles and short texts to analyse the results in order to filter out the interviews relevant to our research question.

The data from the first 20 interviews was transferred to an Excel file and analysed on the basis of the following questions:

- What problems or challenges are seen in logistics or in supply chain management?
- What are the advantages of using the blockchain in the logistics industry?

Due to the fact that not all analysed interviews were able to provide the necessary information to answer our questions, we then transferred and analysed further interviews from our Google search results.

Table 1. Results of the interviews with classification.

Problems/Challenges (current situation)		C	P	Advantages (target situation)		C	P
Class	No	No	No	Class	No	No	No
Process	9	2	7	Process	36	13	23
Trust	5	2	3	Transparency	28	10	18
Conditions	4	2	2	Fraud	8	7	1
Data	2	2	0	Costs	5	0	5
IT-Security	2	2	0	Organization	4	2	2
Fraud	1	1	0	IT-Security	4	3	1
Costs	1	0	1	Collaboration	3	0	3
Standards	1	0	1	Trust	3	2	1

C = Conformance questions; P = Performance questions

We were able to evaluate 35 interviews about current challenges and future expectations and analyse them against the classes shown in Table 1. In a second step, we categorised the answers according to the relevance of process mining as either "performance questions" or "conformance questions". The results show that the processes are very important in this context, both in terms of the current challenges and future expectations. The current problem definition of the interviewees' processes shows a clear focus on the performance of the processes. In addition to the high demands of documenting process steps, the interviewees also mentioned the numerous participants in a process with the associated media disruptions as well as the associated lengthy waiting and idle times. With regard to the future expectations of blockchain-based solutions, the requirements for the processes also predominate. Time plays an important role in the Processes class. The perceived benefits of using blockchain technology are the timely processing of process steps.

4 Case Study

To answer our research questions as defined above, we conduct a case study that, according to Ridder (2017), offers the advantage of detailed description and detailed analysis so that the questions "how" and "why" can be answered more easily [20]. Our case study is suitable for our research topic because a current phenomenon (blockchain) is examined in a real and practical context (MNE) [21]. According to Brüsemeister (2008), a case-by-case study is also helpful if it provides information about a previously under-researched social area aimed at using process mining of blockchain data to improve cross-organizational processes [22].

The company in our case study is Europe's leading manufacturer of semi-trailers and trailers for temperature-controlled freight, general cargo and bulk goods and has an annual production of around 61,000 vehicles with around 6,400 employees. In the 2017/2018 financial year, sales reached over € 2.17 billion [23].

In addition to several production sites in Germany, the MNE also has factories in Lithuania, Turkey, China, Russia and Spain. Sales in Europe are made through the company's own distribution companies in almost every country [24]. Since 2004, the production facility in Germany has produced their own axles with an annual output of approx. 150,000 units [25]. The procurement volume for raw materials, consumables and supplies, for purchased goods and services, amounted to € 1,460.6 million in the financial year [26].

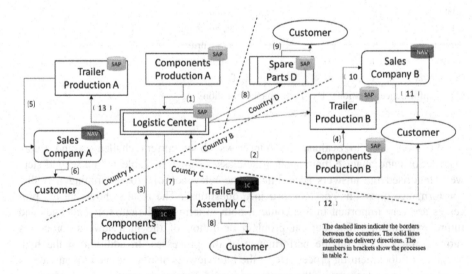

Fig. 2. Supply relationships in the case study.

Figure 2 shows an example of part of the complex performance relationships within the multinational company for five different countries. The headquarters of the multinational is in Country A.

The previous picture shows the factories for the production of components of a trailer (Components Production) and the logistics center (Logistic Center) as a central warehouse for the supply of factories in the group, for example with the Axle from Components Production A. Furthermore, the Factories producing Trailers (Trailer Production) are essential components of value creation. For reasons of customs law, Country C has only a trailer assembly for the assembly of components and kits for trailers. The trade in spare parts at workshops takes place via the spare parts trade (spare parts). In addition to the companies, the IT system used for processing supply relationships is listed.

Table 2. Supply relationships between the entities in the group

No.	Supplying unit	Received unit	IT-System supplier	IT-System receiver	Process
(1)	Components Production A	Logistic Center	SAP-R/3	SAP-R/3	I
(2)	Components Production B	Logistic Center	SAP-R/3	SAP-R/3	S
(3)	Components Production C	Logistic Center	1C	SAP-R/3	S
(4)	Components Production B	Trailer Production B	SAP-R/3	SAP-R/3	S
(5)	Trailer Production A	Sales Company A	SAP-R/3	Navision	S
(6)	Sales Company A	Customer	Navision	Unknown	S
(7)	Logistic Center	Trailer Assembly C	SAP-R/3	1C	S
(8)	Logistic Center	Spare Parts D	SAP-R/3	SAP-R/3	S
(9)	Spare Parts D	Customer	SAP-R/3	Unknown	S
(10)	Trailer Production B	Sales Company B	SAP-R/3	Navision	S
(11)	Sales Company B	Customer	Navision	Unknown	S
(12)	Trailer Assembly C	Customer	1C	Unknown	S
(13)	Logistic Center	Trailer Production A	SAP-R/3	SAP-R/3	I

No.: shows the process step from the Fig. 2/Process: I = Integration; S = Segregation

Table 2 shows the example supply relationships between the companies in the multinational group. The Process column indicates whether the business process between the supplier and recipient IT systems is integrated into a system (I) or whether a media break occurs between the systems (S).

Real integration can only be seen today in the companies with SAP-R/3 in Country A, because this is where the headquarters of the Group is located and where integration efforts are the most advanced. Crossing a border usually results in segregation of processes, even if both companies work with the same SAP R/3 system. Between different IT systems, such as e.g. SAP R/3 and Navision, there is no process integration with interfaces.

Based on the current situation in our case study and the requirements of the analysed interviews, we design a solution based on blockchain technology. Blockchain technology connects the processes that were previously segregated in our solution (see Fig. 3). For example, in the case of a delivery from the SAP R/3 system, the data record is written to the blockchain via a connector and forwarded via a smart contract to the receiving company. The connector then picks up the record and translates it into the local IT system, e.g. Navision. Due to the real-time processing of the blockchain, both the supplying company and the receiving company are always able to provide information about the status of the process.

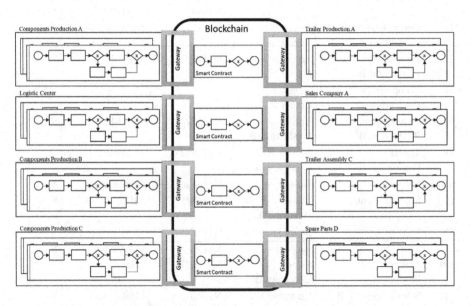

Fig. 3. The blockchain connects the cross-organizational processes.

The blockchain technology can connect processes that were previously separated by media breaks, so that the event logs can be used as part of a data mining process to analyse the processes (see Fig. 4). The blockchain is based on the concept in Fig. 1 applied to the processes in decentralized IT systems such as e.g. SAP R/3 and Navision and database event log implemented. The acceptance of the blockchain solution for all the process participants can reach with a permissioned blockchain. All the process relevant data are secure and can only viewed from the participants with the concrete rights. A permissioned blockchain runs without a consensus algorithms.

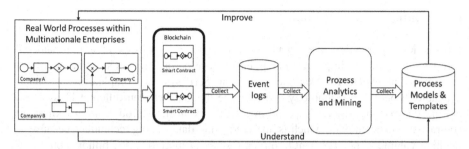

Fig. 4. The integration of the blockchain into process mining, based on Klinger (2018) [28].

5 Discussion of the Results

Our solution is based on blockchain technology and generates event data based on the integration of cross-system processes. It would also be possible to use a relational and distributed database solution. This could take up data from the processes of the distributed systems via interfaces and consolidate it into an overall process. Technically, process mining is also possible using a relational database. However, it should be remembered that in some industries, such as the food industry, pharmaceutical industry and the chemical industry higher requirements of traceability for deliveries is necessary and must be provable. For these increased requirements, the blockchain, because of its immutability, decentralization, and cryptography, could offer a tremendous advantage over a traditional database. In addition, concerns about the integration of third parties, such as easier to clean out suppliers with the blockchain because all parties have access to the same data.

We interviewed experts from industrial practice and process consulting and asked them to complete a standardized questionnaire in order to evaluate our concept. The selection of participants took place via social media contacts in Xing, Linkedin and Facebook with the requirement that the participants have appropriate professional knowledge as well as occupational status. We invited 60 participants to complete the survey in November/December 2018 and received 56 responses by the end of December 2018 - a response rate of 93,3%. Participants are divided into 32 from industry (62,75%), 10 from consulting (19,61%) and nine from other (17,65%). Out of these 56 responses, 35 belonged to a multinational firm. When asked if they could confirm the problem described in our case study regarding process breaks in transitions between different IT systems, 83,67% answered yes. Furthermore, we wanted to know from the participants on the basis of a Likert scale from 1 to 5, how far they regarded our concept as realistic for overcoming the process breaks. The arithmetic mean of the responses is 2,72 with a standard deviation of 0,94. This indicates some uncertainty about a successful implementation. The participants see, among other things, the maturity level of blockchain technology as a hindrance to implementation as well as deficits in the performance of mass data processing. Furthermore, the participants mentioned a lack of risk-taking and the courage to invest in new technologies as a hindrance.

6 Summary and Outlook

Process Mining requires the seamless event data of processes in a multinational company, so as not to improve weak points only at partial process steps in the part of the process in which, for example, due to an SAP system, complete event data is available. Blockchain technology could bridge the gap between different IT systems, closing the existing process gap and generating event data through the use of smart contracts. This event data, which is analysed using data mining techniques because of the high volume of procurement in our case study of more than € 1 billion, could thus serve as the basis for the analysis of cross-company processes. Because of the immutability of the data in the blockchain, this event data could be used for processes in industries with increased requirements for transparency and traceability. By using blockchain technology, the involvement of third parties such as suppliers and downstream value creation stages could succeed. Furthermore, auditors and inspectors could gain access to the blockchain. In the case of cross-border supplies between companies within the group, the tax authority has an important role to play because of transfer pricing rules. The substantive over form principle in tax auditing means an assessment is based on real processes, not on the basis of contracts or agreements. The immutability, cryptography, and chronological order of the data make the blockchain ideal for process mining to control compliance.

References

1. OECD: OECD Transfer Pricing Guidelines for Multinational Enterprises and Tax Administrations (2017). https://dx.doi.org/10.1787/tpg-2017-en
2. Root, F.R.: Entry Strategies for International Markets, 2nd edn. Lexington Books, New York (1994)
3. van der Aalst, W.: Process Mining. Data Science in Action, 2nd edn. Springer, Heidelberg (2016). https://doi.org/10.1007/978-3-662-49851-4
4. Gregor, S., Hevner, A.R.: Positioning and presenting design science research for maximum impact. MIS Q. **37**(2), 337–355 (2013)
5. van der Aalst, W.: Process mining: making knowledge discovery process centric. ACM SIGKDD Explor. Newsl. **13**(2), 45–49 (2011)
6. Tan, Y., Shi, Y., Tang, Q. (eds.): Data Mining and Big Data. Springer, Cham (2018). https://doi.org/10.1007/978-3-319-93803-5
7. Mendling, J., et al.: Blockchains for business process management - challenges and opportunities. ACM Trans. Manage. Inf. Syst. **9**(1), Article 4, 16 pages (2018). https://doi.org/10.1145/3183367
8. Kirchmer, M.: High Performance Through Business Process Management. Strategy Execution in a Digital World, 3rd edn. Springer, New York (2017). https://doi.org/10.1007/978-3-319-51259-4
9. Dumas, M., La Rosa, M., Mendling, J., Reijers, H.A.: Fundamentals of Business Process Management, 2nd edn. Springer, Heidelberg (2018). https://doi.org/10.1007/978-3-662-56509-4

10. Gulden, J., Reinhartz-Berger, I., Schmidt, R., Guerreiro, S., Guédria, W., Bera, P. (eds.): Enterprise, Business-Process and Information Systems Modeling. Springer, Heidelberg (2018). https://doi.org/10.1007/978-3-319-91704-7
11. Al-Ali, H., Damiani, E., Al-Qutayri, M., Abu-Matar, M., Mizouni, R.: Translating BPMN to business rules. In: Ceravolo, P., Guetl, C., Rinderle-Ma, S. (eds.) SIMPDA 2016. LNBIP, vol. 307, pp. 22–36. Springer, Cham (2018). https://doi.org/10.1007/978-3-319-74161-1_2
12. Rebuge, A., Ferreira, D.R.: Business process analysis in health care environments: a methodology based on process mining. Inf. Syst. **37**(2012), 99–116 (2011)
13. Swan, M.: Blockchain, Blueprint for a New Economy. O'Reilly, Sebastopol (2015)
14. Holotiuk, F., Pisani, F., Moormann, J.: The impact of blockchain technology on business models in the payments industry. In: Leimeister, J.M., Brenner, W. (Hrsg.) Proceedings der 13. Internationalen Tagung Wirtschaftsinformatik (WI 2017), St. Gallen (2017)
15. Schmidt, S., Jung, M.: TECHNICAL PAPER. Unibright – The Unified Framework for Blockchain Based Business Integration (2018). https://unibright.io/download/Unibright_Technical_Paper.pdf. Accessed 12 Jan 2018
16. Schmidt, S., et al.: Unibright - The Unified Framework for Blockchain Based Business Integration (2018). https://unibright.io/download/Unibright_Whitepaper.pdf. Accessed 12 Jan 2018
17. Elo, S., Kyngäs, H.: The qualitative content analysis process. J. Adv. Nurs. **62**(1), 107–115 (2007)
18. Rbigui, H., Cho, C.: Purchasing process analysis with process mining of a heavy manufacturing industry. In: The Proceeding of 9th IEEE International Conference of Information and Communication Technology Convergence, 17–19 October 2018, pp. 495–498 (2018)
19. Google Inc., How Search Works. https://www.google.com/intl/ALL/search/howsearchworks/. Accessed 02 Feb 2018
20. Ridder, H.-G.: The theory contribution of case study research designs. Bus. Res. **10**(2), 281–305 (2017)
21. Yin, R.: Case Study Research: Design and Methods. Sage, Thousand Oaks (2002)
22. Brüsemeister, T.: Qualitative Forschung. Ein Überblick, Wiesbaden (2008)
23. Cargobull: 67,000 Vehicles on the Horizon for Latest Objective (2018). https://www.cargobull.com/en/detail_news-563_213_383.html. Accessed 12 Dec 2018
24. Cargobull: Sales Points (2018). https://www.cargobull.com/en/Sales-Points_203_328.html. Accessed 12 Dec 2018
25. Cargobull: 1,000,000th Axle Manufactured (2017). https://www.cargobull.com/en/detail_news-524_213_383.html. Accessed 12 Dec 2018
26. Cargobull: Konzernabschluss zum Geschäftsjahr vom 01.04.2016 bis zum 31.03.2017. Schmitz Cargobull Aktiengesellschaft. Bundesanzeiger (2018)
27. Hof, S.: Process Analytics and Mining. Research Picture (2018). http://www.wi2.fau.de/research/research-projects/pam/. Accessed 12 Jan 2018
28. Klinger, P.: Trustless Cross-Organizational Business Process Integration – Prototyping of a Blockchain based Business Process Management System. Research Picture (2018). http://www.wi2.fau.de/research/research-projects/trustless-cross-organizational-business-process-integration-prototyping-of-a-blockchain-based-business-process-management-system/. Accessed 12 Dec 2018

Modeling the Cashflow Management
of Bike Sharing Industry

Binrui Shen[1], Yu Shan[1], Yunyu Jia[2], Dejun Xie[1(✉)],
and Shengxin Zhu[1(✉)]

[1] Xi'an Jiaotong Liverpool University, Suzhou, China
{Dejun.Xie,Shengxin.Zhu}@xjtlu.edu.cn
[2] University of Liverpool, Liverpool, UK

Abstract. The sharing economy has been widely concerned since its inception and bike sharing is one of the most representative examples. The paper attempts to investigate the cashflow management strategy of bike sharing companies to optimize the overall financial return. The framework of our model is based on the assumption that bike sharing companies may invest operation income in financial market for long-term and short-term earnings. Optimal reserve pool is modeled and estimated using double parameterized compound Poisson distributions. Empirical examples and Monte Carlo analysis are provided for model validation.

Keywords: Sharing economy · Reserve pool · Retention rate · Cashflow · Compound poisson model · Monte Carlo simulation

1 Introduction

In recent decades, the "sharing economy" has grown at an incredible pace, attracting broad attention on a global scale. The success and diffusion of the business model represented by Airbnb and Uber has not only opened up various resources for sharing, but also announced the rise of the sharing economy. It has brought new modes of production, consumption patterns and business operation, which may become the future trend in global economic. Bicycle-sharing, currently the most popular business model in the sharing economy, are one of the most representative examples. The common business practice is that bike-sharing firms charge users a certain amount of guarantee deposit for the bike sharing services to be granted. The deposits, together with other rental service incomes, constitute the main cash inflow for the company to manage as well as to reinvest in various financial instruments and securities, as long as relevant policies and regulations allow. The most important compliance and pledge for the company to fulfill is to make timely refunds of deposits upon customer request. Leading companies in the industry like *Mobike* usually promise to refund within 2 to 7 days after such application is received, but normally the funds would be transferred to the client instantly, if only for best customer satisfaction. In order to achieve timely refunds, companies normally set up a refund reserve (reserve pool) which is undergoing strict monitoring in accordance with the database of existing customers and the expected number and amount of the refund requests. If investment is allowed, the

W. Abramowicz and R. Corchuelo (Eds.): BIS 2019, LNBIP 354, pp. 132–146, 2019.
https://doi.org/10.1007/978-3-030-20482-2_12

investment made on the reserve fund is usually for short term monetary and security market to maintain high liquidity. However, the company may also desire to invest the balance to the reserve account in the long-term capital market expecting to earn a higher return.

The main contribution of this article is to provide a mathematical model to best estimate the monthly amount dedicated to the reserve pool for optimally ensuring quick refunds and maximize the financial returns to the company on the basis that the refund requirement is met. Literature reviews to the field is provided in the following Sect. 2. Our models and methodologies to the problem are presented in Sect. 3. Empirical analysis using historical dataset is given in Sect. 4. Together with Monte Carlo simulation of reserve pool and investment return, this section validates the usefulness of our model. Section 5 summarizes the current work with a discussion on limitations and suggestions for future study.

2 Literature Review

Botsman [2] states that "sharing economy" is an umbrella term with a range of meanings and cannot be given an exact definition. However, some commentators argue that the term is valid to describe normally more democratized marketplaces, even when it's applied to a broader spectrum of services [5]. For example, bike sharing is representative but not an orthodox form of sharing economy according to Schor [14].

Lan et al. [9] points out that sharing economy is of great significance for reflecting the great potential of various social innovations in some important areas to promote the sustainable development of citizens. For instance, bike sharing has numerous perceived benefits such as promoting economic development, improving the urban landscape and enhancing the quality of life [15]. Some studies on business models lend indirect support to the idea that this newly emerging e-business mode will become stably profitable in the future [12]. However, the development of bicycle sharing still encounters many obstacles in China. For example, the high rate of equipment loss caused by moral hazard [1] and management difficulties brought on by the large fleets involved [3] are crucial factors. A few researchers have made rough estimates of the earnings of bike-sharing companies. Ge [7] established a profit model for the Sharing-Bicycle value network model, though the prediction of the deposit investment project only includes two parameters. In contrast, Feng and Ma [6] argue that there are more factors affecting income. Another issue is that the previous studies typically do not include a systematic model to explain the financial sustainability and optimization of bike sharing industry.

In sum, these studies demonstrate the feasibility of this new business model. Existing researches mainly focus on the impact of the sharing economy system in terms of macroscopic social environment and relations. What is lacking is a rigorous mathematical model to explore the financial management of bike-sharing companies in China. The classical compound Poisson risk process (or Cramér – Lundberg model) provides a theoretical layout for companies which have two opposing cash flows: incoming premiums and outgoing claims. The main idea of the model is to find out the probability of bankrupt when the surplus level is less than zero [4]. Value at Risk (VaR) constitutes also a method to measure the risk of investment losses based on probability theory and mathematical statistics [8]. However, VaR only focuses on the statistical characteristics of risk instead of the whole of system risk management. What's more, VaR models are not applicable to scenarios where the data are not stationary [13]. Maasoumi and Racine [11] suggest that information entropy is an appropriate method to measure uncertainty when the probability distribution of the financial market is not clear other than variance. Based on this, Li et al. [10] proposed an information-based risk measurement model. Although this model may measure risk of different investors with less computation, to some extent, the authors did not give any data to prove the effectiveness of the model. Furthermore, an information entropy model of commercial banks was presented to calculate the optimal risk reserve [16].

3 Methodology

3.1 The Reserve Pool Model

Most financial firms would have two dynamic cash flows with opposite directions, cash inflow and outflow, respectively. For simplicity, let $D(t)$ and $W(t)$ be the two random processes representing such two cashflows in a period $(0, t)$, and R be the size of reserve pool. Assume that the frequencies of these two cash flows, $\{M(t); t \geq 0\}$ and $\{N(t); t \geq 0\}$, follow the Poisson processes. All these random variables are independent identically distributed, and independent of other random processes. Then the general risk model for the firms can be formulated as

$$U(t) = R + D(t) - W(t) \tag{1}$$

One primary cash outflow is the reimbursement of customers' guarantee while the opposite cash flow includes the increase in guarantee and customers' service consumption. Note that a prudent financial management of the firm would imply the real time adjustment in the size of reserve pool according to the actual business condition. Now set the unit time of adjustment as one day. If the company does not complete all refund request on any particular day, then it fails to comply with its promise. Let $T_s \geq 1$ stands for the shortest time duration needed to accumulate sufficient fund to meet the refund request, then the fail probability of firm in time period $(0; T_S)$ can be written as: $\varphi(R) = Pr(T < T_S)$. The surplus of firm is then written as

$$U(T_S) = R + D(T_S) - W(T_S).$$

To simply the problem, we first set the $T_S = 1$. Then the risk surplus model becomes

$$U(1) = R + D(1) - W(1)$$
$$\begin{cases} D(1) = M(1)g + \beta I \\ W(1) = N(1)g \end{cases} \tag{2}$$

where g is the amount of guarantee, and I is the daily income and it is reasonable to assume that some daily operating income will be added to the reserve, if revolving fund is not sufficient $(R + M(1)g < N(1)g)$. Considering all refund request should be completed in one day, the probability of failure is given by $\varphi(u) \leq \alpha$, which can be transformed to $P(U(1) \leq 0) \leq \alpha$, expanding to

$$P((R+M(1)g+\beta I - N(1)g)<0) \leq \alpha \tag{3}$$

Because $\{N(t); \ t \geq 0\}$ and $\{M(t); \ t \geq 0\}$ both follows a Poisson distribution respectively, $M(t)g$ and $N(t)g$ also follows a Poisson distribution. Therefore, $[M(1) - N(1)]$ follows a Skellam distribution. And the probability mass function of the Skellam distribution: $S = [M(1) - N(1)]$ between two independent Poisson random variables with parameters λ_1 and λ_2 is given by

$$p(s; \lambda_1, \lambda_2) = Pr\{S = s\} = e^{-(\lambda_1 + \lambda_2)} I_K\left(2\sqrt{\lambda_1 \lambda_2}\right).$$

Then the limiting condition equation can be transformed into

$$P[S < - (\beta I + R)] \leq \alpha \tag{4}$$

where α satisfies the equation $P(S < s_\alpha) = \alpha$, i.e.,

$$s_\alpha = -(\beta I + R) \tag{5}$$

However, Skellam distribution is a dynamical distribution, implying that its $\alpha\%$ value of the bound cannot be found directly. Monte Carlo method can be used to simulate the distribution to estimate s_α (Zhigljavsky, 2004). Generating W random numbers $M_1, M_2, ..., M_W$ and $N_1, N_2, ..., N_W$, the structural probability is reliable as W goes to infinity. We let $W = 100000$ in our simulations. The second step is combining these data to simulate the Skellam distribution S:

$$s_i = M_i - N_i, \quad i = 1, 2 \cdots P \tag{6}$$

Then we sort these data from small to large, and choose the $(100000 \times \alpha)th$ one as the s_α. Then R can be recovered by changing $s_\alpha = -(\beta I + R)$ into

$$R = max(0, -(s_\alpha + \beta I)) \tag{7}$$

Reserve Pool Parameters. The expectation of daily refund frequency λ_1^j for the current month includes two parts. One is the refund to old users. Let the monthly retention rate of old users (p_k^j) be P_1, which is the ratio of the remaining old customers to the total number of customers. Correspondingly, the percentage of departure is $(1 - P_1^j)$. Let F_1^j represent the number of this kind of old users, and U_1^j be the number of total old users in this month, then

$$F_1^j = U_1^j(1 - P_1^j) \tag{8}$$

The other type of refund corresponds to those made by new users in the current month. Similarly, the total refund frequency of this part would be calculated by the retention rate of new users denoted as P_2 and the number of new users (U_2^j) in this month:

$$F_2^j = U_2^j(1 - P_2^j) \tag{9}$$

Combining these two, we obtain the total frequency of refund in one month. Finally, assume that the amount of deposit and withdrawal with respect to one transaction per customer are the same. The parameter λ_1 in this month is then given by

$$\lambda_1^j = \frac{\left\{U_1^j \times \left(1 - P_1^j\right) + U_2^j \times (1 - P_2)\right\}g}{30} \tag{10}$$

The total deposit frequency is the number of new users (U_2^j) in this month. Therefore, the mean daily frequency of deposit λ_2 is

$$\lambda_2^j = \frac{U_2^j \times g}{30} \tag{11}$$

It is noted that the parameter P_2 can be obtained from online research report, while P_1 cannot. In the previous section, the total refund frequency of new users is considered as a variable following the Poisson distribution, which means the duration of new users follows the exponential distribution. The cumulative distribution function is then given by

$$F(t_d) = \begin{cases} 1 - e^{-\lambda_3 t_d} & t_d \geq 0 \\ 0 & t_d \leq 0 \end{cases}, \quad t_d \in N \tag{12}$$

Here $\lambda_3 > 0$ is the parameter of the distribution or the so-called rate parameter. $F(t_d)$ could be considered as the probability that users would stop using firm's service in t_d days, so $1 - F(t_d)$ is the retention rate of users who began using the service t_d days ago.

However, because the property of the exponential distribution, the retention rate $1 - F(t_d)$ would be zero, or equivalently all users would stop using the company's services, when the duration time t_d approaches infinity, which is not a desired property from industry perspective. One alternative distribution is the conditional exponential cumulative distribution function, which makes the ultimate retention rate equals to θ $(0 < \theta < 1)$. Then the new adjusted function is:

$$F(t_d)_{adj} = (1 - \theta)F(t_d) = \begin{cases} (1 - \theta) + (1 - \theta)e^{-\lambda_3 t_d} & t_d \geq 0 \\ 0 & t_d \leq 0 \end{cases}$$

by where $t_d \in N$. θ is a parameter defining the ultimate retention rate parameter. When $\theta = 0$, $F(t_d)_{adj} = F(t_d)$. And $1 - F(t_d)_{adj}$ would approach θ when t_d approaches infinity.

Then we should figure out the parameter λ_3. Let E_i $(i = 1, 2, 3\ldots30)$ stands for the number of customers whose entry time is the i_{th} day of the month, and then these parameters must satisfy

$$E_1 + E_2 + \cdots + E_{30} = U_2$$

From this we estimate that the number of retained users whose entry time is the i_{th} day of the month ago is $(1 - F(30 - i + 1)_{adj})E_i$. Notice that the total number of retained costumers in one month is the sum of the amount of 30 days in that month, we have that

$$\left(1 - F\left(30\right)_{adj}\right)E_1 + \cdots + \left(1 - F\left(1\right)_{adj}\right)E_{30} = P_2 U_2 \qquad (14)$$

To simply, with loss of generality, we may think that $E_1 = E_2 = \ldots = E_{30}$. We could deduce the following equation and then taking $h = e^{-\lambda_3}$, we have

$$30\theta + (1 - \theta)e^{-\lambda_3}\left(1 - \left(e^{-\lambda_3}\right)^{29}\right)/\left(1 - e^{-\lambda_3}\right) = 30P_2$$
$$\lambda_3 = -ln(h).$$

We are now in a position to show the retention rate of the old customers (p_k^j) in this month. Denote the number of users who entered in k_{th} month and did not leave as B_{kj}, then

$$B_{kj} = \left(1 - F\left(30(j - k - 1) + 15\right)_{adj}\right)U_2, 0 < k < j. \qquad (15)$$

By the theorem of conditional probability, we have

$$
\begin{aligned}
p_k^j &= P(t_d > 30(j - k) + 15 | t_d > 30(j - k - 1) + 15), k < j \\
&= \frac{P(t_d > 30(j - k) + 15)}{P(t_d > 30(j - k - 1) + 15)} = \frac{1 - F(30(j - k) + 15)_{adj}}{1 - F(30(j - k - 1) + 15)_{adj}} \\
&= \frac{\theta + (1 - \theta)e^{-\lambda_3(30(j-k) + 15)}}{\theta + (1 - \theta)e^{-\lambda_3(30(j-k-1) + 15)}}
\end{aligned}
\qquad (16)
$$

The total number of retained costumers by this month ($P_1^j U_1^j$) is the sum of retained customers for those who entered in the past several months respectively:

$$B_{1j}p_1^j + B_{2j}p_2^j + \cdots + B_{(j-1)j}P_2 = P_1^j U_1^j \qquad (17)$$

$$U_1^j = U_2^{j-1}P_2 + U_1^{j-1}P_1^j \qquad (18)$$

Combining these two equations, we could obtain:

$$U_1^j = \begin{cases} P_2 U_2^{j-1} & j = 2 \\ P_2 U_2^{j-1} + \sum_{i=1}^{j-2} U_2^i (1 - F(30(j - i - 1) + 15)) & j > 2 \end{cases}, \; j > 1 \qquad (19)$$

It follows that

$$P_1^j = \frac{U_1^{j+1} - U_2^j P_2}{U_1^j}, \quad j > 1 \tag{20}$$

At this point, we have obtained all parameters needed to calculate λ_1 and λ_2.

3.2   Bike-Sharing Company Deposit Benefit

Daily Operation Income. We need first estimate the total usage frequency of rental service in a given month (MUF_j). It is reasonable to approximate the number of monthly active users (MAU_j) to MUF_j. The ratio between the average number of daily unlocking users (N_0^j) and monthly active users is assumed as a proportion of unlocking users denoted by C. That is, if the MAF_j is 100 and C is 0.2, then the average number of daily unlocking users is 20 in this month approximately. In this logic, one can get an effective estimation for MUF_j as

$$MUF_j = 30N_0^j \times C \tag{21}$$

where 30 means there are 30 days in one month. Let p_0 represents the price of a single service set by the company. Then estimation of the daily operation income of the firm is given by

$$I_j = N_0^j \times C \times p_0 \tag{22}$$

As a common business practice, bike sharing companies may provide coupons, special offers, or other kinds of promotions, resulting that p_0 needs to be discounted by a rate of $0 < O < 1$, then a more accurate daily operation income estimation is obtained by

$$I = N_0 C \times p_0 (1 - O) \tag{23}$$

Revenue Reinvestment. We start with the total expected deposit in the current month (D_0^j) which includes deposit from new users (D_1^j) and old users (D_2^j) with the following relation

$$E[D_0^j] = E[D_2^j] + E[D_1^j] \tag{24}$$

Since new users in the month can be divided into retained new users and new users who started but stopped using the firm's service within the same month, we may expand the D_1^j into

$$E[D_1^j] = E[d_1^j] + E[d_2^j] \tag{25}$$

Here, d_1^j and d_2^j stand for the expected amount of daily deposits for such two types of new users respectively. Same logic applies to the old users to obtain

$$E[D_2^j] = E[d_3^j] + E[d_4^j] \tag{26}$$

where d_3^j defines the expected amount of daily deposit of the retained old users in the month and d_4^j of the old users not retained in the same month. In order calculate D_0, it is necessary to have the expectations of deposit contributions of d_i^j (for $i = 1, 2, 3, 4$) by their corresponding service durations (T_i^j) and number of users (n_i^j) in this month.

$$E[d_i^j] = E[T_i^j]E[n_i^j]g \tag{27}$$

Assume that the entry time of new users (T') follows uniform distribution, i.e.,

$$P(T' = 1) = P(T' = 2) = \cdots P(T' = 30) = \tfrac{1}{30} \tag{28}$$

$$E[T'] = 15 \tag{29}$$

So, the number of customers whose entry time is the i_{th} day of the month E_i can be estimated by

$$E_i = P(T' = i)U_2, \ i = 1, 2, 3 \ldots 30. \tag{30}$$

Therefore, the new retained users' service duration is $(30 - T') \sim U(0, 30)$. Since the number of retained new users in this month is $P_2 \times U_2^j$, the expectation of d_1^j could be obtained by

$$E[d_1^j] = \frac{(E[30-T'] \times P_2 \times U_2^j)g}{30} = \frac{(15 \times P_2 \times U_2^j)g}{30} = \frac{(P_2 \times U_2^j)g}{2} \tag{31}$$

And the duration of old retained users is 30 because their deposit is with the firm in the whole month. Similarly, we could get the expectation of d_3:

$$E[d_3^j] = \frac{(30 \times P_1^j \times U_1^j)g}{30} = (P_1^j \times U_1^j)g \tag{32}$$

In addition, it is easy the find the n_2^j and n_4^j by

$$\begin{cases} E[n_2^j] = (1 - P_2)U_2^j \\ E[n_4^j] = (1 - P_1^j)U_1^j \end{cases} \tag{33}$$

The estimation of d_4^j and d_2^j are more complex, since their distributions are adjusted exponential distributions. For old exited users, we could only consider their departure time. So, $E[T_4^j]$ can be achieved

$$E[T_4^j] = \sum_{i=0}^{30} P(T_4^j = i) \times i \tag{34}$$

Here $P(\,T_4^j = i)$ could be written as:

$$(T_4^j = i) = \begin{cases} P(i+1 > T_4^j > i) & i = 0,1,2\cdots 29 \\ P(T_4^j \ge 30) & i = 30 \end{cases}$$
$$= \begin{cases} P(30j+i+1 > t_d > 30j+i | t_d > 30j) & i = 0,1,2,\cdots 29 \\ P(T_4^j \ge 30) & i = 30 \end{cases}$$

And the expansion of it is

$$P(T_4^j = i) = \begin{cases} \dfrac{\sum_{k=1}^{j-1}((H+1)_{adj}-F(H+1)_{adj})B_{kj}}{(H-i_{adj})U_1^j} & i = 0,1\cdots 29 \\ 1 - P(0 < T_4^j < 30) & i = 30 \end{cases} \tag{35}$$

Here H is defined as

$$H = 30(j-k-1) + E[T'] + i.$$

Combining these equations, we obtain

$$E[T_4^j] = \sum_{i=0}^{29} P(T_4^j = i) \times i + (1 - P(0 < T_4^j < 30)) \times 30.$$

With respect to d_2^j, both entry time and departure time should be considered. Denote the entry time of new users as T_2^i. For $i = 1, ...,30$, we could deduce $E[T_2^i]$ and $E[T_2]$ as

$$E[T_2^i] = \sum_{j=2}^{i-1}((F(j)_{adj} - F(j-1)_{adj})(j-1)) + (1 - F(j)_{adj}) \times i. \tag{36}$$
$$E[T_2] = \sum_{i=1}^{30} E[T_2^i]P(T' = i) = \frac{\sum_{i=1}^{30}E[T_2^i]}{30}$$

Now we obtain a preliminary estimation of D_0^j as

$$E[D_0^j] = E[d_1^j] + E[d_2^j] + E[d_3^j] + E[d_4^j] \tag{37}$$

However, considering the various promotions and third-party allowances, the total deposit D_T^j needs to be discounted by $0 < A < 1$:

$$D_T^j = (1 - A)D_0^j \tag{38}$$

Finally, let K_j be the monthly investment revenue. Assuming that monthly interest rates of monetary fund and long-term investment be r_1 and r_2, respectively, the expectation of K_j is given by

$$K_j = R_j \times r_1 + \left(D_T^j - R_1^j\right) \times r_2.$$

4 Empirical Results

This model would be tested through a case study of *Mobike* using the data ranging from July 2016 to June 2017. Most of the data used are from publicly accessible internet data platforms, such as Quest mobile, Trust Data and I-Research. As sampling estimations were typically used in the data reporting process, most of the available data, including *MAUj* and *U2*, are not perfectly accurate. To mitigate such bias, official statistical database from China Internet Network Information Center are used for calibration. By comparison, those indirect data, such as ratio data, are more precise. The following Table 1 provides a summary of the monthly operating data of *Mobike* used in the test.

It should be noted that certain variables such as the discount ratio (O) and the level of risk (α) are not publicly available. In addition, the values may be various for different firms. Therefore, we appeal to the Monte-Carlo approaches to explore the financial significance of the variables.

Table 1. Mobike operating data

Time	16-Jul	16-Aug	16-Sep	16-Oct	16-Nov	16-Dec
U_2	193.55	683.63	1474.72	1921.57	2395.48	4170.24
MAU	384.5	832.05	2190.449	3696.39	5686.81	9793.224
	17-Jan	17-Feb	17-Mar	17-Apr	17-May	17-Jun
U_2	3859.24	7300.89	10451.53	12690.2	9720.23	8451.45
MAU	11519.36	19196.53	32957.10	45330.36	48329.08	47446.52
Mobike Ratio Data						
P_2		0.7	C	0.5	P_1	1

The unit of measurement for these data is number of people reduced by factor of 1,000. The data in red shows the progress of its order of magnitudes. Purple highlights indicate data affected by cooperation between Mobike and Wechat.

4.1 Daily Income

The following Table 2 simulates the average daily operating income for the company *Mobike* in *j*th month based on different size of *O*.

The table shows that the average daily operating income has increased dramatically in 2017, far exceeding that of the previous year. In addition, increasing sales promotion has a significant positive impact on daily income. On the contrary, the amount of monthly active users and the unlocking rate will decrease correspondingly, which may have a negative impact on income.

Table 2. Average daily income

Time	16-Jul	16-Aug	16-Sep	16-Oct	16-Nov	16-Dec
[0.4]	55368	119816	315424	532281	818901	1410244
[0.3]	64596	139785	367995	620994	955385	1645261
[0.2]	73824	159754	420566	709708	1091868	1880299
	17-Jan	17-Feb	17-Mar	17-Apr	17-May	17-Jun
[0.4]	1658788	2764300	4745823	6527572	6959388	6832299
[0.3]	1935253	3225017	5536793	7615501	8119286	7971015
[0.2]	2211718	3685734	6327764	8703430	9279184	9109732

The values in "[]" represent the size of the special offers discount rate. Red highlights show the progress of its order of magnitudes.

4.2 Reserve Pool Parameters

The number of old users and the retention rate of old users per month are estimated based on the size of in the corresponding month and the results are presented by Table 3.

We observe that the size of θ and the number of old users is positively correlated. This implies that the company can increase the number of their customers by increasing the parameter θ. It is also shown that the company obtains more loyal customers as time

Table 3. Old user data

Number of Old Users						
Time	16-Jul	16-Aug	16-Sep	16-Oct	16-Nov	16-Dec
[0.45]	154.46	557.40	1303.37	2211.06	3310.48	5202.05
[0.55]	161.57	615.54	1494.02	2586.04	3921.11	6226.99
[0.65]	168.67	673.69	1684.66	2961.03	4531.74	7251.93
	17-Jan	17-Feb	17-Mar	17-Apr	17-May	17-Jun
[0.45]	6944.26	10233.86	14939.46	20651.13	25025.49	28828.57
[0.55]	8354.11	12373.04	18123.36	25103.82	30450.11	35098.26
[0.65]	9763.97	14512.22	21307.26	29556.50	35874.73	41367.96
Retention Rate of Old Users						
Time	16-Jul	16-Aug	16-Sep	16-Oct	16-Nov	16-Dec
[0.45]	0	0.5105	0.4863	0.6643	0.7388	0.6895
[0.55]	0	0.8479	0.7500	0.8306	0.8678	0.8435
[0.65]	0	1.1569	0.9683	0.9592	0.9641	0.9560
	17-Jan	17-Feb	17-Mar	17-Apr	17-May	17-Jun
[0.45]	0.8155	0.7377	0.7449	0.7877	0.8823	0.9155
[0.55]	0.9077	0.8693	0.8734	0.8950	0.9419	0.9583
[0.65]	0.9738	0.9628	0.9649	0.9702	0.9835	0.9882

The unit of measurement for U_i^f is number of people reduced by factor of 1,000. The values in "[]" represent the size of θ. Green highlights signify unreasonable data.

goes on, and the growth of loyal old customers could increase the retention rate of old users. In addition, some simulated results are unreasonable when unreasonable values are chosen for the parameter θ. In actuality, the retention rate of old users in any given month should be smaller than 1 and larger than the retention rate of new users, since the retention rate always increases as time goes on. Therefore, we assume that $\theta = 0.55$ for the purposes of calculating other variables in the next subsection (Fig. 1).

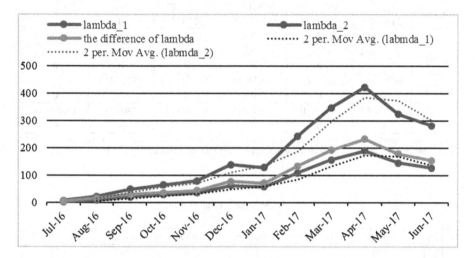

Fig. 1. The Figure shows that λ_1 and λ_2 increase from July 2016 to April 2017, and peak in April 2017, after which they begin to decrease. From business perspective, λ_1 predicts the number of exited users and λ_2 the added users. As such, λ_2 is always larger than λ_1 implies that the number of the customers increases indefinitely in time.

4.3   Reserve Pool and Investment Return Simulation

As an example of Monte Carlo simulation, one million random numbers are created (for visualization purposes, the unit of λ reduced by factor 1,000) after input of the two parameters, λ_1 (70 thousand times) and λ_2 (125 thousand times) of the Poisson distribution. Then the risk that the reserve pool is not enough to cover the refund requests is graphically shown by the simulated graphics of Skellam distributions. As the company's risk level parameter and the capital turnover coefficient β are not publicly known, the results of R are based on simulated values of α, β.

From the Table 4, we observe that if β is big enough, R will be zero. This suggests that if a company has a large enough daily income and fast enough fund turnover, it would have no need of a reserve pool. With regards to α, a smaller value of it would bring a larger size of reserve pool. The financial interpretation of this result is that a company has more available income for high return projects should predict a lower risk. The predicted monthly expected amount of the total deposit (D_T) of the firm and the funds invested in long-term projects are given by Table 5.

Table 4. Reserve pool

Time	16-Jul	16-Aug	16-Sep	16-Oct	16-Nov	16-Dec
[5%]	2287272 **672362**	6302340 **2807704**	10557028 **1357140**	11666783 **0**	12566843 **0**	19617113 **0**
[1%]	3184272 **1569362**	7498340 **4003704**	12351028 **3151140**	13759783 **0**	14958843 **0**	22607113 **0**
[0.5%]	3782272 **2167362**	8993340 **5498704**	14444028 **5244140**	15852783 **327917**	17350843 **0**	25896113 **0**
	17-Jan	17-Feb	17-Mar	17-Apr	17-May	17-Jun
[5%]	15711664 **0**	29178641 **0**	33866040 **0**	34030115 **0**	13240457 **0**	6393151 **0**
[1%]	18701664 **0**	33065641.8 **0**	38650040 **0**	39412115.3 **0**	17725457 **0**	10579151 **0**
[0.5%]	21990664 **0**	37550641 **0**	44032040 **0**	44794115 **0**	22808457 **0**	15064151 **0**

Data in bold signifies $\beta=0.5$ and non-bold signifies $\beta=0.1$. The values in "[]" stand for the risk level α. Red highlights show the progress of its order of magnitudes and blue ones indicate the rate has become zero.

Table 5. Mobike's total deposit

Time	16-Jul	16-Aug	16-Sep	16-Oct	16-Nov	16-Dec
[0.5]	255	241	210	292	382	647
[0.55]	257	250	213	297	390	660
[0.6]	261	301	291	443	597	1036
	17-Jan	17-Feb	17-Mar	17-Apr	17-May	17-Jun
[0.5]	1321	2871	3386	4533	5750	5439
[0.55]	1935	3225	5536	7615	811	7971
[0.6]	2211	3685	6327	8703	927	9109

The values in "[]" represent θ.

5   Conclusion

This article analyzes the cashflow management of bike-sharing firms. The main revenue source of bike sharing companies is service charge and returnable deposits from customers. In order to put forward a reasonable and effective model to analyze the financial returns of the bike sharing companies, risk models following compounded Poisson distributions are proposed and empirically tested. Assuming that the accumulated deposit and service revenue can be divided into refund reserve part and the part possible for long term investment, the key idea of the model is comprised of the size estimation of the reserve pool, the optimal solution of which has been achieved via Monte Carlo methods. To our best knowledge, this article provides the first systematic framework to mathematically model the cashflow management of bike sharing industry.

In addition, the innovative approaches contained in the current paper are applicable to other similar situations. For instance, the special form of the compound Poisson distribution in this article can be used to determine the optimal size of the business reserve for other similar industries. Furthermore, an adjusted exponential cumulative distribution function, shown as effective for the current study, could be used to estimate the retention rate for other industries also.

The model has some limitations and a number of aspects could be further improved. The first limitation arises from the lack of raw data from the industry. Second, the mathematical methods used to process the data seem rather complex at this stage. Moreover, the estimates of long-term investments in our analysis may be over-conservative, and industry specific knowledge may be helpful to provide even well-founded forecasts. All such limitations are hopefully to be addressed in future studies.

Acknowledgement. The authors are grateful to the valuable comments from the anonymous reviewers, which helped to improve the earlier version of the paper. The research is supported by Natural Science Foundation of China (NSFC.11501044, 1181339), Jiangsu Science and Technology Basic Research Programme (BK20171237), Key Program Special Fund in XJTLU (KSF-E-21), Research Development Fund of XJTLU (RDF-2017-02-23), and partially supported by NSFC (No.11571002,11571047,11671049,11671051,61672003).

References

1. Bart, M.E., Clinch, G.: Scale effects in capital markets- based accounting Michael Ahillen, Derlie Mateobabiano, and Jonathan Corcoran. Dynamics of bike sharing in washington, dc and brisbane, australia: Implications for policy and planning. Int. J. Sustain. Transp. **10**(5), 150106050055009, 2016 (2001)
2. Botsman, R.: The sharing economy lacks a shared definition (2013)
3. Boudiche, M.: The determinants of profit forecast by tunisian companies. Asian Econ. Financ. Rev. **3**(9), 1180–1194 (2013)
4. Delbaen, F., Haezendonck, J.: Classical risk theory in an economic environment. Insur. Math. Econ. **6**(2), 85–116 (2006)
5. Ertz, M., Durif, F., Arcand, M.: Collaborative consumption or the rise of the two-sided consumer. Int. J. Bus. Manage. **4**(6), 195–209 (2016)
6. Feng, R.X., Ma, Y.: Innovation of business model under sharing economy - take ofo for example. Jiangsu Bus. Theor. **19**, 28–29 (2017)
7. Ge, W.J.: Discussion on the construction and profit model of the sharing-bicycle value network model: take moby and ofo for example. China Bus. Trade **15**, 174–176 (2017)
8. Jorion, P.: Value At Risk: The New Benchmark for Managing Financial Risk. Risk Management. McGraw-Hill, New York (2006)
9. Lan, J., Ma, Y., Zhu, D., Mangalagiu, D., Thornton, T.F.: Enabling value co-creation in the sharing economy: the case of mobike. Sustainability **9**(9), 1504 (2017)
10. Li, Y.J., Li, X.S., Jiang, Y.X.: Study on information entropy as a solution to measure risk in financial market. Oper. Res. Manage. Sci. **16**(5), 111–116 (2007)
11. Maasoumi, E., Racine, J.: Entropy and predictability of stock market returns. J. Econometrics **107**(1), 291–312 (2002)
12. Magali, D., Alexer, O., Yves, P.: E-business model design, classification, and measurements. Thunderbird Int. Bus. Rev. **44**(1), 5–23 (2010)

13. McNeil, A., Frey, R., Embrechts, P.: Quantitative risk management: concepts. Tech. Tools **101**, 10 (2005)
14. Schor, J.: Debating the sharing economy (2014)
15. Shaheen, A., Guzman, S., Zhang, H.: Bikesharing in Europe, the Americas, and Asia: past, present, and future. Transp. Res. Rec. **2143**, 159–167 (2010)
16. Yan, T.H., Feng, Q.S., Xu, J.: Study on measurement model of risk reserve for commercial banks. J. Quant. Tech. Econ. **18**(4), 84–87 (2001)

Predicting Material Requirements in the Automotive Industry Using Data Mining

Tobias Widmer$^{(\boxtimes)}$, Achim Klein, Philipp Wachter,
and Sebastian Meyl

University of Hohenheim, Stuttgart, Germany
{tobias.widmer,achim.klein,philipp.wachter,
sebastian.mey}@uni-hohenheim.de

Abstract. Advanced capabilities in artificial intelligence pave the way for improving the prediction of material requirements in automotive industry applications. Due to uncertainty of demand, it is essential to understand how historical data on customer orders can effectively be used to predict the quantities of parts with long lead times. For determining the accuracy of these predications, we propose a novel data mining technique. Our experimental evaluation uses a unique, real-world data set. Throughout the experiments, the proposed technique achieves high accuracy of up to 98%. Our research contributes to the emerging field of data-driven decision support in the automotive industry.

Keywords: Predictive manufacturing · Material requirements planning ·
Data mining · Artificial intelligence · Automotive industry

1 Introduction

Predictive manufacturing has become a major challenge to the industrial production sector [1]. Manufacturers integrate business information systems into their production environment to create competitive advantages and to enhance efficiency and productivity [2]. These information systems increasingly use artificial intelligence for planning and controlling manufacturing operations [3]. In particular, the automotive industry is progressively adopting methods from artificial intelligence research in a wide range of industrial applications. For instance, Audi has developed intelligent Big Data capabilities to optimize their production and sales processes [4].

Cars are subject of increasing individualization, which exemplifies in the high number of possible variants. This complexity puts a burden on supply chain management in the automotive industry and calls for intelligent business information systems, which integrate supply and demand [5, 6]. For instance, BMW offers more than 10^{32} car variants, out of which several thousands are in fact ordered by customers [7]. Due to the high variance in products given globally distributed production plants, car manufacturers use planned orders based on forecasts to optimize their material requirements planning. As manufacturers transit from build-to-stock to build-to-order strategies, planning processes are reorganized by implementing advanced planning systems such as predictive manufacturing. In particular, quantities of car parts with long lead times must

© Springer Nature Switzerland AG 2019
W. Abramowicz and R. Corchuelo (Eds.): BIS 2019, LNBIP 354, pp. 147–161, 2019.
https://doi.org/10.1007/978-3-030-20482-2_13

be predicted accurately to prevent shortages and excess stock, respectively. Predictive manufacturing systems provide tools and methods to process historical data about customized orders into information that can explain planning uncertainties and support managers in making more informed decisions. These decisions typically concern strategies for planning material requirements along the entire supply chain.

To facilitate an efficient production in the presence of long lead time suppliers, manufacturers depend on accurate estimates about the material requirements for production. The increasing number of available options and option combinations for vehicle equipment entails highly complex and interdependent parts requirements lists (PRL) that are necessary to build a vehicle. Due to uncertainties emerging from suppliers with long lead times, however, manufacturers do not know in advance the exact quantity of the parts and components needed at each production plant [8]. While manufacturers use historical customer orders to estimate future sales and analytical high-level models for production planning, they have not yet exploited the full potential of their data to predict fine-grained material requirements. Therefore, we contribute a technique that exploits a unique dataset of fully specified vehicle orders with all product options and required material parts for predicting the material requirements of parts with long lead times.

Given the incomplete vehicle specifications of estimated future customer orders, it is essential to understand how historical orders can be used more effectively to improve the prediction for parts with potentially long lead times. We aim to enhance this understanding by proposing a data mining technique for predicting the quantities of parts with long lead times. These parts must be ordered at a time where the associated customer orders are not yet available. Therefore, we base our prediction on historical customer orders. We represent these orders as vectors in which each element corresponds to the frequency of a product option ordered by a set of customers. First, we exploit the concept of cosine similarity for quantifying the similarity between orders of different customer sets [9]. Then, we select the most similar set of customer orders and use the associated set of known required parts as predicted parts for the estimated set of future orders. Finally, we quantify the prediction quality of our approach using accuracy defined as the ratio between the predicted quantity and the actual quantity of parts.

To validate our proposed technique, we carried out a set of experiments using a unique data set, covering real-world purchase orders placed at an international automobile manufacturer during a fixed period of time. These orders contain information about the specific combinations of product options ordered by customers and associated required material parts during a fixed production cycle.

We calculated the accuracy of the prediction for customer order groups of varying sizes and uncertainty levels. We find that larger customer order groups yield higher accuracy across different uncertainty levels. Specifically, we find that our proposed technique achieves an accuracy between 88% and 98% throughout all experiments. This finding is consistent with the growing trend toward modularization in the automotive production industry [10]. As digital technology platforms emerge in smart production environments, car manufacturers can begin exploiting digitalized infrastructures to design and control innovative components of higher levels of standardization [11]. Our findings help explain to what degree the increasing modularization and standardization of components impact the prediction of the requirements for parts with long lead times.

The remainder of this paper is organized as follows. The next section discusses prior research. In Sect. 3, we present the proposed data mining technique for predicting material requirements. In Sect. 4, we report on the experimental evaluation and discuss our findings. Finally, we provide our conclusion in Sect. 5.

2 Prior Research

We discuss prior research on (1) predicting material requirements in industrial production under demand uncertainty, and (2) data mining approaches for estimating the similarity between different sets of product orders.

2.1 Predicting Material Requirements

Manufacturing industries use material requirements planning (MRP) systems for inventory management and for planning and forecasting the quantities of parts and components required for production. Since their first implementation in the 1960s, MRP systems evolved into MRP II and later into enterprise resource planning (ERP) systems. MRP systems use master and transaction data as input. While master data include information about structure and variants of each component, transaction data are created when a customer places an order [12]. In the presence of demand uncertainty, planning systems typically follow either a supply-oriented or a demand-oriented approach.

First, in **supply-oriented approaches**, manufacturers estimate the required quantities by optimizing a given objective function subject to production capacity, storage, and market constraints [12]. Gupta and Maranas [13] develop a stochastic model for minimizing the total cost of a multi-product and multi-site supply chain under uncertain demand. They solve the objective function using optimization methods in two stages. First, all manufacturing decisions are made before the demand is known. Second, inventory levels, supply policies, safety stock deficits and customer shortages are determined after the demand is already known while taking the quantities produced in the first stage into account. The main difference to our approach is that Gupta and Maranas analytically solve a model on the level of different products while we use a fully data-based approach for predicting material parts requirements of products. Whereas Gupta and Maranas focus on total production and logistics cost, we implicitly optimize cost by predicting material requirements for reducing potential over- or underproduction.

Second, **demand-oriented approaches** for production planning focus on forecasting future demand and adjusting the production accordingly. Common techniques used for this purpose include moving averages and exponential smoothing based on historical customer orders [12]. Zorgdrager et al. [14] compare various regression and statistical models to forecast the material demand for aircraft non-routine maintenance. They find that the exponential moving average model offers the best tradeoff between forecast errors and robustness over time.

Lee et al. [15] model uncertain demand using fuzzy logic theory. They integrate triangular fuzzy numbers in a part-period balancing lot-sizing algorithm to determine the optimal lot size under uncertain demand. Chih-Ting Du and Wolfe [16] propose a decision support system to determine the optimal ordering quantity for materials and the safety stock. The system utilizes fuzzy logic controllers and neural networks and takes variables such as the ordering costs, inventory carry costs and uncertainty into account. The role of the neural networks is to increase the fault tolerance of the system and increase rule evaluation performance by learning and replacing imprecise and complicated fuzzy if-then rules. In contrast to Lee et al., we model uncertainty of demand in terms of ordered vehicle configurations in a simpler way by randomly removing product options from individual orders in our dataset.

Steuer et al. [17] predict the total demand for new automotive spare parts in three steps. First, they cluster automotive parts with similar product life cycle curves using k-medoids clustering and chi-square distances. The optimal number of clusters is determined by calculating the Dunn index and the silhouette width of the clusters for various number of clusters. Second, their approach identifies common features that products in the same cluster share. Third, they use a classification model to match new parts to clusters by estimating the feature similarity between them. They find that among all evaluated algorithms, support vector machines achieve the highest accuracy of 68.4%. We extend the cluster method of Steuer et al. by integrating the different option combinations of the vehicles ordered by customers. Whereas Steuer et al. focuses on spare parts only, we consider the material requirements for the complete production process.

In summary, existing approaches do not internalize real-world customer orders that include all possible option configurations. We contribute a novel data mining technique for predicting fine-grained material parts requirements given uncertain demand about vehicle option configurations. To this end, our approach can be used to complement existing approaches for fine-tuning the prediction of material parts.

2.2 Data Mining Approaches

Data-based prediction of required material parts in a production supply chain relates to analyzing historical parts requirements in transaction data for a certain product demand pattern. As such, predicting parts requirements involves comparing imprecisely specified current demand with closest known demand pattern for which the required parts are known. Thus, we face two problems: (1) forming clusters in historical transaction data (containing product options and required parts) to model potential product demand with respect to certain product options, and (2) measuring closeness of imprecise current demand vs. historical demand patterns. Because we tackle both problems on the basis of the vector space model, we first motivate and review the model and specific approaches for the two outlined problems.

The vector space model is a well-established model in the fields of data mining and text mining [18]. This model has been widely used for pattern matching and in particular for text retrieval and text classification [19]. In the scope of this work, we are interested in comparing and matching patterns of imprecise current demand with fully specified historical demand, for which required parts are known. Thus, the quantities of order details (e.g., ten times product type A, five times product option B) are used as elements of a vector. While the vector-based approach received little attention for predicting parts requirements in prior research, we contribute to existing literature by transferring previous findings from using the vector space model in text mining to demand prediction for car production.

A major challenge in interpreting demand patterns as vectors in vector space is the assumption of linear independence of the dimensions of a vector space. It seems obvious that the dimensions offered by product types and product options are not fully independent. However, the quantification of text as vectors by interpreting the words of the vocabulary as dimensions and counting word occurrences in a vector's elements also clearly violates the independence assumption. The reason is that words in a sentence or a full text depend on each other. The same is true for product options, when cars are configured by customers (e.g., demand for certain luxury product options might be correlated). Despite the obvious violation of independence, text mining research has shown that text classification approaches still achieve high accuracy [20]. These findings also apply to our research as reported in Sect. 4.2 and discussed in Sect. 4.3.

Another challenge includes the imprecise formulation of demand patterns that must be represented as vectors. Imprecise or uncertain knowledge means that some quantities might not be known exactly. Furthermore, the quantities for some options might be completely unknown. Thus, the corresponding vector elements are zero, which leads to sparsity in the vector. To this respect, text classification research provides evidence for high performance despite sparse vectors, referring to the application of Support Vector Machines [21]. Our data mining technique is different because it is based on cosine similarities at its core. However, our results indicate high predictive performance.

Several clustering approaches are available for addressing the problem of forming clusters of historical transactions to create synthetic demand patterns. A possible approach consists in selecting transaction data randomly to form a cluster. Another option is choosing transactions based on similarity. In this case clustering algorithms from the field of unsupervised machine learning can be used. A prominent example for such an algorithm is k-means clustering [22]. K-means clustering partitions transaction data in the vector space by iteratively forming k clusters around centroid vectors with a minimum sum of squared distances of all other vectors with respect to the centroid cluster. Apart from this algorithm, research in data mining examines approaches based on hierarchy, fuzzy theory, distribution, density, graph theory, grids, fractal theory, and other models [23].

A number of approaches for addressing the problem of closeness of vectors have been proposed. Examples include the Minkowski distance, Euclidian distance, cosine similarity, Pearson correlation distance, and the Mahalanobis distance [23]. In the context of text mining, the most common approach is the cosine similarity. Cosine similarity measures the angle between two vectors in the same vector space. Uncertain demand, represented as vector with elements counting product options, is compared to fully specified demand vectors by the angles between the vectors. It is then assumed that the demand vector with the smallest angle is the most similar to the uncertain demand vector.

We address both problems in vector space, i.e., forming clusters and measuring closeness. Thus, we evaluate the suitability and performance (i.e., prediction quality) of our approach and contribute to the transfer of knowledge in text-related research in information systems (e.g., [24]) for predicting material requirements.

3 Data Mining Technique

We describe our data mining technique for predicting material parts by providing formal notations, illustrating its use in an example, and defining an accuracy measure. The proposed technique predicts quantities for parts with long times based on historical customer orders.

3.1 Measuring Similarity of Customer Order Groups

A single customer order describes a fully customized vehicle as ordered by a customer (e.g., through a web-based car configurator). A typical customer order includes a car configuration such as car model, engine type, navigation system, electric exterior mirrors, sunroof, and so on. Each customer order is accompanied by a parts requirements list (PRL). This list is used by the production plant to assemble the vehicle.

Now suppose the manufacturer wants to predict the quantity of parts required to produce a set of cars X given a basic configuration of product options. To achieve this, all historical customer orders are divided into groups randomly. All groups are sized equally by a pre-determined size. Let $G = \{g_1, \ldots, g_n\}$ be the set of all customer order groups. Further, let $O = \{o_1, \ldots, o_m\}$ be the set of distinct options present in G and X. Each group and also X are then represented by an m-dimensional vector $\vec{o_g}$. Further, let $f(g, o)$ denote the frequency of option $o \in O$ in group $g \in G$. Then, the vector representation is given by

$$\vec{o_g} = (f(g, o_1), \ldots, f(g, o_m)) \tag{1}$$

The vector representation assumes linear independence of the dimensions, which may not hold, but still the approach has achieved good results in other fields [20]. Once the vectors for all groups are formed, we measure the similarity between the set X of cars to produce and a historic group of cars $g \in G$ by calculating the cosine of the associated angles. We use cosine similarity because of its simplicity and effectiveness to get an initial validation of our data mining technique. The cosine similarity between vectors $\vec{o_g}$ and $\vec{o_X}$ can be derived using the Euclidean dot product formula,

$$S(\vec{o_g}, \vec{o_X}) := \cos(\theta) = \frac{\vec{o_g} \cdot \vec{o_X}}{\vec{o_g} \cdot \vec{o_X}} \tag{2}$$

Because each dimension within the vectors $\vec{o_g}$ and $\vec{o_X}$ equals the frequency of a distinct option in the corresponding groups and these frequencies cannot be negative, the cosine similarity is bounded in the interval $[0, 1]$. Thus, the closer S gets to 1, the more similar are X and g. If $S = 1$, X and g are said to be identical. In other words, we use the required parts to produce cars in group $g \in G$ with highest associated similarity value regarding X as prediction for required parts in X.

3.2   Illustrative Example

We provide a simple example to illustrate our data mining technique for estimating the parts requirements based on historical customer orders. Suppose a manufacturer seeks to forecast the number of parts required for producing a set X of 10 cars out of which 8 will have option o_1 (e.g., navigation system) and 4 will have option o_2 (e.g., sunroof). All other options are unknown to the manufacturer at this point in time. Implicitly, the frequency of unknown options is assumed to be zero. Hence, the vector representation for set X is given by

$$\overrightarrow{o_X} = (f(X, o_1), f(X, o_2)) = (8, 4) \tag{3}$$

To predict the number of parts in the presence of uncertainty about the final configuration, we divide the complete set of customer orders into 10 random groups of size 10 respectively; that is, $G = \{g_1, \ldots, g_{10}\}$. Each of these groups contain a set of distinct options. For instance, suppose customer order group g_1 consists of 10 cars out of which 9 are configured with option o_1 (i.e., navigation system), 7 are configured with option o_2 (i.e., sunroof), and 3 are configured with option o_3 (e.g., electric exterior mirrors). The vector representation of this group is then given by

$$\overrightarrow{o_{g_1}} = (f(g_1, o_1), f(g_1, o_2), f(g_1, o_3)) = (9, 7, 3) \tag{4}$$

Likewise, all other groups $\{g_2, \ldots, g_{10}\}$ are represented by vectors containing the frequency of product options over all orders of a group. Using cosine similarity, our data mining technique now discovers the group that is most similar to set X:

$$S(\overrightarrow{o_{g_1}}, \overrightarrow{o_X}) = \frac{(9, 7, 3) \cdot (8, 4, 0)}{(9, 7, 3) \cdot (8, 4, 0)} \approx 0.9483 \tag{5}$$

The cosine similarity between set X and the remaining groups $\{g_2, \ldots, g_{10}\}$ is calculated analogously. Suppose group g_1 is closest to X according to cosine similarity; that is, among all cosine similarities, 0.9483 is closest to 1. Thus, we use PRL of group g_1 as prediction for the PRL of set X.

3.3   Measuring the Accuracy of Predictions of Parts Requirements

Each individual customer order is associated with a unique parts requirements list $PRL = \{i_1, i_2, \ldots, i_N\}$, where N is the number of unique parts required to produce the vehicle, and $i_{l \in \{1, \ldots, N\}}$ denotes the quantity of each part. For example, $PRL = \{12, 3, 5\}$ means that part 1 is required twelve times, part 2 three times, and part 3 five times.

We use an accuracy measure to quantify the quality of our prediction of required parts as follows. First, we subtract the quantity of each part in the predicted PRL from the respective quantity of that part in the benchmark PRL. Then, we aggregate the absolute differences in quantities and divide the resulting value by the total quantity of parts occurring in the benchmark PRL. Let $PRL_{Benchmark} = \{I_1, I_2, \ldots, I_K\}$ and $PRL_{Prediction} = \{J_1, J_2, \ldots, J_K\}$ denote the benchmark PRL and the predicted PRL, respectively, where K is the total number of unique parts in the union of both lists. Then, the difference between $PRL_{Benchmark}$ and $PRL_{Prediction}$ is given by

$$D := PRL_{Benchmark} - PRL_{Prediction} = \sum_{k=1}^{K} |I_k - J_k| \qquad (6)$$

Given this notation, the accuracy A is

$$A = 1 - \frac{D}{\sum_{k=1}^{K} I_k} \qquad (7)$$

Accuracy A gives the percentage of correctly predicted quantities within the benchmark PRL. For instance, if $A = 0.97$, the predicted PRL deviates by 3% from the benchmark PRL in the quantities of parts.

4   Evaluation

This section reports an experimental evaluation of the proposed data mining technique. We describe the setup, report the empirical results, and discuss the findings.

4.1   Experimental Setup

Our experiments used a unique data set of 47,499 actual orders received by a car manufacturer within a given time period. These orders contain fully customized car orders (i.e., including all configured options), associated with the specific PRL for each vehicle. For instance, in a random group of 20 orders, 9 vehicles were ordered with rear-view camera, 11 with active parking assist, 10 with cruise control, 2 with traffic sign recognition, and so on. The complete data set contained 55 different options for customers to choose from. Figure 1 shows an excerpt of the frequency of the configured options in this group taken from our unique data set.

We consider a scenario where the manufacturer does not know the exact option configurations for future orders. For example, the manufacturer estimates that from within 20 future orders, 5 orders contain a rear-view camera, 10 an active parking assist, and 12 a cruise control. At this point in time, the manufacturer does not have more information concerning all other potential options. In the presence of incomplete information, the manufacturer now wants to predict the quantities of those parts with long lead times that are required to produce these 20 vehicles.

To predict the PRL in this scenario, we randomly selected groups of varying sizes from {20, 100, 200, 500} as benchmark groups. Then, we systematically removed varying sets of options from these groups. By removing these sets of options, we mimic the incomplete option estimate provided by the sales manager. Next, we applied our data mining technique to identify the group in the historic order data set that is most similar to the benchmark group. Finally, we compared the aggregated PRL of the most similar group to that of the benchmark group. Figure 2 depicts the flow chart of the proposed data mining technique for predicting the PRL.

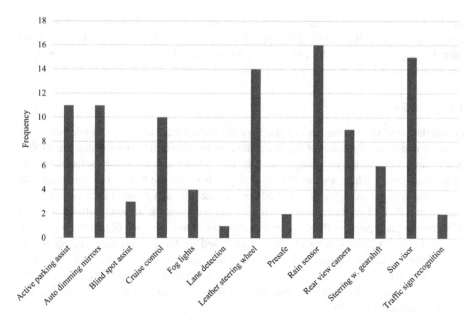

Fig. 1. Frequency of configured options in a random group of 20 orders (excerpt).

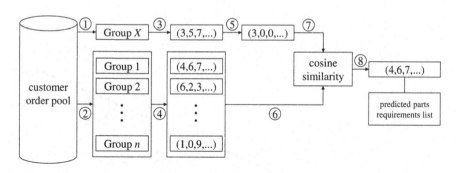

Fig. 2. Flow chart of the proposed data mining technique.

In step 1, we randomly select group X as the benchmark group, for which the PRL is known. Then, the historic order set is randomly divided into n distinct groups of equal size (step 2). Group X and groups 1 to n are then vectorized using the frequency of each option in the associated group (steps 3 and 4). In step 5, we define three uncertainty levels of the ordered vehicle configuration: low, medium, and high. For this purpose, we randomly remove a varying number of options from group X by setting the associated frequency to zero. When 13 out of 55 total options are removed, the level of uncertainty is said to be low (i.e., 23.6%). When 26 of 55 options are removed, the level of uncertainty is said to be medium (i.e., 47.3%). Finally, when 39 of 55 options are removed, the level of uncertainty is said to be high (i.e., 70.9%). In steps 6 and 7, we determine the cosine similarity between group X and groups 1 to n to identify the group that is most similar to the "stripped" group X. Once the most similar group has been found, we compare the associated PRL to the PRL of the original group X to estimate the accuracy of the predicted part requirements.

4.2   Results

To validate our data mining technique, we calculated the accuracy of our prediction (of required parts) as a function of the associated cosine similarity. We divided the 47,499 orders into 475 groups of group size 100. One group was randomly selected as the benchmark group. Then, we calculated all angles between the benchmark group and the remaining groups. Next, we determined the accuracy of each group's PRL compared to the original PRL of the benchmark group. Figure 3 depicts the accuracy obtained for all 475 groups. Each point in the diagram corresponds to the accuracy obtained for a single group. The red line illustrates the trend line based on linear regression. As shown in Fig. 3, the accuracy of the prediction increases as the cosine similarity increases. Notice that increasing cosine similarities result in decreasing angles between vectors. This result implies that our data mining technique is valid and can be applied to the problem studied in this work.

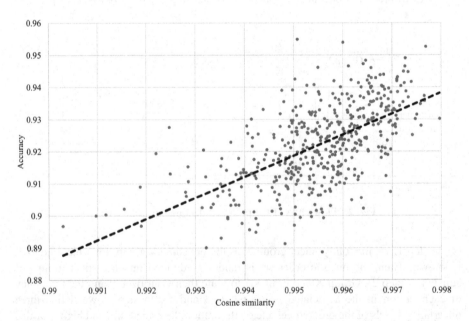

Fig. 3. Accuracy as a function of cosine similarity and linear regression trend line (red) (Color figure online).

After having successfully validated our approach, we now report on the results obtained in our simulation study. Table 1 presents the results of the simulation. For each group size 20, 100, 200, and 500, we calculated the accuracy of the PRL for the scenarios "none", "low", "medium", and "high." Here, scenario "none" corresponds to a benchmark group that contains all options (no uncertainty). Thus, if that exact group were contained in the data set, the accuracy would be 100%.

Table 1. Accuracy for varying group sizes and uncertainty levels.

Group size	Accuracy for uncertainty level			
	0% (none)	23.6% (low)	47.3% (medium)	70.9% (high)
20	91.98	88.16	88.14	88.91
100	93.02	92.92	91.60	91.27
200	97.30	96.50	96.18	96.18
500	97.73	97.41	97.02	97.02

Figure 4 depicts the results graphically. For group size 20 (blue line), the accuracy is decreasing for increasing uncertainty level, reaching 88.14% at uncertainty level medium. Then, the accuracy increases up to 88.91% for uncertainty level high. When a group contains 100 orders (yellow line), the accuracy decreases for all uncertainty levels with its highest value of 93.02% down to its lowest value of 91.27%. For group size 200 (green line), the accuracy decreases from 97.3% to 96.18% for all uncertainty levels. Finally, when 500 orders are grouped (grey line), the accuracy also decreases for all uncertainty levels, falling from 97.73% to 97.02%.

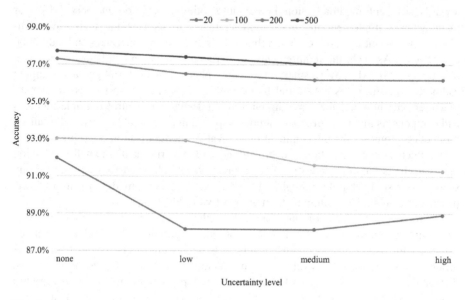

Fig. 4. Accuracy for different uncertainty levels at varying group sizes (Color figure online).

4.3 Discussion

Our experiments demonstrate the impact of uncertainty about future customer orders on the accuracy of predicting the material requirements for production in automotive industry applications. Our findings provide evidence for the efficacy of the proposed data mining technique to predict the quantities of parts with long lead times based on a large data set of historical customer orders. In the following paragraphs, we discuss the insights that can be obtained from our research.

First, we find that the proposed cosine similarity measure suits well for predicting material requirements. As the cosine similarity of vectors increases, the accuracy increases also (see Fig. 3). This finding implies that the frequencies of options within a historical customer order group correlate with the future requirements of parts subject to long lead times. The fact that the smallest observed cosine similarity between vectors is approximately 0.99 indicates that different order groups exhibit similar requirement lists for long lead time parts. In consequence, the quantities required to produce these vehicles can be predicted with an accuracy of close to 96%. This result indicates that potential violations of the assumption of linearly independent dimensions in the underlying vector space are not too detrimental to the accuracy results achieved by our technique.

Second, we find that larger customer order groups entail higher accuracy. As more customer orders are pooled, the quantities of parts and components required to produce these cars converge to those quantities in the associated benchmark group. This finding implies that the quantity of parts required for production becomes invariant as group sizes increase. In other words, individual parts and components are re-used by manufacturers for producing different car variants. This finding is consistent with the trend towards stringent modularization in the automotive production industry [10]. Car manufacturers implement modularization strategies to manage the increasing complexity and variant diversity of their vehicles by standardizing interfaces and individual components. As such, the requirements for mass customization can be achieved more effectively [25]. Moreover, smart production plants benefit from pervasive digital technology platforms as the central focus of the firm's innovation process. Car manufacturers can now use the same digital tools to design and control multiple modules and components that were dispersed among supplies in the past [11]. Hence, digitalized production promotes the development of innovative modularization concepts which in turn influences the prediction and procurement of material at distributed production plants in the automotive industry. The accuracy values obtained for increasing group sizes in our study thus help explain to what extent product modularity impacts the prediction of material requirements in automotive production.

Third, one advantage of our approach is that it can deal with high levels of uncertainty about the demand of possible option configurations. We find that accuracy decreases for increasing uncertainty levels. It is interesting to observe that for bigger group sizes the uncertainty level barely impacts the accuracy of the prediction. That is, if many customer orders are pooled, the number of parts and components required to produce this group of vehicles virtually matches that of the associated benchmark group. This finding corroborates that car manufacturers pursue a sustainable platform strategy for managing the complexity of their product variants. To this end, the transformation toward digitalized production encourages the integration of data-driven analytics into business information systems to advance current prediction methods in automotive industry applications.

5 Conclusion

The contribution of this research is a data mining technique for predicting the requirements of parts with long lead times in the automotive industry. To evaluate our approach, we used a unique data set containing actual customer orders received by an international car manufacturer. In a first step, our approach incorporates the concept of cosine similarity to discover similar customer order groups within the data set. Then, we aggregate the quantities of the required parts and components for producing the vehicles within these different groups. Finally, we calculate the accuracy of our prediction relative to a predefined benchmark group. We find that increasing group sizes result in higher accuracy across all uncertainty levels. As car manufacturers continue to optimize product modularization using digital platform technologies, standardized parts and components for producing cars facilitate an improved prediction of material requirements even in the presence of uncertainty concerning future customer orders.

From a managerial perspective, our study can support supply chain managers in making more informed decisions about choosing the appropriate customer group size for predicting the demand in parts and components with long lead times. Because larger group sizes imply higher accuracy, managers can pool heterogenous estimates about future customer orders based on production capacity. At the same time, managers can focus on small sets of equipment options when forecasting material requirements because varying uncertainty levels show little impact on accuracy.

Future research can be pursued in four directions. First, our experimental evaluation could be extended by implementing k-means clustering on the data set [9, 22]. Unlike the random group formation used in our technique, the k-means clustering algorithm divides the data set into k clusters relative to the nearest mean. For benchmarking purposes, the accuracy obtained by k-means clustering could then be compared to the accuracy achieved in our study. Second, while the cosine similarity measure suits well to obtain high accuracy, other measures of similarity such as k-median or k-means + + algorithms could be used for comparing customer order groups. Third, for comparing parts requirements lists, different weights could be placed at different quantities of components. This change could provide further insights of how economic factors such as price and economies of scale affect parts requirement predictions. Fourth, future research can assess the applicability of our approach to other industries in which large quantities of parts are needed and customers can individualize the products.

Acknowledgements. This work has been partially supported by the Federal Ministry of Economic Affairs and Energy under grant ZF4541001ED8. We would like to thank Hansjörg Tutsch for his valuable comments on earlier versions of this paper. We also thank Lyubomir Kirilov for helping to develop and improve parts of this paper.

References

1. Lee, J., Lapira, E., Yang, S., Kao, A.: Recent advances and trends in predictive manufacturing systems in big data environment. Manuf. Lett. **1**(1), 38–41 (2013)
2. Ghabri, R., Hirmer, P., Mitschang, B.: A hybrid approach to implement data driven optimization into production environments. In: Abramowicz, W., Paschke, A. (eds.) BIS 2018. LNBIP, vol. 320, pp. 3–14. Springer, Cham (2018)
3. Renzi, C., Leali, F., Cavazzuti, M., Andrisano, A.O.: A review on artificial intelligence applications to the optimal design of dedicated and reconfigurable manufacturing systems. Int. J. Adv. Manuf. Technol. **72**, 403–418 (2014)
4. Dremel, C., Herterich, M.M., Wulf, J., Waizmann, J.-C., Brenner, W.: How AUDI AG established big data analytics in its digital transformation. MIS Quart. Exec. **16**(2), 81–100 (2017)
5. Leukel, J., Jacob, A., Karaenke, P., Kirn, S., Klein, A.: Individualization of goods and services: towards a logistics knowledge infrastructure for agile supply chains. In: Proceedings of the AAAI Spring Symposium (2011)
6. Widmer, T., Premm, M., Kirn, S.: A formalization of multiagent organizations in business information systems. In: Abramowicz, W., Alt, R., Franczyk, B. (eds.) BIS 2016. LNBIP, vol. 255, pp. 265–276. Springer, Cham (2016)
7. Meyr, H.: Supply chain planning in the German automotive industry. OR Spectr. **26**, 447–470 (2004)
8. Lee, H.L.: Aligning supply chain strategies with product uncertainties. Calif. Manag. Rev. **44**(3), 105–119 (2002)
9. Baeza-Yates, R., Ribeiro-Neto, B.: Modern Information Retrieval, 2nd edn. ACM Press, New York (1999)
10. Takeishi, A., Fujimoto, T.: Modularization in the auto industry: interlinked multiple hierarchies of product, production and supplier systems. Int. J. Automot. Technol. Manage. **1**(4), 379–396 (2001)
11. Yoo, Y., Boland, R.J., Lyytinen, K., Majchrzak, A.: Organizing for innovation in the digitized World. Organ. Sci. **23**(5), 1398–1408 (2012)
12. Kurbel, K.E.: MRP: material requirements planning. In: Swamidass, P.M. (ed.) Enterprise Resource Planning and Supply Chain Management, pp. 19–60. Springer, Heidelberg (2013)
13. Gupta, A., Maranas, C.D.: Managing demand uncertainty in supply chain planning. Comput. Chem. Eng. **27**(8–9), 1219–1227 (2003)
14. Zorgdrager, M., Curran, R., Verhagen, W., Boesten, B., Water, C.: A predictive method for the estimation of material demand for aircraft non-routine maintenance. In: 20th ISPE International Conference on Concurrent Engineering (2013)
15. Lee, Y.Y., Kramer, B.A., Hwang, C.L.: Part-period balancing with uncertainty: a fuzzy sets theory approach. Int. J. Prod. Res. **28**(10), 1771–1778 (1990)
16. Du Chih-Ting, T., Wolfe, P.M.: Building an active material requirements planning system. Int. J. Prod. Res. **38**(2), 241–252 (2000)
17. Steuer, D., Korevaar, P., Hutterer, V., Fromm, H.: A similarity-based approach for the all-time demand prediction of new automotive spare parts. In: 51st Hawaii International Conference on System Sciences (HICSS 2018), pp. 1525–1532. Waikoloa Village (2018)
18. Salton, G., Wong, A., Yang, C.S.: A vector space model for automatic indexing. Commun. ACM **18**(11), 613–620 (1975)
19. Manning, C.D., Ragahvan, P., Schutze, H.: An Introduction to Information Retrieval. Cambridge University Press, Cambridge (2009). Online ed

20. McCallum, A., Nigam, K.: A comparison of event models for naive Bayes text classification. In: 15th National Conference on Artificial Intelligence (AAAI 1998): Workshop on Learning for Text Categorization, pp. 41–48, Madison (1998)
21. Joachims, T.: Text categorization with support vector machines: learning with many relevant features. In: Nédellec, C., Rouveirol, C. (eds.) ECML 1998. LNCS (LNAI), vol. 1398, pp. 137–142. Springer, Heidelberg (1998)
22. MacQueen, J.: Some methods for classification and analysis of multivariate observations. In: 5th Berkeley Symposium on Mathematical Statistics and Probability, vol. 1, pp. 281–297. University of California Press, Berkeley (1967)
23. Xu, D., Tian, Y.: A comprehensive survey of clustering algorithms. Annals Data Sci. **2**(2), 165–193 (2015)
24. Riekert, M., Leukel, J., Klein, A.: Online media sentiment: understanding machine learning-based classifiers. In: 24th European Conference on Information Systems (ECIS 2016), Istanbul (2016)
25. Tu, Q., Vonderembse, M.A., Ragu-Nathan, T.S., Ragu-Nathan, B.: Measuring modularity-based manufacturing practices and their impact on mass customization capability: a customer-driven perspective. Decis. Sci. **35**(2), 147–168 (2004)

Are Similar Cases Treated Similarly?
A Comparison Between Process Workers

Mark Pijnenburg[1,2]([✉]) [iD] and Wojtek Kowalczyk[1] [iD]

[1] Leiden Institute of Advanced Computer Science,
Leiden University, Niels Bohrweg 1, Leiden, The Netherlands
{m.g.f.pijnenburg,w.j.kowalczyk}@liacs.leidenuniv.nl
[2] Netherlands Tax and Customs Administration,
Croeselaan 14, Utrecht, The Netherlands

Abstract. In processes involving human professional judgment (e.g., in Knowledge Intensive processes) it is not easy to verify if similar cases receive similar treatment. In these processes there is a risk of dissimilar treatment as human process workers may develop their individual experiences and convictions or change their behavior due to changes in workload or season. Awareness of dissimilar treatment of similar cases may prevent disputes, inefficiencies, or non-compliance with regulations that require similar treatment of similar cases. In this article two procedures are presented for testing in an objective (statistical) way if different groups of process workers treat similar cases in a similar way. The testing is based on splitting the event log of a process in parts corresponding to the different (groups of) process workers and analyzing the sequences of events in each part. The two procedures are demonstrated on an example using synthetic data and on a real life event log.

Keywords: Process mining · Knowledge intensive processes ·
Case management · Statistical auditing · Statistical testing

1 Introduction

Knowledge Intensive (KI) processes present an expanding topic in the field of Business Process Management (BPM), see for instance the paper of Marin et al. [9]. KI processes differ from classical processes by the presence of human process workers whose domain knowledge has an important influence on the next steps in the process. KI processes occur typically in organizations that complete complex tasks.

The human component in KI processes makes the flow of activities less predictive and raises the question whether similar cases are treated similarly. Differences in treatment of similar cases may result from differences in the expertise, workload, and opinions of the human process workers, especially when the same process is executed at two or more different sites. For example, when similar cases are presented to two process workers, they may process them in different ways.

© Springer Nature Switzerland AG 2019
W. Abramowicz and R. Corchuelo (Eds.): BIS 2019, LNBIP 354, pp. 162–176, 2019.
https://doi.org/10.1007/978-3-030-20482-2_14

Awareness of dissimilar treatment of similar cases by different groups of process workers (e.g., at different locations) is important for businesses, not only because a uniform treatment is usually preferred with a view on efficiency, but also since the lack of uniformity may create disputes between customers and the company. For governmental organizations similar treatment of similar cases is even more important, as similar treatment is often demanded by law or policy. As such, the topic is of interest for auditors as well. Auditors, besides the classical task of checking financial statements, are also expected to check compliance with regulations and the law.

Verifying similarity of treatments for two or more (groups of) process workers is complicated by two facts. First, the characteristics ('attributes') of cases presented to process workers may differ from process worker to process worker. For instance, when two process workers are employed in different regions, one should take into account the regional differences of the cases presented to these process workers. Second, human professionals often have the possibility (and are expected) to gather additional information about the cases they treat. Although this (unstructured) information may be stored in a case management system, it is usually not registered in the event log of the process. This 'data incompleteness' introduces an additional stochastic component when modelling the decisions of a process worker. Hence statistical techniques are required to make measure similarity of treatments, see Sect. 3 for more details.

A simple example may illustrate the issue of similar treatment. Consider a Knowledge Intensive service process of a manufacturer of laptops and printers, as shown in Fig. 1. The service process supports customers whose device breaks down within the warranty period. A defect device that comes in, initially starts two activities that are processed in parallel: registration (V_1) and sending a confirmation (V_2). Part of the registration process is the recording of the *device type* (possible values: 'laptop' and 'printer') and the original purchase price (*price*). After registration and confirmation, a human process worker considers the defect device and has three options: return the original purchase price (V_3), send a

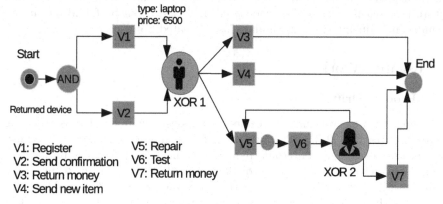

Fig. 1. Simple process model of a service process of a laptop and printer manufacturer.

new device (V_4), or send the defect device for repair (V_5). The choice of option depends on an assessment of the actual defect, the recorded values, and possibly additional information gathered by contacting the customer. When the repair option is chosen (V_5), the device will be send to an external repair shop and the device will be tested (V_6) afterwards. Depending on these test results, a second process worker will decide on the next activity; a good test result will lead to ship the repaired device back to the customer. Otherwise the process worker may choose to send the device again for repair or to return the money to the customer (V_7).

Traces in the event log of this process may look like this:

- 244 Printer · Start V1 V2 XOR1 V5 V6 XOR2 V5 V6 XOR2 V7 End
- 69 Printer · Start V2 V1 XOR1 V3 End
- 224 Laptop · Start V1 V2 XOR1 V4 End
- 1082 Laptop · Start V1 V2 XOR1 V5 V6 XOR2 V7 End
- 67 Printer · Start V1 V2 XOR1 V4 End

where the first number is the purchase price of the device.

The service process is deployed at various regions. The manufacturer is interested in knowing whether similar defects receive the same treatment at each region to deal with a rise of complaints claiming that customers in region A usually receive a new item quickly, while customers in region B have to wait for a repair. The service process of this manufacturer will serve as an example throughout this paper.

The research into similar treatment is motivated by two real-life examples where knowledge of similar treatment of similar cases is found to be important. One is the debt-collection process of a tax administration where it is expected that debtors in various regions are treated in a uniform way, see Sect. 4.2 for more details. The other is the treatment of patients in two wards of the same hospital according to the same protocol.

The paper is organized as follows. Section 2 reviews related work. Section 3 describes two testing procedures. Section 4 demonstrates these procedures on the 'warranty'-example and the tax data set. We end with conclusions and future research in Sect. 5. The code used in Sect. 4.1 can be found on github: github.com/PijnenburgMark/Similar_Cases_Treated_Similarly/.

2 Related Work

2.1 Process Drift

Detecting different variants of business processes is a topic that is receiving increasing attention [2,3,8,10]. Most works focus on detecting changes in a business process over time, but some papers also mention differences by location (e.g., departments), like the paper of Pauwels and Calders [11], or mention the more general applicability of their method, like the paper of Bolt et al. [2].

One of the earliest papers in the field of process drift is the paper of Bose et al. [3]. The method described in this paper extracts features from an event log and compares the values of these features at two different points in time. The features demonstrated in the paper only allow for finding 'structural changes', i.e., changes in the process model.

The paper of Ostovar et al. [10] differs from the paper of Bose et al. [3] by using a technique for automated discovery of process trees and considering changes in these trees instead of using features extracted from event logs. Moreover, root cause analysis is supported by providing natural language statements to explain the change behind the drift.

In the paper of Pauwels and Calders [11] the concept of process drift is applied to the data set of the BPI challenge 2018 [4]. Besides showing concrete results, the authors add a new model-based approach relying on Dynamical Bayesian Networks. Their approach is strengthened by providing some nice aides for detecting process drift visually.

The paper of Maaradji et al. [8] adds a formal statistical test to the comparison between different time points and thus adds objectivity. Moreover the detection of process drift is determined by looking at the frequencies of sequences of activities, and is thus capable of detecting structural changes as well as changes in frequencies of process paths.

In our approach, similarly to Maaradji et al. ([8]), we also look at the activities themselves instead of derived features and we also apply a formal statistical test. However, there are several differences. From the contextual viewpoint there are two main differences. First, we consider differences in process execution between groups of users instead of differences in time and second, we focus on the human decision making in the process and thus leaving aside the activities that are triggered automatically by the workflow management software. As a consequence of the latter, we have to consider fewer differences, making the comparison simpler and the statistical testing more powerful. From the technical point of view, there are two differences as well. First we take into account the attributes of the cases that enter the process and second, we consider the frequently occurring situation where some or all decisions in the process are independent of each other. If this assumption holds, the χ^2-test becomes much more powerful and easy to use, because of the mathematical property that the sum of independent χ^2-distributions is again a χ^2-distribution.

2.2 Sequence Analysis

Event logs form the basis of process mining and are in essence a number of sequences of activities. For this reason it is not surprising that techniques from sequence analysis (also known as 'sequential pattern analysis' or 'sequential pattern mining') are applicable in process mining. One of the techniques from sequence analysis that is applied frequently in this paper, is Pearson's χ^2-test to test the probabilities of going from one activity to the next against a theoretical model, see the book of Bakeman and Gottman [1]. We apply this test repeatedly under the null hypothesis of no differences in treatment between (groups of) process workers. As described by Bakeman and Gottman [1], Pearson's χ^2-test can be also applied to sequences of events, although the number of possible sequences (and thus the number of observations needed to fulfill the assumptions of the test) grows exponentially in the sequence length. We overcome this problem by making an assumption of independence of decisions (Sect. 3.1) or by focusing on the most frequent traces (Sect. 3.2).

3   Testing for Similar Treatment of Similar Cases

When testing similar treatment of similar cases for two groups of process workers, a first step is finding the points in the process where human process workers decide on the next activity. We will call these points 'decision points'. In a process model these are 'XOR-junctions' (eXclusive OR-junctions), i.e., places where the flow of activities can take two or more directions.

We define 'similar treatment of similar cases' as the situation where at each decision point holds that two similar cases (measured by having the same attributes) have the same probability distribution over all possible follow-up activities.

In case only one decision point is present, the situation is relatively simple and the approach is described in Sect. 3.1. When multiple decision points are present, two situations can occur: (a) the decisions at the decision points can be considered independent of each other (frequently called a 'Markov assumption' in sequence analysis). In this situation the comparing of two groups of process workers can be done for each decision point individually and the results can later be combined. This case is addresses in Sect. 3.1. (b) the decisions cannot be considered independent, in other words the decision at one decision point is influenced by the previous decisions (i.e., it matters for the decision what path a case took through the process model to reach the decision point). The problem that arises in this situation is that the number of possible combinations of decisions increases exponentially in the number of decision points and hence a lot of traces in the event log are needed if a naive approach is taken. In Sect. 3.2 we present a heuristic that requires less data, by assuming that a subset of possible traces accounts for most observed traces in the event log. An assumption which is often reasonable in practice.

3.1   Approach 1: Independence of Decisions

A Test for a Single Decision Point. If we compare two groups of users, say from location L_1 and L_2, and we focus for the moment on *one* decision point, a table like Table 1 forms the basis for analysis (see also Fig. 2). In the first column of Table 1 'Cluster of case', we see that all cases are grouped based on their attributes into K clusters. This clustering is done to make groups of similar cases. Then in the next column we see the decision that was made at the decision point, i.e., the next activity for the case. Then follow two columns indicating the count data of cases for both groups of users (L_1 and L_2). See Table 7 for two examples of filled out tables.

Table 1. Three-way contingency table needed to apply the χ^2-test for one decision point for two groups of process workers.

Cluster of case	Output activity	Group of process workers	
		L_1	L_2
x = 1	V^1	$n_1^{1,1}$	$n_2^{1,1}$
	$\vdots$	$\vdots$	$\vdots$
	V^q	$n_1^{1,q}$	$n_2^{1,q}$
x = 2	V^1	$n_1^{2,1}$	$n_2^{2,1}$
	$\vdots$	$\vdots$	$\vdots$
	V^q	$n_1^{2,q}$	$n_2^{2,q}$
$\vdots$	$\vdots$	$\vdots$	$\vdots$
x = K	V^1	$n_1^{K,1}$	$n_2^{K,1}$
	$\vdots$	$\vdots$	$\vdots$
	V^q	$n_1^{K,q}$	$n_2^{K,q}$

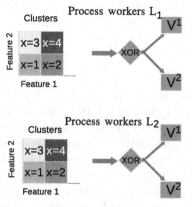

Fig. 2. Figure illustrating the elements of the table at the left

If we consider a subtable of Table 1 that belongs to one cluster, we can apply the well known two sample χ^2-test (see for instance the book by Kanji [6]) to test whether the differences between the two user groups can be attributed to chance, or that there is a reason to reject the hypothesis of similar treatment. If we want to apply the χ^2-test to the whole table (i.e., for all clusters simultaneously), we can take the approach as described for instance by Hald [5] (page 746). Note that Table 1 contains only two locations. This is only for demonstrative purposes as the χ^2-test works equally well for any number of user groups.

Mathematically the test for one decision point comes down to the following. To test the null hypothesis of similar treatment of similar cases, we apply the (two sample) χ^2-test. So we calculate,

$$\chi^2_{XORj} = \sum_{i \in \text{all cells}} \frac{(O_i - E_i)^2}{E_i} = \sum_{x,V,L} \frac{(n_L^{x,V} - E_L^{x,V})^2}{E_L^{x,V}}, \qquad (1)$$

where $E_L^{x,V}$ denotes the expected number of counts. These expected numbers are estimated assuming no differences between the groups of process workers ($\mathcal{H}_0$), so,

$$E_L^{x,V} = \frac{n_\bullet^{x,V} n_L^{x\bullet}}{n_\bullet^{x\bullet}}, \qquad (2)$$

where,

$$n_\bullet^{x,V} = \sum_L n_L^{x,V}, \qquad n_L^{x,\bullet} = \sum_V n_L^{x,V}, \qquad n_\bullet^{x\bullet} = \sum_{V,L} n_L^{x,V}. \qquad (3)$$

For one cluster (e.g., $x = 1$), the degrees of freedom are equal to $(q-1)(t-1)$ [5] (where t is the number of groups of process workers, and q the number of next activities). Since there are K clusters, we have,

$$df_{XORj} = K(q-1)(t-1). \tag{4}$$

The statistic (1) is χ^2-distributed with degrees of freedom given in (4) if the assumptions of the Pearson's χ^2-test are fulfilled.

Experiments have demonstrated that the assumptions of Pearson's χ^2-test are sufficiently fulfilled if *no more than 20% of the expected counts are less than 5 and all individual expected counts are 1 or greater*, see the book by Yates et al. [12]. One way of meeting this condition is to find a balance in the number of clusters K, such that K is small enough for each cluster to contain enough cases, while simultaneously keeping K large enough to make sure that cases that are considered as dissimilar by experts are in different clusters. See Sect. 4 for the values used in the experiments. If after choosing an appropriate number of clusters the assumptions of Pearson's test are still violated, we propose to remove rows from Table 1 with infrequent row sums (i.e., activities that are seldom chosen by any groups of process workers) and apply Eq. (1) on the reduced table. Since the frequency of these activities is low for all process workers, the effect of removing will be small in practice. Of course, the degrees of freedom have to be adjusted. If l rows are omitted, then the number of degrees of freedom is given by:

$$df_{XORj}^{adj} = K(q-1)(t-1) - l(t-1). \tag{5}$$

By introducing two groups of process workers L_1 and L_2 and comparing the model of a decision point for these two groups, we must be cautious that possible differences are only caused by differences in treatment of the groups of process workers L_1 and L_2 and *not* by differences in the attributes of the cases that are presented to group L_1 and group L_2. Otherwise we would erroneously conclude there is dissimilar treatment of similar cases, while in reality there is dissimilar treatment of *dissimilar* cases. In part, these differences in the cases of each group are prevented by the clustering the cases into K clusters based on the known attributes. However, typically not all attributes are recorded. If one suspects the existence of an attribute that has an influence on the decision taken by the process workers and this feature is also related to the group of users a case is assigned to, one must, within each cluster of cases, distribute cases randomly over the groups of process workers. This way one ensures the same probability distribution for the groups and one can safely apply the tests.

Multiple Decision Points that Are Independent. If there are multiple decision points in the process and we can assume that the decisions by the human process workers are independent from previous decisions in the process, we can easily extend our χ^2-test. When the statistics χ_{XORi}^2 have been computed for the individual decision points, a test of similar treatment for the whole process (i.e., all m decision points) can be constructed by adding the individual statistics as well as the degrees of freedom:

$$\chi^2_{overall} = \chi^2_{XOR1} + \cdots + \chi^2_{XORm}, \qquad (6)$$

with

$$df_{overall} = df^{adj}_{XOR1} + \cdots + df^{adj}_{XORm}. \qquad (7)$$

The final statistic $\chi^2_{overall}$ is χ^2-distributed with $df_{overall}$ degrees of freedom since the sum of independent χ^2-statistics is again χ^2-distributed with the degrees of freedom equal to the sum of the original ones. The independence of the individual statistics is ensured by the Markov assumption.

The value of the statistic $\chi^2_{overall}$ leads to a p-value indicating the probability that the hypothesis $\mathcal{H}_0$ is due to the randomness in the sample. A value lower than 0.05 is generally accepted as a rejection of $\mathcal{H}_0$ and is thus evident of dissimilar treatment of similar cases.

The complete test procedure in case of independent decision points is summarized in Test Procedure 1.

Test Procedure 1. Procedure to test on similar treatment by two or more groups of process workers when decision points are independent

Input : Event logs
Output: A p-value
1 Find the decision points in the process;
2 **for** *each decision point* **do**
3   Find relevant attributes;
4   Cluster the cases of all event logs into K clusters based on the attributes (perform the clustering on the collective data of all user groups);
5   Construct a contingency table like Table 1;
6   Check the assumptions of Pearson's χ^2-test and possibly remove rows with low row sums (adjust degrees of freedom accordingly);
7   Use equation (1) to compute the test statistic for this decision point, and equation (5) for the degrees of freedom;
8 **end**
9 Apply equation (6) and (7) and read off the p-value from standard tables of the χ^2-distribution.

3.2   Approach 2: Frequent Paths

For some processes the Markov assumption will not hold. A straightforward generalization in this case is to cluster cases and consider for each cluster all possible traces and test if some traces occur significantly more frequently for one group of process workers. Although this approach is in line with the approach taken in the previous section, it has the drawback that a lot of data is needed in order to satisfy the underlying assumptions of Pearson's χ^2-test. Namely, the number of possible traces grows exponentially in the number of decision points.

For this reason we will work out an approach that does not take into account *all* traces, but only the most frequent ones (i.e., 'typical paths') for each cluster of similar cases. Then the observed frequencies of these traces for each group of process workers are compared with the expected frequencies (based on the average frequencies over all process workers) using the χ^2-test.

Technically, the frequent paths approach comes down to first clustering cases based on known features, and then finding the most frequent traces for each cluster. If the event log contains many traces, special algorithms can be applied to find the most frequent traces, like the SPADE algorithm described by Zaki [13].

Subsequently, a table like Table 2 is created that contains the frequencies of these frequent paths for all groups of process workers. Then Eq. (8) can be applied to each frequent trace s in cluster x,

$$\chi^2 = \sum_{i \in \text{all cells}} \frac{(O_i - E_i)^2}{E_i} = \sum_{x,s,L} \frac{(n_L^{x,s} - E_L^{x,s})^2}{E_L^{x,s}}, \tag{8}$$

where $E_L^{x,s}$ is estimated by,

$$E_L^{x,s} = \frac{n_\bullet^{x,s} n_L^{x\bullet}}{n_\bullet^{x\bullet}}, \tag{9}$$

and,

$$n_\bullet^{x,s} = \sum_L n_L^{x,s}, \qquad n_L^{x,\bullet} = \sum_s n_L^{x,s}, \qquad n_\bullet^{x\bullet} = \sum_{s,L} n_L^{x,s}. \tag{10}$$

The degrees of freedom of the statistic in Eq. (8) equals:

$$df = (N - K)(t - 1), \tag{11}$$

where N is the number of rows in the contingency table (Table 2), K the number of clusters and t the number of groups of process workers. This can be seen by realizing that for each cluster i the degrees of freedom equals $(s_i - 1)(t - 1)$ [5], where s_i is the number of frequent paths for cases in cluster i. The Test Procedure for the Frequent Path approach is presented in Test Procedure 2.

Test Procedure 2. Procedure to test on similar treatment by two or more groups of process workers when decision points are dependent and most traces belong to a few frequently occurring traces

Input : Event log
Output: A p-value
1 Cluster all cases based on their features into K clusters;
2 Determine for each cluster the most frequent traces in the event log, for instance by applying a frequent sequence algorithm like SPADE. Discard all other traces;
3 Make a contingency table like Table 2;
4 Use equation (8) to compute the test statistic, and equation (11) for the degrees of freedom;
5 Find the p-value from standard tables of the χ^2-distribution.

Table 2. Three-way contingency table needed to apply the χ^2-test for all decision points for two groups of process workers when the assumption of independence of decision points does not hold.

		Group of process workers	
Cluster of case	Frequent trace	L_1	L_2
x = 1	$V^1V^3V^5$	$n_1^{1,1}$	$n_2^{1,1}$
	$\vdots$	$\vdots$	$\vdots$
	$V^2V^3V^4$	$n_1^{1,r}$	$n_2^{1,r}$
x = 2	$\vdots$	$\vdots$	$\vdots$
$\vdots$	$\vdots$	$\vdots$	$\vdots$
x = K	V^1V^5	$n_1^{K,1}$	$n_2^{K,1}$
	$\vdots$	$\vdots$	$\vdots$
	$V^1V^2V^3V^5$	$n_1^{K,s}$	$n_2^{K,s}$

4 Experiments

4.1 Synthetic Data

In this section we will apply the test procedures to the example mentioned in the introduction. The code used for this can be found on github: github.com/PijnenburgMark/Similar_Cases_Treated_Similarly/.

We used the process model of Fig. 1 to generate four event logs. Each event log consists of 500 traces and the value of two meaningful features for each case (price and type). The four event logs differ in the behavior of the two decision points and in the distribution of the two features, see Table 3.

Table 6 specifies the behavior I and behavior II of the decision points as well as the distributions of the input cases. For instance we see that under behavior I, a printer that is returned with an original purchase price of 200 has the probability of 0.2 for being send for repair, while this probability under behavior II is 0.6. Moreover we see that the behavior of decision point 2 (XOR 2) depends on the number of times the device has been send for repair before (with two options: one time or more than one time) as well as the result of the test that is performed

Table 3. Differences of the four generated event logs

		XOR-junctions	
		Behavior I	Behavior II
Features	Distribution I	Event log 1	Event log 2
	Distribution II	Event log 3	Event log 4

after the repair (activity V_6 in Fig. 1). Note that if the results of the test are positive, the device is ready and thus there is a probability of one that the process trace will be ended. In our example we set the probability that the test of V_6 gives a positive result to 0.7.

Clearly, event logs 1 and 3 have the same process (i.e., no dissimilar treatment), but differ in the input (event log 1 mainly cheap printers, event log 3 more laptops). In contrast event logs 2 and 4 are generated with other process parameters. In these latter two event logs the process workers have a preference for the repair option instead of sending a new item or returning the money.

We will compare the four event logs pairwise. Large p-values (over 0.05) are expected when comparing event logs that have the same behavior (i.e., event log 1 with event log 3, and event log 2 with event log 4), while low p-value are expected if we compare event logs with different behavior such as event log 1 and event log 2.

The p-values resulting of applying Test Procedure 1 are shown in Table 4. The application of this test procedure is justified because the decision points are independent. The results are as expected, i.e., the test procedure is able to clearly distinguish dissimilar treatment from differences in the input. For feature selection we used a standard feature selection method from python's scikit learn package based on the ANOVA F-value, and for the clustering we applied a standard Gaussian Mixture clustering algorithm from the same package. The number of clusters was chosen $K = \#$ cases$/100$, where '# cases' are the number of cases that go through the decision point in the event log (for both groups).

We also applied Test Procedure 2, resulting in the p-values of Table 5. These values demonstrate that our algorithm properly captures relevant properties of this synthetic data set. In applying Test Procedure 2 we used again Gaussian Mixture clustering. The number of clusters was chosen $K = \#$ traces $/250$, where '# traces' are the number of traces of the combined event logs for both group of process workers.

4.2   Tax Debt Collection Data

For demonstration purposes, we applied Test Procedure 1 to real data of the Netherlands Tax and Customs Administration (NTCA). In the year 2013 a debt

Table 4. p-values resulting from applying Test Procedure 1 to all pairs of the four event logs. We used 10 clusters.

1st Event Log		2nd Event Log			
		Event log 1	Event log 2	Event log 3	Event log 4
	Event log 1	1.000000	0.000000	0.975989	0.000000
	Event log 2	0.000000	1.000000	0.000000	0.057141
	Event log 3	0.975989	0.000000	1.000000	0.000000
	Event log 4	0.000000	0.057141	0.000000	1.000000

Table 5. p-values resulting from applying Test Procedure 2 to all pairs of the four event logs. We used 4 clusters.

1st Event Log	2nd Event Log			
	Event log 1	Event log 2	Event log 3	Event log 4
Event log 1	1.000000	0.000000	0.982920	0.000000
Event log 2	0.000000	1.000000	0.000000	0.662541
Event log 3	0.982920	0.000000	1.000000	0.000000
Event log 4	0.000000	0.662541	0.000000	1.000000

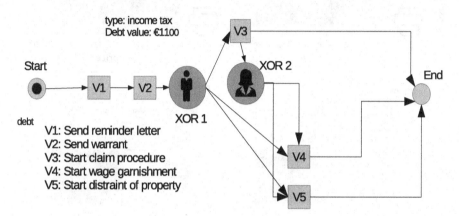

Fig. 3. Part of the debt collection process of the NTCA in 2013.

collection process was in place, a part of which is sketched in Fig. 3. As soon as a debt is over its due date, automatically a reminder letter is sent. When no payment has taken place, a legal notice (warrant) is sent that allows for legal actions afterwards. Usually, there are several legal actions possible and the choice is determined by a human process worker (XOR 1). The legal actions in this part of the process are: a special claim procedure that allows to take money of a savings account (V_3), wage garnishment (V_4), and distraint of property, e.g., a car (V_5). When the special claim procedure V_3 does not lead to payment of the debt, a second human process worker (XOR 2) can decide for actions V_4 or V_5.

We have compared event logs from two locations. We took 1000 debts from each location in the period 2013. We took into account two features of each debt: the debt value and the tax type and restricted ourselves to two tax types. The debt value has been used for constructing the clusters for both decision points, while the tax type played a role for the clustering of XOR 1 only. The contingency tables for both decision points are displayed in Table 7.

The value of the χ^2-statistic for the first decision point is 59.247 (8 degrees of freedom), while 1.785 (2 degrees of freedom) for the second decision point. For the combination of the two decision point we find, under the assumption

Table 6. Worked out example: two types of behavior of the XOR junctions and two distributions of the input cases.

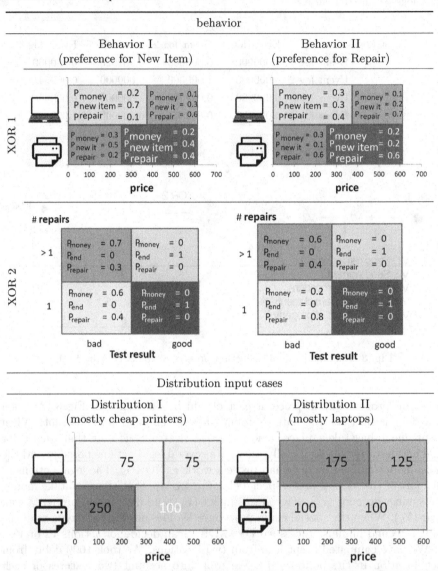

of independent decisions, a value of 61.032 (10 degrees of freedom) which corresponds to a p-value of $2.31 \cdot 10^{-9}$. The low p-value indicates that, under this model, we should reject the hypothesis of similar treatment of similar cases. Due to privacy reasons we did not use some important features that could lead to different results.

Table 7. Contingency tables belonging to the decision points XOR 1 and XOR 2 of the tax collection data. Note we were not able to analyze 77 of the 2000 cases.

XOR 1		location	
Cluster	Activity	A	B
1	V_3	16	14
1	V_4	42	48
1	V_5	21	17
2	V_3	1	2
2	V_4	16	39
2	V_5	64	35
3	V_3	381	379
3	V_4	25	59
3	V_5	80	90
4	V_3	217	213
4	V_4	8	47
4	V_5	59	50
	Total	930	993

XOR 2		location	
Cluster	Activity	A	B
1	V_4	17	17
1	V_5	64	53
2	V_4	8	9
2	V_5	42	24
	Total	131	103

5   Conclusion and Future Research

The paper started by noting that the human factor in Knowledge Intensive processes raises the question of similar treatment by several groups of process workers. The two test procedures that are proposed in Sect. 3 are able to test similarly treatment as is demonstrated in Sect. 4 on an artificial example of a service process of a hardware manufacturer and a debt collection process of a tax administration.

The two test procedures presented in this paper are based on rather elementary statistical techniques. We choose deliberately to solve the problem with elementary techniques as simple methods display the nature of the problem most clearly. Moreover an elementary approach allows to communicate clearly with business experts. Besides an elementary approach allows for easy extensions to meet more particular needs. For instance most organizations may tolerate a certain small level of dissimilar treatment and the test may be extended to take this into account. However, applying some recently developed algorithms may have some advantages as well. In particular recurrent neural networks can be used successfully for modelling complex sequential data [7]. Unfortunately these networks are 'black-box models' and provide little insight into the nature of detected differences in the behaviour of groups of process workers.

Finally note that other interpretations of 'similar treatment' are possible and might be appropriate in some settings. In this paper similar treatment has been defined in terms of probabilities of going to the next activity in the process model. However similar treatment can also be defined for instance in terms of the amount of time that is spend on each case, or as the level of expertise that is involved in each case. Tests based on these metrics have not been explored yet, but may be the subject of further research.

References

1. Bakeman, R., Gottman, J.M.: Observing Interaction: An Introduction to Sequential Analysis. Cambridge University Press, New York (1997)
2. Bolt, A., van der Aalst, W.M.P., de Leoni, M.: Finding process variants in event logs. In: Panetto, H., et al. (eds.) OTM 2017. LNCS, vol. 10573, pp. 45–52. Springer, Cham (2017)
3. Bose, R.J.C., Van Der Aalst, W.M., Zliobaite, I., Pechenizkiy, M.: Dealing with concept drifts in process mining. IEEE Trans. Neural Netw. Learn.Syst. **25**(1), 154–171 (2014)
4. van Dongen, B.F.; Borchert, F.: BPI challenge 2018 (2018). https://doi.org/10.4121/uuid:3301445f-95e8-4ff0-98a4-901f1f204972
5. Hald, A.: Statistical theory with engineering applications. In: Statistical Theory with Engineering Applications. Wiley (1952)
6. Kanji, G.K.: 100 Statistical Tests, 3rd edn. Sage Publications, London (2006)
7. Lipton, Z.C., Berkowitz, J., Elkan, C.: A critical review of recurrent neural networks for sequence learning (2015). arXiv preprint arXiv:1506.00019
8. Maaradji, A., Dumas, M., La Rosa, M., Ostovar, A.: Fast and accurate business process drift detection. In: Motahari-Nezhad, H.R., Recker, J., Weidlich, M. (eds.) BPM 2015. LNCS, vol. 9253, pp. 406–422. Springer, Cham (2015). https://doi.org/10.1007/978-3-319-23063-4_27
9. Marin, M.A., Hauder, M., Matthes, F.: Case management: an evaluation of existing approaches for knowledge-intensive processes. In: Reichert, M., Reijers, H.A. (eds.) BPM 2015. LNBIP, vol. 256, pp. 5–16. Springer, Cham (2016). https://doi.org/10.1007/978-3-319-42887-1_1
10. Ostovar, A., Leemans, S.J., La Rosa, M.: Robust drift characterization from event streams of business processes. Internal Report (2018)
11. Pauwels, S., Calders, T.: Detecting and explaining drifts in yearly grant applications (2018). arXiv preprint arXiv:1809.05650
12. Yates, D., Moore, D., McCabe, G.: The Practice of Statistics, 1st edn. W.H. Freeman, New York (1999)
13. Zaki, M.J.: Spade: an efficient algorithm for mining frequent sequences. Mach. Learn. **42**(1–2), 31–60 (2001)

Mining Labor Market Requirements Using Distributional Semantic Models and Deep Learning

Dmitriy Botov[1]([✉]), Julius Klenin[1], Andrey Melnikov[2], Yuri Dmitrin[1], Ivan Nikolaev[1], and Mikhail Vinel[1]

[1] Chelyabinsk State University, 129, Bratiev Kashirinykh Street, 454001 Chelyabinsk, Russia
dmbotov@gmail.com
[2] Ugra Research Institute of Information Technologies, 151 Mira str., Khanty-Mansiysk, Russia

Abstract. This article describes a new method for analyzing labor market requirements by matching job listings from online recruitment platforms with professional standards to weigh the importance of particular professional functions and requirements and enrich the general concepts of professional standards using real labor market requirements. Our approach aims to combat the gap between professional standards and reality of fast changing requirements in developing branches of economy. First, we determine professions for each job description, using the multi-label classifier based on convolutional neural networks. Secondly, we solve the task of concept matching between job descriptions and standards for the respective professions by applying distributional semantic models. In this task, the average word2vec model achieved the best performance among other vector space models. Finally, we experiment with expanding general vocabulary of professional standards with the most frequent unigrams and bigrams occurring in matching job descriptions. Performance evaluation is carried out on a representative corpus of job listings and professional standards in the field of IT.

Keywords: Natural language processing ·
Distributional semantic model · Deep learning ·
Convolutional neural networks · Multilabel classification ·
Semantic similarity · Information extraction ·
Labor market requirements · Professional standards

1 Introduction

Nowadays, during the transition to a digital economy, leading industries, such as IT develop more and more rapidly, with technology life cycle being reduced to 1–2 years. Demands of IT companies are constantly changing, while the shortage of qualified personnel is only growing. The concept of lifelong learning, which

© Springer Nature Switzerland AG 2019
W. Abramowicz and R. Corchuelo (Eds.): BIS 2019, LNBIP 354, pp. 177–190, 2019.
https://doi.org/10.1007/978-3-030-20482-2_15

emerged in the 20th century, becomes increasingly more relevant in the new information age [1]-after all, what was learned only yesterday, today has already lost its relevance. At the same time, requirements of educational and professional standards are too general and do not capture a complete picture of what knowledge and skills are the most relevant, focusing instead on determining which competencies will ensure the successful development of a particular region and the country as a whole. All of these factors force the developers of curricula, online courses, and programs for advanced training and refresher courses to regularly update educational content and ensure the relevance of the learning outcomes. Job listings contain labor market demands, which can be efficiently extracted and analyzed using various data mining approaches, like the ones in [2–6]. However, Russian language and the federal Russian professional standards themselves have certain specifics, which are yet to be investigated in existing research. Thus, we know of no research into the efficiency of various distributive language models and machine learning algorithms in semantic analysis of such texts, which would take into account these specifics. The goal of this paper is the development of such approach; able to mine the actual labor market requirements by comparing them with professional standards, using IT industry as an example. In order to reach this goal, we define the following tasks:

- use the job description to determine if it matches one or several professions according to the classification described in the professional standards for IT industry;
- match specific requirements and duties from job descriptions with the functions and requirements of the standard;
- determine the most significant functions and requirements of the standard;
- enrich the general vocabulary of the professional standard with key concepts from the job descriptions.

To do so, we apply both classic machine learning models and distributive semantic based deep learning algorithms to this task. Evaluation of these models and algorithms is carried out on a representative corpus of job listings and professional standards.

2   Related Work

In [2], authors present their job search engine, which, among other filters, allows user to filter listings by certain skills. Here, skills are acquired by extracting information from social networks, then processed by removing unimportant words and lemmatizing the rest with regard for common word pairs. In order to produce ranked lists of job advertisements, the system weights job description by averaging the weights of each skill in every job description, which in turn are calculated by using TF-IDF and the probability of occurrence for any given skill, based on the title of a job description. While the system does show promising practical results, no scientific evaluation is present to support the claims to quality. Authors of [3] perform an analysis of currently most demanded jobs in several regions, based on the online job listings and a O*NET database of occupational requirements. Their method is based on evaluating similarities between LSA vectors of O*NET descriptions and job listings, in order to map the latter onto the former. Once again, a paper shows the application for an approach, but does not test the quality of the approach itself. Another analytical application is described in [4]. Here, authors apply LSA to the descriptions of a number of online job listings, with the goal of acquiring so-called ideal types of employees: especially effective or successful combinations of skills, which are in high demand. The paper describes various groups of professions, extracted for different numbers of LSA classes, as well as providing the analysis of the ideal skills for each class. A task similar to ours is shown in [5], where authors present their NER-based system, capable of matching skills from job listings onto the skill ontology, generated using Wikipedia. First step is taxonomy generation, using seed phrases from skill descriptions to retrieve matching categories from Wikipedia articles. The second step is skill taggingmatching skill description to one of the surface forms in taxonomy. The paper presents a proper evaluation of the approach with taxonomy having a quality of 83 F1 points, and tagging having a quality of 75.5 F1 points. Authors of [6] presented the evaluation of various approaches for soft skill extraction. The task was specified as a binary classification of text fragments as either containing a description of a soft skill or not. From a variety of classification methods, trained and tested on top of the word2vec vectors for text documents, the best results were achieved by a LSTM neural network, trained on the unmodified texts.

3   Method

3.1   Conceptual Model

Conceptual model (Fig. 1) of the domain can be represented as a directed graph $G = (V, E)$, where V is a set of vertices describing the basic domain concepts, and E is a set of edges defining asymmetric semantic relations between vertices. The set $V = F, R$ is divided into two subsets: Flabor functions/actions and Rrequirements to knowledge and skills, education, or work experience. The set of relationships (edges of the graph) $E = Include, Require, Match$, includes three subsets describing the possible types of relations between concepts:

- $Include \subset F \times F$ is a part-whole relation between generalized labor functions, position, and specific labor actions;
- $Require \subset F \times R$ is a relation between the required knowledge, skill, experience, or education on one side and labor functions or position on the other;
- $Match \subset S \times J$ is a semantic correspondence relation between the common functions and requirements of professional standards: $S = F_s \cup R_s$ and similar specific actions and requirements of job description: $J = F_j \cup R_j$.

3.2   Algorithm

Include and Require relations can be determined by analyzing the structure of professional standards and job descriptions from online recruitment systems and composing a number of simple rules. To be able to mine the actual requirements of the

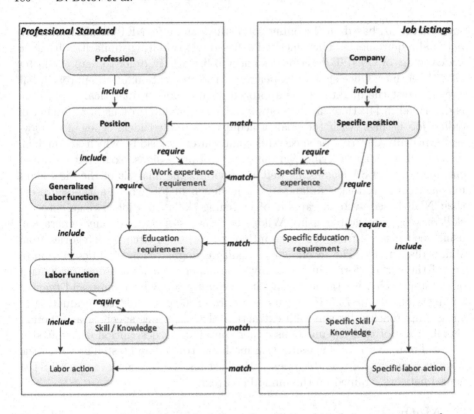

Fig. 1. A conceptual model, which defines the relationships between concepts from professional standards (S) and job descriptions (J)

labor market, we propose to discover Match relations by matching job descriptions published by employers on the Internet with the general requirements of professional standards for relevant professions. We can then determine the significance of certain functions and requirements listed in the standards, based on the frequency of references to them occurring in job listings. Finally, we propose enrichment of the general concepts from professional standards with the most frequent keywords from job descriptions. Thus, the structure and content of professional standards, in a way, acts as a framework, with the particular requirements of the labor market being added on to it, complementing the picture of professional field, placing emphasis on what is currently the highest priority from the point of view of industrys employers. A detailed algorithm is presented in Fig. 2.

Preprocessing. We have tested our models using different combinations of preprocessing steps, as preprocessing does destroy or modify original information. Thus we have the following list of preprocessing steps, some of which we have skipped, if that led to improvement for a given model:

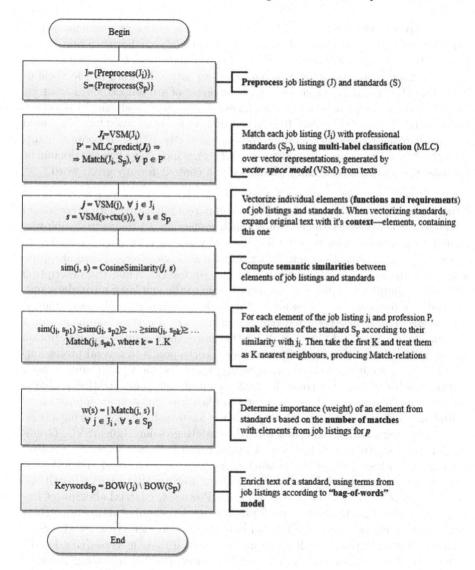

Fig. 2. Algorithm for mining the requirements of the labor market by matching job listings and professional standards

- multiline documents are joined into a single line;
- any symbol that is not alphanumeric, whitespace, or selected special character is removed;
- sequence is tokenized and each token is lemmatized (where possible);
- lemmatized tokens are appended with their POS-tag;
- stopwords are removed (conjunctions, prepositions, pronouns).

Additionally, for the case of character N-grams, we have replaced word tokenization with separation into character N-grams.

Vector Space Models. Machine learning models in our experiments utilize various vector representations of texts (vector models): one-hot encoding, which represents words and documents with a binary, zero-filled vector, with a one being assigned to the position, corresponding to a specific word in the vocabulary; character N-grams, which are a continuous set of groupings of N characters; TF-IDF (Term Frequency Inverse Document Frequency), a classical frequency-based weighting scheme used often in various NLP tasks; word2vec [7], based on a shallow neural network, trained to connect words with their context; para-graph2vec [8] (sometimes referred to as doc2vec), which considers the document itself as well as the surrounding words, to be a context for any given word.

Multi-label Classification. In order to compare a job listing with professional standards, it is necessary to determine which profession it belongs to. Employers formulate positions titles with high degree of variability, so it is impossible to achieve high quality with a simple set of rules or a gazetteer. There are quite a few examples where title is related to one profession, while the description points to a completely different one, usually due to the employer misunderstanding the requirements of professional standards or mixing up similarly named professions (e.g. employers often call the position of a system administrator a system programmer and vice versa). Furthermore, since a single job description can include a description that simultaneously corresponds to several professions (e.g. a system analyst and a software testing specialist, or a programmer and a software architect), we can treat this task as a multilabel classification problem. We use a both a classic algorithms for classification and a more sophisticated neural network classifier. In our research we have tested the following classifiers, as implemented in scikit-learn [9]: LogisticRegression, LinearSVC, GradientBoostingClassifier, RandomForestClassifier, MLPClassifier (neural network model) and others. As for multi-label strategies, we have used the following implementations, provided in scikit-multilearn library [10]: OneVsRestClassifier, BinaryRelevance, ClassifierChain, LabelPowerset, MatrixLabelSpaceClusterer. Text classification can be performed using deep neural networks, such as convolutional neural networks [11] and long short-term memory networks. Such architectures show traditionally high quality in such tasks as classifying intents in conversational systems or performing sentiment analysis including Russian-language datasets [12]. In this paper we experiment with a CNN-based classifier.

Matching Concepts. In order to determine the most significant concepts in professional standards, we have to match each concept from a job description to their closest concepts from a professional standard. It is worth noting that at this stage we basically perform short text analysis, unlike during the previous steps. To improve matching accuracy, we propose expansion of the concepts text using the text of the labor function and the generalized labor function that contains it in conceptual graph. Text vectorization is performed using one of the basic distributional semantic models: word2vec trained on a large body of job listings and professional standards. This model shows the best results in competitions on semantic similarity and word sense extraction in Russian [13,14]. To measure semantic similarity itself of vector representations itself, we use a cosine similarity measure.

4 Evaluation Methods and Results

4.1 Text Collections

For various parts of our experiments, we have assemble 4 datasets, which are detailed in Table 1. Datasets and experimental results are placed in the repository: https://github.com/master8/vacanciesanalysis. First, to train word and sentence embedding models, we have assembled a large corpus of job listings (Job653K) for positions in the IT industry. To do so we collected some of the listings from the online recruitment platforms: headhunter and superjob. To ensure that collected documents are similarly up-to-date, we have only retrieved postings from the past 5 years. Secondly, we have also collected a collection of 40 professional standards (Std40Label) from category 06 - Communication, information and communication technologies. Before training, we have labelled elements of these standards as either generalized labor functions, concrete labor functions, labor activities, knowledge/skills requirements, work experience requirements, or education requirements. In order to train and evaluate multilabel classification, a separate corpus of job listings was prepared, covering 20 professions (Job20Label), labelled manually by experts. The details of Job20Label are presented in Table 2. Details of label counts for different classes are presented in

Table 1. Dataset specifications

Dataset	Document count	Label count (number of professions)	Concept count	Token count	Unique token count (vocabulary size)
Job653K	653K	-	3.6M	130M	200K
Std40Label	40	40	26.6K	574K	3.7K
Job20Label	4652	20	24K	195K	9K
Job120Concepts	98	4	120	754	425

Table 2. Job20Label dataset detalization

Element type	Element count
Knowledge	8985
Activities	5058
Skills	3803
Posts	3270
Education and training requirements	1916
Requirements for practical experience	1345

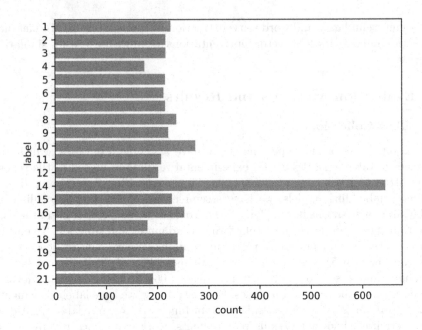

Fig. 3. Label counts for different professions in Job20Label

image 3. The class labelled with "14" is the programmer class the most overlapping class in the dataset. Finally, from the test dataset of job listings, we have produced a separate collection of 120 elements (labor actions, knowledge/skills requirements) for 4 professions to serve as an assessment for pairwise matching methods for concepts from job descriptions and standards. (Job120Concepts) Experts assessed the relevance between each element of a job description and 5 most semantically similar (using cosine similarity) elements of professional standards.

Neural Language Models Setup. We used word2vec and paragraph2vec (doc2vec) implementation from the gensim library as our neural language models. Table 3 presents the training parameters for them.

Table 3. Training parameters for neural language models

Model	Architecture	Dimensionality	Min. occurrence	Epoches
Paragraph2vec	PV-DM	200	3	5
Paragraph2vec	PV-DBOW	200	3	5
Word2vec	skip-gram	300	3	5
Word2vec	CBOW	300	3	5

Multilabel Classification Evaluation and Results. Our solution to the job description classification task, was to train a variety of multilabel classifiers using the models and approaches mentioned in Sect. 3. Specifically, the full texts of job descriptions were preprocessed and used in training, including the position title. In our evaluation we compute both micro- and macro-averaged versions of F1 score. The main difference between them is in the way they treat results of individual label classification, with micro approach using results of all labels together, calculating a total precision and recall across all of them, while macro approach calculates precision and recall for each label individually, averaging them afterwards. This results in macro approach being more descriptive of overall quality and micro being more fare when datasets contain imbalanced classes. To reduce the dependency of our results on chance, we perform 5-fold cross-validation, every time splitting dataset randomly 7 to 3, train and test subset respectively. Impact of overfitting should be minimal, since class counts in Job20Label are balanced. Table 4 presents the best classifiers for each of the vector space models. In case of linear classifiers, it can be noted that different

Table 4. Results of multi-label profession classification task

Multi-labeling Classifier/Strategy	Base classifier	Vector representation	F1-micro	F1-macro
One vs Rest classifier	Logistic regression	TF-IDF	0.8384	0.8349
		Avg. Word2Vec (CBOW)	0.8522	0.8515
		Doc2Vec (DBOW)	0.6091	0.6039
LabelPowerset	Logistic regression	TF-IDF	**0.8767**	**0.8782**
		Avg. Word2Vec (CBOW)	0.8662	0.8682
		Doc2Vec (DBOW)	0.6182	0.6176
	LinearSVC	TF-IDF	0.8529	0.8572
		Avg. Word2Vec (CBOW)	0.8251	0.8294
	MLP classifier	TF-IDF	0.8326	0.8308
		Avg. Word2Vec (CBOW)	0.8364	0.8390
Convolutional Neural Networks (CNN)		Char N-grams with punct	0.8730	0.8764
		Char N-grams w/o punct	0.8592	0.8643
		Word Emb with lemm	0.8734	0.8796
		Word Emb w/o lemm	**0.8900**	**0.8876**

Table 5. Hyperparameters of CNN classifier

Embedding	Vocabulary size depending on the corpus
	Output size of embedding-layer 300
	Spatial dropout 0.2
CNN	Vocabulary size depending on the corpus
	Output size of embedding-layer 300
	Spatial dropout 0.2
Dense	Number of convolution filters 1024
	Kernel size 3
	Activation function RELU
	GlobalMaxPooling
Loss function: binary cross-entropy	
Optimizer: adam with default values	
The average number of learning epochs: 20	

multi-label strategies, cause different vector representations to have the best results. Among the basic classification models, Logistic Regression consistently achieved the best results. For instance, in the case of OneVsRestClassifier strategy, the best result was achieved using averaged word2vec, as was the case with LabelPowerset and TF-IDF. Paragraph2vec (Doc2vec) proved to be significantly worse than both averaged word2vec and TF-IDF. However, the best result for either version of F1, was achieved by a more sophisticated CNN classifier, using one-hot word embeddings, on data, preprocessed without lemmatization. It is worth mentioning that the word2vec pretrained on large Internet corpora (wiki, Russian National Corpus (RNC), Araneum) performed significantly worse, in contrast to word2vec models we have trained on our Job653K dataset. CNN architecture, which had the best performance in this experiment, used Keras framework implementation. Details of the its hyperparameters, are presented in Table 5. To analyze effects of the training sample size on the performance of the classifier, we plotted training curves for the best classifiers (Fig. 4). It should be noted that CNN already starts leading in F1-macro just at 50 examples per class (out of 20 classes). Analyzing classifier errors, it is worth noting that linear classifiers using word2vec and CNN are more likely to make mistakes on longer job descriptions, where most text is not a description of a position, but rather a description of the company. The quality of classification could be improved by developing rules to remove this kind of information from text (Table 6).

Concept Matching Evaluation and Results. We treat concept matching as a task of finding semantic similarity of paraphrases with paraphrases being individual concepts from the Job120Concepts corpus and 4 professional standards. These standards correspond to professions defined for original job description via multilabel classification. While we match job concepts directly to standard

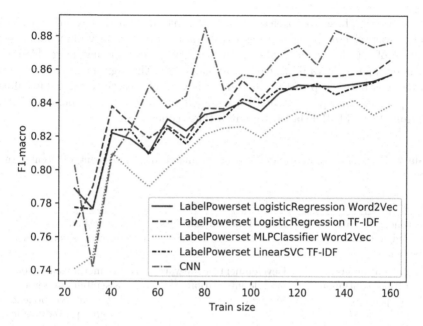

Fig. 4. Learning curves for the best multi-label classifiers (F1-macro/train size per class)

concepts, we do make note of the entire labor function description for each matched standard concept. Relevance of each matched pair of concepts, as well as relevance to the standard concepts containing labor function was determined and labeled by experts. The resulting labelled sequence of selected standard concepts was evaluated using traditional metric used for ranking tasks: mean average precision (MAP) for 1, 3 and 5 concepts, ordered by their predicted semantic similarity. Matching job concepts to concepts of all standards from Std40Label

Table 6. Results of concept matching task

Matching method	Distribution semantic model	Matching labor function			Matching labor action/req.		
		MAP@1	MAP@3	MAP@5	MAP@1	MAP@3	MAP@5
Matching with all professional standards (w/o multilabel classification)	TF-IDF	0.6734	0.6944	0.6687	0.4167	0.5278	0.4750
	Avg. word2vec (CBOW)	0.9167	0.8472	0.8333	0.8315	0.7639	0.7583
Matching with one professional standards (using multilabel classification)	TF-IDF	0.8333	0.8611	0.8365	0.6250	0.6667	0.6583
	Avg. word2vec (CBOW)	**0.9967**	**0.9722**	**0.9500**	**0.8333**	**0.8889**	**0.8167**

with no filtering based on profession, acts as a baseline for this experiment. This baseline was selected in order to assess the impact preliminary multi-label classification of job descriptions has on the quality of concept matching. TF-IDFs The poor quality of ranking can be explained by the significant difference of vocabularies between professional standards and job descriptions. Table 7 illustrates pairs of concepts matched by averaged word2vec, despite being different in terminology. TF-IDF failed to match these concepts.

Table 7. Pairs of job and standard concepts, matched despite using different terminology

Job description	Profession standards	
Job concepts: skill, knowledge, action, experience	Labor function concepts	Labor action, requirements concepts
Build a continuous integration, continuous delivery process	Development of software module integration procedures	Apply methods and tools for assembling modules and software components, developing procedures for deploying software ...
Work in the project team (developers, analysts, key users)	Requirements Development and Software Design	Distribution of tasks between programmers in accordance with the technical specifications
Communication with the customer and the team of Backend programmers	Requirements Development and Software Design	Communicate with stakeholders
Providing professional consulting services for customers	Software requirements analysis	Evaluate and justify recommended solutions.
Refinement of the existing functionality in accordance with user requests	Functional check and software code refactoring	Making changes to the program code to eliminate the detected defects

Keyword Extraction Results. We enrich professional standards using the most frequent terms from job descriptions. To do so, we have constructed a vocabulary from the texts of job descriptions marked with the label of the relevant profession, with terminology from professional standards excluded from this vocabulary. We then used a term frequency (TF) analysis and the analysis of TF-IDF weights of vocabularies for each class to generate word clouds. Word clouds visualization is implemented using WordCloud [15]. The results of the experiments are presented in the Fig. 5.

Fig. 5. Learning curves for the best multi-label classifiers (F1-macro/train size per class)

5   Conclusion and Future Work

In this paper, we achieve our goal of creating a method for mining the actual requirements of the labor market, based on a matching concepts between job listings and professional standards which allows us to deal with the gap, existing between some professional standards and ever-developing reality. In our first step, the multi-label classification of jobs by profession, the best results were achieved by a model based on a convolutional neural network trained on one-hot word embeddings of unlemmatized documents. Slightly worse was the performance of a classic logistic regression classifier, trained on TF-IDF vectors, while using LabelPowerset multi-label strategy. In the task of matching individual concepts between job descriptions and professional standards, average word2vec was in a significant lead, when compared to others. We also demonstrated a simple approach to enriching the vocabulary of a professional standard with the most frequent terms from the corresponding vacancies with the option of result visualization. In the future, we plan to continue improving our approach in following ways:

- Explore other distributional semantic models, such as fasttext, as well as topic modeling with additive regularization (ARTM);
- Experimentally evaluate other vector space models for individual concepts, while taking into account the context (structural links in the conceptual model);
- Develop a methodology for determining learning outcomes to be used in creation and updating of the educational programs, relevant to the requirements of the labor market;
- Implement a interactive software interface for labor market analysis and results visualization.

Acknowledgments. Research has been supported by the RFBR grant No. 18-47-860013 r_a Intelligent system for the formation of educational programs based on neural network models of natural language to meet the requirements of the digital economy.

We are grateful to the students and lecturers of Chelyabinsk State University for help in preparing and marking data, as well as in conducting experiments. We are grateful to the head and IT-specialists of the Intersvyaz company (is74.ru) who provided the necessary computational platform for the experiments.

References

1. Gorshkov, M.K., Kliucharev, G.A.: Nepreryvnoe obrazovanie v kontekste modern-izatsii. [Continuing education in the context of modernization]. Moscow: IS RAN, FGNU TsSI, p. 232 (2011)
2. Muthyala, R., Wood, S., Jin, Y., Qin, Y., Gao, H., Rai, A.: Data-driven job search engine using skills and company attribute filters. In: 2017 IEEE International Conference on Data Mining Workshops (ICDMW) (2017)
3. Karakatsanis, I., et al.: Data mining approach to monitoring the requirements of the job market: a case study. Inf. Syst. **65**, 16 (2017)
4. Mller, O., Schmiedel, T., Gorbacheva, E., Brocke, J.V.: Towards a typology of business process management professionals: identifying patterns of competences through latent semantic analysis. Enterp. Inf. Syst. **10**, 5080 (2014)
5. Zhao, M., Javed, F., Jacob, F., McNair, M.: SKILL: a system for skill identification and normalization. In: Proceedings of the Twenty-Seventh Conference on Innovative Applications of Artificial Intelligence, pp. 4012–4018, January 2015
6. Sayfullina, L., Malmi, E., Kannala, J.: Learning representations for soft skill matching. In: van der Aalst, W.M.P., et al. (eds.) AIST 2018. LNCS, vol. 11179, pp. 141–152. Springer, Cham (2018). https://doi.org/10.1007/978-3-030-11027-7_15
7. Mikolov, T., Chen, K., Corrado, G., Dean, J.: Efficient estimation of word representations in vector space. arXiv preprint arXiv:1301.3781 (2013)
8. Le, Q., Mikolov, T.: Distributed representations of sentences and documents. In: Proceedings of the 31st International Conference on Machine Learning. Beijing, China, JMLR: W&CP 2014, vol. 32, pp. 1188–1196 (2014)
9. Pedregosa, F., et al.: Scikit-learn: machine learning in Python. J. Mach. Learn. Res. **12**(Oct), 2825–2830 (2011)
10. Szymaski, P., Kajdanowicz, T.: A scikit-based Python environment for performing multi-label classification. arXiv preprint arXiv:1702.01460 (2017)
11. Kim, Y.: Convolutional neural networks for sentence classification. In: Proceedings of the 2014 Conference on Empirical Methods in Natural Language Processing (EMNLP), Doha, Qatar, 1746–1751 (2014)
12. Arkhipenko, K., Kozlov, I., Trofimovich, J., Skorniakov, K., Gomzin, A., Turdakov, D.: Comparison of neural network architectures for sentiment analysis of Russian tweets. In: Computational Linguistics and Intellectual Technologies: Proceedings of the International Conference Dialogue 2016, pp. 50–58. RGGU, Moscow (2016)
13. Panchenko, A., Loukachevitch, N., Ustalov, D., Paperno, D., Meyer, C., Konstantinova., N.: RUSSE: the first workshop on Russian semantic similarity. In: Computational Linguistics and Intellectual Technologies: Proceedings of the International Conference Dialogue 2015, vol. 2, pp. 89–105. RGGU, Moscow (2015)
14. Panchenko, A., et al.: a shared task on word sense induction for the Russian language. In: Computational Linguistics and Intellectual Technologies: Proceedings of the International Conference Dialogue 2015, pp. 547–564. RGGU, Moscow (2018)
15. WordCloud for Python Documentation. https://amueller.github.io/word_cloud/. Accessed 29 Nov 2018

Enhancing Supply Chain Risk Management by Applying Machine Learning to Identify Risks

Ahmad Pajam Hassan[(✉)]

Department of Computing Science, Oldenburg University, Oldenburg, Germany
ahmad.pajam.hassan@uni-oldenburg.de

Abstract. Supply chain risks negatively affect the success of an OEM in automotive industry. Finding relevant information for supply chain risk management (SCRM) is a critical task. This investigation utilizes machine learning to find risk within textual documents. It contributes to the supply chain management (SCM) by designing (i) a conceptual model for supply risk identification in textual data. This addresses the requirement to see the direct connection between data analytics and SCM. (ii) An experiment in which a prototype is evaluated contributes the requirement to have more empirical insight in the interdisciplinary field of data analytics in SCRM.

Keywords: Supply chain risk management · NLP · Data analytics · Machine learning · Risk identification

1 Introduction

1.1 Situation and Research Gaps

Supply chain management (SCM) is the management of material, information and financial flows within a network of organizations. It has the goal to produce and deliver products or services between participants of the network, e.g. suppliers, manufacturers, logistics providers, wholesalers, distributors and retailers [1]. Recent investigations focused on the synergy between SCM and advanced analytics techniques, also known as big data, business analytics, predictive analytics or data science. They help researchers and practitioners to understand how the technology could impact the design of SCM and identify research gaps [2–5]. Upcoming studies have to combine fundamentals of SCM with technology within the field of advanced analytics to show direct connections between them [5] and validate new concepts empirically [6]. To address these requirements this investigation will demonstrate how a concrete implementation of a machine learning workflow could improve SCM, in particular supply chain risk management (SCRM). SCRM is "characterized by a cross-company orientation aiming at the identification and reduction of risks not only at the company level, but rather focusing on the entire supply chain" [7, p. 243)]. Supply risks are defined as "an individual's perception of the total potential loss associated with the disruption of supply of a particular purchased item from a particular supplier" [8, p. 36]. The entire

W. Abramowicz and R. Corchuelo (Eds.): BIS 2019, LNBIP 354, pp. 191–205, 2019.
https://doi.org/10.1007/978-3-030-20482-2_16

SCRM consists of the process steps risk identification, risk assessment, risk mitigation, risk monitoring [9].

1.2   Problem and Delimitation

This investigation focuses on applying machine learning to the risk identification process which is part of the SCRM as subdomain of SCM. The challenge is to build a classification model with a learning algorithm. The trained classification model has to identify potential supply risks for automotive supply chains in text documents (e.g. news articles). Using the example of the massive fire at Meridian Lightweight Technologies on May 02, 2018, an online news article mentions that the production of the automotive supplier is negatively affected by explosions [10].

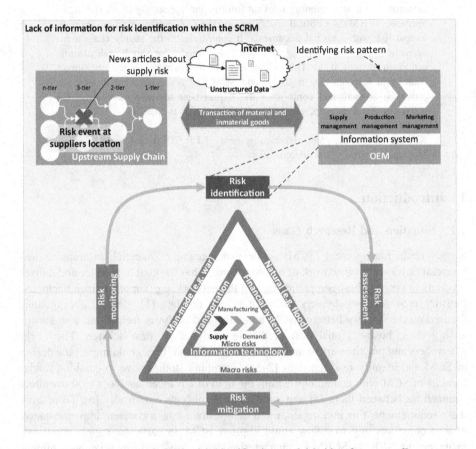

Fig. 1. Integrated supply risk identification model [with reference to 6]

In this context a machine trained classifier digitalize the human task to explore all available news and inspect them for potential supply risks. The machine accelerate and speed up this procedure by process massive amount of data. Figure 1 illustrates on the one hand a conceptual SCRM model [6] in which relevant aspects are accentuated. These are risk identification within the risk cycle as well as supply risks within a micro risk subsystem. It shows an extract of an upstream supply chain, where a risk event occurs, which affects a supplier within the network of organizations connected to an OEM. In situations like that, news articles referencing the risk event outside of the enterprise information system (in [10] an explosion within the property of a supplier). It is necessary to classify risk events and integrate them into the information system of an OEM. This investigation is focused on the first part – identifying risk within unstructured data, in particular textual documents. The boundaries of this investigation are the requirements for the integration into a real enterprise information system.

1.3   Goals and Research Questions

The goals of this investigation are (i) designing a conceptual model for risk identification algorithm, (ii) the implementation of an instance of it, (iii) its evaluation by means of an experiment for measuring the performance. To achieve these objectives the following research questions are structuring this paper and have to be answered:

- RQ1: How does the conceptual design of an artefact that classifies supply risk within text documents look like?
- RQ2: Could an implementation of the conceptual model technically prove that it is possible to enhance the risk identification process with online news?
- RQ3: How does the model perform at classifying the supply chain risk?

2   Research Methodology

Design science in information systems research aims to develop and study artefacts within the field information system. An artefact could be classified as system design, method, notation, algorithm, guideline, requirements, pattern or metrics [11]. In this investigation three different artefact types will be created.

- A conceptual model of the risk identification workflow.
- An implementation of the conceptual model as proof of concept.
- An evaluation layout which will be used as design for an experiment to measure performance and optimize parameters.

To achieve these artefacts the design process is based on the method framework for design science research [12]. Figure 2 summarizes the research design applied in this investigation. In the introduction the situation and its problems are explicated. Based on a review of recent studies, requirements and research gaps within the SCM are identified. In General, artefacts are made by humans with the goal of bridging knowledge gaps and solving problems for several disciplines [11]. In particular the SCRM tackles the problem of identifying supply chain risks. The goal of this investigation is to apply machine learning for classifying text documents which indicate supply chain disruptions.

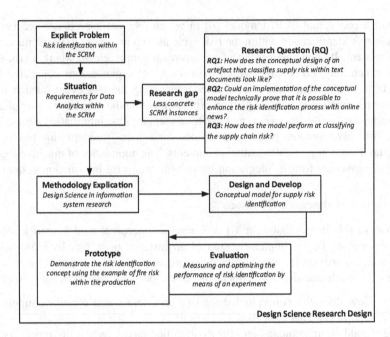

Fig. 2. Research design to develop a supply risk identification concept

Hereby the value of machine learning for SCRM is demonstrated by using the example of fire incidences within the automotive industry. In relation to this specific case the last section evaluates the designed and developed artefacts by performance measuring of the results of the classification experiment.

3 Text Classification - A Field for Applying Machine Learning

Several disciplines (e.g. database, data mining, and information retrieval) have widely studied the problem of machine based classification [13] and prediction [42]. Text classification is a common task within the field of text mining. Based on a labeled training set a classification model will be trained. By applying the built classification model to new documents, unlabeled documents will be classified into one or more pre-defined classes [14]. The implementation of the classification task is realized by machine learning algorithms. Previous studies have shown support vector machines (SVM) are suitable to assign text documents to predefined (risk) classes [15]. In comparison to machine learning approaches the main difference to text mining is the occurrence of semantic information within the textual data. The sense of a word depends on multiple factors, e.g. negation or ambiguity [16]. Recent investigations focus on bridging limitations within existing classification systems or designing new ones. They are focused on three aspects to improve the state of art [17]. The first one is high dimensionality of the vector space; the second aspect is handling the data sparseness. Both are addressed by applying feature selection algorithms. The feature selection will heuristically reduce the dimensionality of a vector [18, 19]. This can be achieved with the selection of features by wrappers or by applying filters [20]. The third aspect is the combination of different classifiers. In a multiple classifier system the assumption is that the combination of several classifiers will improve the accuracy rate in comparison to an individual classifier [21, 26]. Based on the studies from [13, 22–24] state of the art text classification elements could be identified [17]. The following elements are the fundamentals in the design of the supply risk analytics concept.

1. Pre-processing
2. Vector space model
3. Dimensionality reduction
    (a) Feature selection
    (b) Feature projection
4. Training method of a classification function
5. Evaluation measurement technique
6. Performance measures
7. Dataset(s)
8. Description of domain-specific difficulties

4   Conceptual Model to Identify Supply Risks Within Textual Data

In this section the design of the conceptual model for the supply risk identification system will be presented. Figure 3 shows that it contains 14 steps, referenced alphabetically. In Sect. 4.1 the steps from (a) to (f) will be explained on an abstract level. Section 4.2 to 4.4 are described in detail to allow a high traceability of the calculations form (d), (e) and (h). In Sect. 5.1 are (g) and (j) and in Sect. 5.2 (i), (k) and (l) are described. Section 5.3 include (m) and (n).

4.1   Preliminary

(a) A risk manager as expert for supply risks classifies news documents within a textual data set in a way that the unlabeled data becomes annotated with adequate labels. In this investigation the supply risk manager determine if a supply risk label for fire incidents should be assigned to a document or not. The expert answers the question, if the news article contains indicators for a supply risk or not. (b) If the corpus is completed, (c) in the content extraction step metadata such as html or xml code will be filtered. (d) In the term extraction step, terms will be extracted after a document is transformed into the vector space model. In vector space model each document becomes a vector of words [31]. A word or token by document matrix A represents the occurrences of a token in a document, i.e. $A = (a_{ik})$ where a is the weight of a token i in a document k. Since documents don't include all words, A is a spare matrix with high dimensionality [22]. (e) To reduce dimensionality, stop words[1] will be removed and word stemming[2] will be executed. In the phase of term selection there are several steps. The basic idea is to weight terms and to remove them in case of less information value.

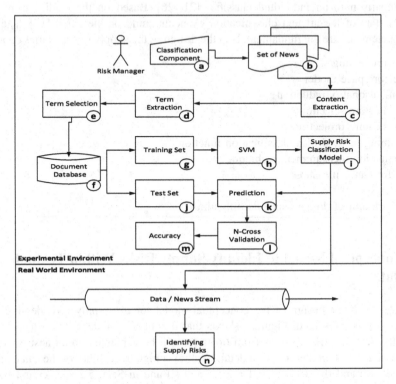

Fig. 3. Design of the conceptual model

4.2 Term Relevancy (d)

In this investigation several weighting options will be used for determining the weight a_{ik} of word i in document k. The first one is the *word frequency weight*, which is defined as $a_{ik} = f_{ik}$ and where f_{ik} is the frequency of term i in document k. Against this measure stands the assumption that the more often a word occurs in a risk indicating document, the more relevant it is to its risk class. But the more often it appears in

---

[1] Frequent words that carry irrelevant information (i.e. pronouns, prepositions, conjunctions, etc.) [22]. In https://github.com/stopwords-iso/stopwords-de/blob/master/raw/stop-words-german.txt (accessed 09.08.2018).

[2] Stemming is the process of removing suffixes from a term to generate word stems [27].

documents with other class associations, the less it discriminates between documents. This leads to the extended *tf-idf weight*, which is defined as:

$$a_{ik} = f_{ik} * \log\left(\frac{N}{n_i}\right) \tag{1}$$

and where N is the number of documents in a given data set and n_i is the overall count of a term i [32]. Extending the frequency weight by the inverse proportion of documents within the corpus which include the term at least ones, still disregards the fact that in short documents, a single word has stronger relevancy to its risk indicating topic than in a large one. The *tfc weighting* takes the length of a document into account and normalizes it [32]. There are numerous modifications, e.g. [33] softens up the weight of f_{ik} by assigning the common logarithm to it. This led to:

$$a_{ik} = \frac{\log(f_{ik} + 1) * \log\left(\frac{M}{m_i}\right)}{\sqrt{\sum_{j=1}^{M}\left[\log\left(f_{jk} + 1\right) * \log\left(\frac{N}{n_i}\right)\right]^2}} \tag{2),)}$$

where M is the number of terms in the given corpus after reducing its dimensionality, in particular after the removal of stop words and word stemming. n_i is the overall count of a term i in the document corpus. Another relevancy measurement is *entropy weighting*, based on [34] it is defined as:

$$a_{ik} = \log(f_{ik} + 1) * \left(1 + u_{ij}\right), \text{With } u_{ij} = \frac{1}{\log(N)}\sum_{j=1}^{N}\left[\frac{f_{ij}}{n_i}\log\left(\frac{f_{ij}}{n_i}\right)\right], \tag{3}$$

where u_{ij} is the entropy or the average uncertainty of word i. If the word is equally distributed over all documents u becomes -1 and 0 if the word occurs only in one document.

4.3   Feature Selection (e)

After determining the weights for a_{ik} to improve the effectiveness of classifying supply risks, feature selection will be applied to increase effectiveness of categorization and reduce computational complexity. The basic idea is to remove words from documents which are rare. This will be done by document frequency thresholding and information gain. *Document frequency thresholding* means to remove words which occur less than or equal to a predefined threshold in the whole document corpus. Based on [22] the core assumption is that these rare words are non-informative for a risk category prediction or have no influence in global performance. Based on the knowledge that different feature selection approaches have a comparable performance [35], it is coherent to not devote much effort to many variations on this process step. For double-checking the influences of feature selection approaches on risk identification, the *information gain* will be used as criterium for the goodness of a feature as well. It measures the number of bits of information obtained for category prediction by knowing the presence or absence of a term in a document [35]. It is defined as:

$$IG(w) = -\sum_{j=1}^{K} P(c_j) \log P(c_j) + P(w) \sum_{j=1}^{K} P(c_j|w) \log P(c_j|w) + P(\overline{w}) \sum_{j=1}^{K} P(c_j|\overline{w}) \log P(c_j|\overline{w})$$

(1)

Where c_j are possible supply risk classes.

4.4 Machine Learning Method (h)

By integrating dimension reduction and classification, support vector machines (SVMs) have been widely studied in the context of text classification problems [25, 36]. The SVM separates multi-classification problems in a series of dichotomous classification problems [22]. It assigns to vector d either -1 or 1 by the following equation:

$$s = w^T \phi(d) + b = \sum_{i=1}^{N} \alpha_i y_i \, K(d, d_i) + b, \text{With } y = \begin{cases} 1 & if\ s > s \\ -1 & otherwise \end{cases},$$

(5)

Where $\{d_i\}_{i=1}^{N}$ is the set of training vectors, $\{y_i\}_{i=1}^{N}$ is the corresponding classes ($y_i = \{-1; 1\}$ and $K(d_i, d_j)$ is denoted a kernel.

5 Experimental Results

In this section the experiment and its results will be presented. In the experiment several specifications of the kernel function are implemented. First the data set will be discussed, before the context of the experiment and finally the prototyping and its results will be presented. The experiment includes a prototype with a RBF kernel. Also the exploration of an appropriate penalty parameter and the kernel parameters are part of this section.

5.1 Data Set (g) (j)

There are several data sets for classification tasks, e.g. *20 Newsgroup, ACM, CSTR, Dmoz Computers 500, Dmoz Health 500, Dmoz Science 500, Dmoz Sports 500, Enron Top20, FBIS, Hitech, Irish Sentiment, La1s, La2s, LATimes, Multi Domain Sentiment, New 3, NFS, Ohscal, Ohsumed-400, Opinosis, Reuters-21578, Review Polarity, SpamAssassin, SpamTrec-3000, SyskillWebert, TDT2T30, WAP, WebACE, WebKB* [28–30]. In addition, other data sets are available in the internet.[3,4,5,6] However all of the reviewed data sets are not suitable for the specific task of supply risk identification

---

[3] http://ana.cachopo.org/datasets-for-single-label-text-categorization.

[4] http://www.cad.zju.edu.cn/home/dengcai/Data/TextData.html.

[5] http://sites.labic.icmc.usp.br/text.

[6] http://qwone.com/~jason/20Newsgroups/.

within an automotive supply chain. Due to the limited amount of problem specific data sets in this field, it is necessary to manually create a labeled data set. In general the domain specificity problem is a commonly known in the field of machine learning and in particular in text mining. Taking this into account is an important requirement for the design and development process [17]. The manually created text corpus contains over 741 text documents from which 299 are labeled as fire risk. The labeling process is examined by two risk managers which work in the field of SCRM in automotive industry.

5.2 Technical Implementation – An Instance of the Designed Artefact (i) (k) (l)

Feature Construction. The labeled data set is transformed appropriately by applying OpenNLP SimpleTokenizer[7]. After transforming the documents to bag of words, stop words[8] will be removed. To stem the words the snowball stemmer library[9,10] is used.

Document Frequency Thresholding and Term Weighting. The minimum document frequency is specified by 0.5%. When a term occurs less than the defined threshold, the feature is filtered. Term weighting is computed by the relative term frequency.

Matrix Specification and Document Database Split. The matrix includes 1,518 columns and 740 rows. Each row represents a preprocessed document. 299 documents are classified as supply risks and labeled as fire based supply chain disruptions. The portioning of the processed documents is examined by stratified sampling[11]. According to the performed split, 70% of the documents are training data and 30% test data.

Train Classifier (i). Taking the knowledge from [15] into account, SVMs are suitable to assign text documents to predefined classes. The LibSVM library in WEAK 3.7 is used to build a supply risk classifier to identify fire incidence within automotive supply chains within a news corpus with real news items, labeled by a human and double-checked by a second expert. The SVM (type: C-SVC) implementation based on [37] with a RBF kernel $K(x, \gamma) = e^{-\gamma \|x-v\|^2}$. The radial based function nonlinearly maps samples into a higher dimensional space. It has less hyperparameters in comparison to a polynomial kernel. To identify supply risks, the SVM is initially parameterized with $C = 1$ and $\gamma = 0.01$.

Parameter Optimization. Giving a predefined step size, all possible parameters between two predefined boundaries will be iteratively assigned to parameter C ($C_i = \sum_{i=1}^{10} C * 10$) and the gamma parameter γ ($\gamma_j = \sum_{j=1}^{100} \gamma * 0.01$). By this grid layout the

---

[7] https://opennlp.apache.org/docs/1.8.0/manual/opennlp.html#tools.tokenizer.

[8] https://github.com/stopwords-iso/stopwords-de/blob/master/stopwords-de.txt.

[9] https://github.com/snowballstem.

[10] http://snowballstem.org/algorithms/german/stemmer.html.

[11] The distribution of the supply risk is approximately retained in training and test set.

approximately best parameter specification for penalty parameter C and for γ – the inverse of the radius of influence of samples determined by the support vectors parameter [40] – will be determined.

Cross Validation. To avoid overfitting and validate the performance, the split stratified sampling is repeated for each parametrization setting of the kernel function. By splitting the data set in 5 different sets, five different classifiers will be trained. Taking [41] into account, the 5-fold cross validation step generalizes the performance of the supply risk classifier.

5.3 Evaluation (m) (n)

In this section the best performance out of 5,000 trained classification models will be presented in general and the best performing classifier will be presented in detail.

Parameter Optimization. As exhaustive search strategy, a grid search layout is designed for parameter optimization. In combination with the 5-fold cross-validation a total of 5,000 supply risk models are trained and evaluated. Figure 4 shows the search grid where the red area highlights good performance. The lowest measured accuracy performance is 0.59595 and is highlighted green. The maximum Measurement accuracy is 0.96486 with gamma 0.02 (x-axis) and C 21 (y-axis).

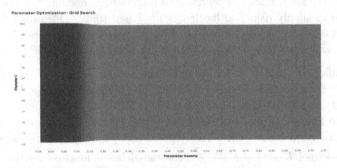

Fig. 4. Results from search grid parameter optimization (Color figure online)

Table 1 shows the cross validation results of the best measured performance. The error count ranges from 4 to 7.

Table 1. 5-fold cross validation

	Error in %	Size of test set	Error count
fold 0	2.702702703	148	4
fold 1	4.054054054	148	6
fold 2	4.72972973	148	7
fold 3	3.378378378	148	5
fold 4	4.054054054	148	6

In Table 2 the rows represent the actual class and the columns the prediction of the class, in particular the fire incidents.

Table 2. Confusion matrix

	Other	Fire incident
Other	418	23
Fire incident	5	294

Table 3 shows the accuracy statistic of the experiment the calculated parameters are defined in [22], where true positives (tp) are the number of documents which are correctly assigned to the risk class, false positives (fp) are the number of documents incorrectly assigned to the risk category, true negatives (tn) are the number of documents correctly rejected from risk class and false negatives (fn) are the number of documents incorrectly rejected from the risk class. The precision is the likelihood that a randomly chosen document is correctly classified as risk. The recall is quotient of the number of TP divided by FP plus FN.

Table 3. Accuracy statistics

	TP	FP	TN	FN	Recall	Precision
Other	418	5	294	23	0.9478	0.9882
Fire incident	294	23	418	5	0.9833	0.9274

Table 4 represents the F-measure as quality metrics. It is calculated by $F = 2 * \frac{precision*recall}{precision + recall}$ and has a value of 0.9545 and an accuracy value of 0.9622. Since the F-measure didn't take the true negatives into account, Cohen's kappa is calculated. Cohen's kappa represents the possibility of the agreement occurring by chance. According to [38], values less than 0.4 are critical and values over 0.75 represents a strong inter-rater reliability performance. With reference to [39] the performance of the trained risk classifier is almost perfect with a value of 0.9222.

Table 4. Accuracy matrix

	F-measure	Accuracy	Cohen's kappa
Other	0.967592593		
Fire incident	0.954545455		
Overall		0.962162162	0.922185

6 Conclusion

In this section, first the evaluation and the contributions of this investigation will be presented. Second the limitations of the designed artefact will be discussed and an outlook regarding upcoming research questions is given.

6.1 Evaluation and Contribution

RQ1. Section 4 presents the conceptual design of the risk identification artefact. It addresses the classification of supply risk using the example of fire based supply disruptions referenced in textual data. The conceptual design is on an abstract level. It should be understood as guideline for other researchers and practitioners.

RQ2. In Sect. 5 an instance of the conceptual model is implemented. It is the evidence that machine enhance the risk identification process by classifying supply risk indicator in online news or other textual documents. The implantation addresses the knowledge gap which [5] identify. They request a demonstration of direct relations between supply chain management and big data [5]. This investigation contributes knowledge to the mentioned knowledge gap by showing that 294 different supply risk cases could be identified successfully.

RQ3. In Sect. 5 the prototype is evaluated within an examined experiment. Based on state of the art methods, the evaluation subsection presents formal parameters which measure the performance of the prototype by classifying supply chain risk explosion. On one hand 0.9274 precision proves that it is possible to identify supply risk and shows that potential for optimization exists. On the other hand 0.9833 recall demonstrates a strong performance. Overall the F1 measurement in combination with Cohen's kappa shows a good result. Regarding that, this investigation addresses the identified knowledge gap from [6]. Ho et al. request more empirical evidence within the interdisciplinary filed of data analytics in SCRM. This investigation contributes empirical insights to the mentioned knowledge gap.

6.2 Limitations and Outlook

From a technical perspective the mentioned potential in the subsection before should be explored in further researches. Regarding that, the question should be answered how to reduce false positives within the identification process. However, the evaluation step of the classification model shows good results regarding the examined labor experiment. Upcoming investigations should take this research as proof of concept and experiment with different classification algorithms or change the parametrization of the kernel functions. It will also be helpful for researchers and practitioners to demonstrate that the concept is portable and could be applied to other risks or even opportunities.

Furthermore upcoming investigations have to focus on technical and domain specific integration aspects. One aspect is to integrate the designed and developed concept into an integrative SCRM concept. For an effective SCRM, it is crucial that more than one process step within SCRM is considered [6]. Applying machine learning to the risk identification process is only the first step. Next would be to explore possible applications in other risk management process steps. Based on the requirement of an integrated SCRM concept, the classified risks have to assign to the risk assessment process.

The other aspect addresses the integration into the entire business information system to provide actionable insights to operative decision making processes within the automotive industry. To bridge its limitation, it would be necessary to explore the requirements of real SCM environments and integrating the designed artefact in it. The contribution of this investigation is a proof of concept that design and develop an artefact for risk identification. Further investigations have to explore industrial requirements to integrate the artefact to IT architecture in automotive industry and help practitioners identify their supply risks. Also a possible transfer to other industries should be investigated. By changing the training data and considering the industry requirements, the designed and developed concept should be easily transferred to other industry sectors.

References

1. Tang, C.S.: Perspectives in supply chain risk management. Int. J. Prod. Econ. **103**, 451–488 (2006)
2. Tiwari, S., Wee, H.M., Daryanto, Y.: Big data analytics in supply chain management between 2010 and 2016: insights to industries. Comput. Ind. Eng. **115**, 319–330 (2018)
3. Wang, G., Gunasekaran, A., Ngai, E.W.T., Papadopoulos, T.: Big data analytics in logistics and supply chain management: certain investigations for research and applications. Int. J. Prod. Econ. **176**, 98–110 (2016)
4. Gunasekaran, A., et al.: Big data and predictive analytics for supply chain and organizational performance. J. Bus. Res. **70**, 308–317 (2017)
5. Waller, M.A., Fawcett, S.E.: Data science, predictive analytics, and big data: a revolution that will transform supply chain design and management. J. Bus. Logistics **34**(2), 77–84 (2013)
6. Ho, W., Zheng, T., Yildiz, H., Talluri, S.: Supply chain risk management: a literature review. Int. J. Prod. Econ. **53**(16), 5031–5069 (2015)
7. Thun, J., Hoenig, D.: An empirical analysis of supply chain risk management in the German automotive industry. Int. J. Prod. Econ. **131**, 242–249 (2011)
8. Ellis, S.C., Henry, R.M., Shockley, J.: Buyer perceptions of supply disruption risk: a behavioral view and empirical assessment. J. Oper. Manag. **28**, 34–46 (2010)
9. ISO 31000:2018: Risk management – guidelines
10. Walsworth, J.: Fire, explosions rip through instrument panel plant in Michigan. In: automotivenews.com by Crain Communications Inc., 2 May 2018. http://www.autonews.com/article/20180502/OEM10/180509951/fire-magnesium-supplier-general-motors-michigan. Accessed 30 July 2018
11. Offermann, P., Blom, S., Schönherr, M., Bub, U.: Artifact types in information systems design science – a literature review. In: Winter, R., Zhao, J.L., Aier, S. (eds.) DESRIST 2010. LNCS, vol. 6105, pp. 77–92. Springer, Heidelberg (2010). https://doi.org/10.1007/978-3-642-13335-0_6

12. Johannesson, P., Perjons, E.: A method framework for design science research. In: An Introduction to Design Science, pp. 75–89. Springer, Cham (2014). https://doi.org/10.1007/978-3-319-10632-8_4

13. Aggarwal, C.C., Zhai, C.: A survey of text classification algorithms. In: Aggarwal, C., Zhai, C. (eds.) Mining Text Data, pp. 163–222. Springer, Boston (2012). https://doi.org/10.1007/978-1-4614-3223-4_6

14. Liu, A.B.: Web data mining: exploring hyperlinks, contents, and usage data. Springer (2007)

15. Abbasi, A., Chen, H.: CyberGate: a system and design framework for text analysis of computer-mediated communication. MIS Q. 32(4), 811–837 (2008)

16. Altınel, B., Ganiz, M. C., Diri, B.: A corpus-based semantic kernel for text classification by using meaning values of terms. In: Engineering Applications of Artificial Intelligence, vol. 43, pp. 54–66 (2015)

17. Mironczuk, M.M., Protasiewicz, J.: A recent overview of the state-of-the-art elements of text classification. Expert Syst. Appl. **106**, 36–54 (2018)

18. Chen, J., Huang, H., Tian, S., Qu, Y.: Feature selection for text classification with naive Bayes. Expert Syst. Appl. **36**(3), 5432–5435 (2009)

19. Forman, G.: An extensive empirical study of feature selection metrics for text classification. J. Mach. Learn. Res. **3**, 1289–1305 (2003)

20. John, G.H., Kohavi, R., Pfleger, K.: Irrelevant features and the subset selection problem. In: Proceedings of the 11th International Conference on machine learning, pp. 121–129. Morgan Kaufmann, San Francisco (1994)

21. Zhu, J., Xie, Q., Zheng, K.: An improved early detection method of type-2 diabetes mellitus using multiple classifier system. Inf. Sci. **292**, 1–14 (2015)

22. Aas, K., Eikvil, L.: Text categorisation: a survey. Technical report. Norwegian Computing Center (1999)

23. Guzella, T.S., Caminhas, W.M.: A review of machine learning approaches to spam filtering. Expert Syst. Appl. **36**(7), 10206–10222 (2009)

24. Adeva, J.G., Atxa, J.P., Carrillo, M.U., Zengotitabengoa, E.A.: Automatic text classification to support systematic reviews in medicine. Expert Syst. Appl. **41**(4, Part 1), 1498–1508 (2014)

25. Altınel, B., Ganiz, M.C., Diri, B.: A novel higher-order semantic kernel. In: Proceedings of the IEEE 10th International Conference on Electronics Computer and Computation (ICECCO), pp. 216–219 (2013)

26. Ali, R., Lee, S., Chung, T.C.: Accurate multi-criteria decision making methodology for recommending machine learning algorithm. Expert Syst. Appl. **71**, 257–278 (2017)

27. Porter, M.F.: An algorithm for suffix stripping. Program **14**(3), 130–137 (1980)

28. Pinheiro, R.H.W., Cavalcanti, G.D.C., Tsang, I.R.: Combining dissimilarity spaces for text categorization. Inf. Sci. **406–407**, 87–101 (2017)

29. Sabbah, T., et al.: Modified frequency-based term weighting schemes for text classification. Appl. Soft Comput. **58**, 193–206 (2017)

30. Cardoso-Cachopo, A.: Improving methods for single-label text categorization. PdD Thesis, Instituto Superior Tecnico, Universidade Tecnica de Lisboa (2007)

31. Slaton, G., McGill, M.J.: An Introduction to Modern Information Retrieval. McGraw-Hill, New York (1983)

32. Slaton, G., Buckley, C.: Term weighting approaches in automatic text retrieval. Inf. Process. Manage. **24**(5), 513–523 (1988)

33. Buckley, C., Slaton, G., Allan, J., Singhal, A.: Automatic query expansion using SMART: TREC 3. In: Proceedings of the 3rd Text Retrieval Conference, NIST (1994)

34. Dumais, S.T.: Improving the retrieval of information from external sources. Behav. Res. Methods Instrum. Comput. **23**(2), 229–236 (1991)

35. Yang, Y., Pedersen, J.O.: A comparative study on feature selection in text categorization. ICML **97**, 412–420 (1997)
36. Joachims, T.: Text categorization with support vector machines: learning with many relevant features. In: Nédellec, C., Rouveirol, C. (eds.) ECML 1998. LNCS, vol. 1398, pp. 137–142. Springer, Heidelberg (1998). https://doi.org/10.1007/BFb0026683
37. Chang, C.-C., Lin, C.-J.: LIBSVM: a library for support vector machines. ACM Trans. Intell. Syst. Technol. **2**(3), 2:27:1–27:27 (2011)
38. Greve, W., Wentura, D.: Wissenschaftliche Beobachtung. Eine Einführung. Psychologie Verlags Union, München (1997)
39. Landis, J.R., Koch, G.G.: The measurement of observer agreement for categorical data. Biometrics **33**(1), 159–174 (1977)
40. Hsu, C.-W., Chang, C.-C., Lin, C.-J.: A practical guide to support vector classification. Technical report, Department of Computer Science, National Taiwan University (2003)
41. Bergstra, J., Bengio, Y.: Random search for hyper-parameter optimization. J. Mach. Learn. Res. **13**, 281–305 (2012)
42. Candanedo, I.S., Nieves, E.H., González, S.R., Martín, M.T.S., Briones, A.G.: Machine learning predictive model for industry 4.0. In: Uden, L., Hadzima, B., Ting, I.-H. (eds.) KMO 2018. CCIS, vol. 877, pp. 501–510. Springer, Cham (2018). https://doi.org/10.1007/978-3-319-95204-8_42

Deep Neural Networks for Driver Identification Using Accelerometer Signals from Smartphones

Sara Hernández Sánchez[1]([⊠]) [ID], Rubén Fernández Pozo[2] [ID],
and Luis Alfonso Hernández Gómez[1] [ID]

[1] Department of Signals, Systems and Radiocommunications, Universidad
Politécnica de Madrid, 28040 Madrid, Spain
{sara.hernandez.sanchez, luisalfonso.hernandez}@upm.es
[2] Group of Biometry, Biosignals, Security, and Smart Mobility, Universidad
Politécnica de Madrid, 28040 Madrid, Spain
ruben.fernandez@upm.es

Abstract. With the evolution of the onboard communications services and the applications of ride-sharing, there is a growing need to identify the driver. This identification, within a given driver set, helps in tasks of antitheft, autonomous driving, fleet management systems or automobile insurance. The object of this paper is to identify a driver in the least invasive way possible, using the smartphone that the driver carries inside the vehicle in a free position, and using the minimum number of sensors, only with the tri-axial accelerometer signals from the smartphone. For this purpose, different Deep Neural Networks have been tested, such as the ResNet-50 model and Recurrent Neural Networks. For the training, temporal signals of the accelerometers have been transformed as images. The accuracies obtained have been 69.92% and 90.31% at top-1 and top-5 driver level respectively, for a group of 25 drivers. These results outperform works in the state of the art, which can even utilize more signals (like GPS- Global Positioning System- measurement data) or extra-equipment (like the Controller Area-Network of the vehicle).

Keywords: Driving identification · Smartphone · Accelerometers ·
Deep learning · Neural networks · Fine-tuning

1 Introduction

Driver identification, understood as the driver recognition within a set of given drivers, is a growing need in many areas of work. For instance, for fleet management companies, it is important to know who is driving and how is he/she driving. Also, companies that offer transportation services with a professional driver, such as Uber or Cabify, need to confirm if the driver registering the journey is really him/her. In many of these applications, once you have selected a trip, you can know the driver, and when the trip is over, users can make assessments about the assigned driver. In applications of ride-sharing services such as BlaBlaCar, reliability of the drivers is a great concern for users. In all these type of services it may be important to detect drivers who shared

W. Abramowicz and R. Corchuelo (Eds.): BIS 2019, LNBIP 354, pp. 206–220, 2019.
https://doi.org/10.1007/978-3-030-20482-2_17

their registration with other unauthorized people. Besides, the survey conducted by the [1], showed that one of the most frequent problems in BlaBlaCar is a poor quality of rides.

The purpose of this work is to contribute to the identification of the driver within a given set, in the least invasive way possible from the point of view of necessary hardware or software equipment, using only the smartphone that the driver carries inside the vehicle, and with the minimum number of sensors (only the tri-axial accelerometer signals). With the developed method, based on the data from the smartphone, the identification of drivers can be applied to multiple fields, from fleet management systems to insurance companies, among others. Most of works related to the driver identification need to have access to the sensors of the vehicle's Controller Area-Network (CAN) or to install extra equipment for obtaining specific signals. Those that employ the smartphone as capture device, use the accelerometers plus other signals like gyroscope or GPS. Our work utilizes only the accelerometer signals for reasons of battery saving, avoiding signals such as GPS, and to avoid sensors that are not included in all current-market smartphones, such as the gyroscope.

In order to carry out the driver identification, different Neural Networks have been trained. The signals of accelerometers captured through smartphones have been transformed into images, to take advantage of already pre-trained networks for object recognizing, such as the Residual Neural Networks (ResNets). The final results obtained shown an accuracy of 69.92% and 90.31% at the top-1 and top-5 levels respectively, for a total set of 25 drivers. That is to say, in 69.92% of the journeys, the driver of the trip is the one who has obtained the highest probability, among the group of 25 drivers, using the identification networks. And in 90.31% of the journeys, the driver is at least among the 5 drivers who have obtained most probability, also among the total set of 25 drivers.

Paper has been organized as follows. In Sect. 2 we present an extensive analysis of the state of the art in driver identification, summarizing in a table the most relevant works for the research community in this area. In Sect. 3, the procedure carried out to achieve the identification is described. Finally, Sect. 4 shows the results obtained, and we discuss the conclusions and possible future lines.

2 Related Work

Both driver identification and verification are areas of great interest for driving applications and for on-board communications services, as well as for safety and control tasks. Recognizing a driver and his behavior is widely in demand.

The characterization of the driver defines its behavior, for instance if a driver is aggressive or not or what kind of actions he/she performs while driving. There are many works related to driver characterization, for example in [2], where it is emphasizes the importance of improving road safety and how many of the accidents are caused by human errors. To do this, they propose a system to characterize the behavior of the driver that consists of capturing an image while driving. This image is used to analyze both the driver's pose and the contextual information by exploring the skin-like region. The skin-like regions are first extracted with a Gaussian Mixture Model

(GMM) algorithm of skin images, and then they are sent to a Convolutional Network that will determine the action of the driver (for example if he/she is manipulating the cell phone, eating, with hands on the steering wheel, etc.).

In [3], they propose an unsupervised model of Deep Learning to study behavior and risk patterns using the Global Positioning System (GPS) signal. For this they use Autoencoders and Self-organized Maps to extract the features and classify the behavior of the driver. The speed or overspeed patterns are divided into 4 clusters: none, mild, moderate and strong. Acceleration, braking (deceleration) or turning in 3 clusters: mild, moderate or strong.

Also in [4], authors try to characterize the behavior using four sensors of the smartphone (the accelerometer, the linear acceleration, the magnetometer and the gyroscope); and they compare four Machine Learning algorithms (Artificial Neural Networks (ANN), Support Vector Machines (SVM), Random Forest (RF) and Bayesian Network (BN)). The characterization is carried out by detecting seven types of driving events: aggressive braking (or deceleration), aggressive acceleration, aggressive left turns, aggressive right turns, aggressive left lane changes, aggressive right lane changes and non-aggressive maneuvers.

The number of research works on driver characterization is very extensive and is closely related to driver identification, since in many cases both use the detection and classification of maneuvers for these purposes. There are also numerous works on driver identification; for example in [5] where they make use of three different databases. One of them is made out of datasets coming from the CAN bus of the car, other from smartphones inside the vehicles through an application, and the last one is a public database. The database of smartphones collects GPS information (speed, latitude, longitude, altitude, course, etc.), plus information from accelerometers and gyroscopes. The CAN bus database also collects recorded video signals while driving. In the public database there is also physiological information with sensors for the conductance of the skin, the temperature or the ECG (electrocardiogram). They define three behaviors of the driver: normal, aggressive and drowsy. They apply their model with different Machine Learning algorithms to the three databases and one of the conclusions they draw is that data from the brake pedal or the steering wheel are very relevant to identify the driver.

Driver identification can be very valuable both in tasks of antitheft, autonomous driving or car insurance. Works like [6] have studied a dataset of 14 drivers for two years. In their experiments, they have seen an important relationship between the mean and the maximum acceleration within the acceleration maneuvers of a driver, and that there are large variations between these ratios for different drivers.

Also [7] uses the controller software when accelerating and decelerating to extract spectral features that are modeled through a GMM. Or in [8], they affirm that it is possible to distinguish the identity of a driver, within a set of them, from the information of the sensors of a single turn.

Driver identification is becoming very relevant for shared car services. These services, such as those offered by Uber or Cabify, work through an application that can be downloaded on smartphones. With this application customers can request transport vehicles with a driver. Most of these applications have a driver assessment system after completing the journey, so that the customer has information about the driver and the vehicle before requesting the trip and he/she can discard or select another request. However, in countries such as China, there are cases of Uber drivers [9] registering false trips or using multiple accounts to receive bonuses from the transport company. Also in the United States there have been cases of false drivers of Uber [10], as well as other types of problems [11], in which the identification and verification of the driver is very important.

The driver identification allows knowing who is driving the vehicle or how much time it spends on the road or in a certain place (Telematics). Companies dedicated to fleet management, which need to know who is driving the vehicle to provide personalized information, often they use methods that require physical devices in the vehicle, such as card readers. With the evolution of Deep Learning techniques, we try to identify or verify the driver directly with the signals recorded during their journeys. For example, [8] proposes a method to predict the identity of the driver within a given set of individuals. Works as [12], in addition to identifying the driver, try to distinguish if the signal of the captured sensors corresponds to the driver's or passenger's device.

Table 1 shows a summary of the most relevant works in this field. Those that try to classify among a greater number of drivers, such as in [13] that identify between 38 drivers, make use of the GPS signal of the smartphone. In our work, we avoid such use due to the GPS requires a considerable increase of battery. Or in [14], which have employed a set of 217 families, utilizes In-Vehicle Data Recorder (IVDR) to measure the driver's actions and the vehicle's movement performance. In principle, they do not need to install a device in the vehicle; however they need information about the position of the car, as well as the most significant events. They also define four categories for trips: home to home (trips that start and end in the area around home), home to other (they start at home to a more distant place), other to home (the opposite of the previous one), and other to other (those start and end from locations distant from the home). The place can be a variable too determinant to identify a driver, since normally drivers do the same route, so it would be positive to be independent of this variable. Also [8] studies the identification in a set of 64 drivers, however although they use 10 cars, all of them are the same model. Besides, the vehicles had to be modified to store all sensor readings, from the different electronic systems in the car. Others like [15] or [16] need the sensors of the CAN-bus of the vehicle.

Some works such as [17] or [18] obtain the database using a driving simulator, so they are not real driving journeys. Works like [19] make use of the accelerometers and the GPS of the smartphone, in order to identify among 10 school bus drivers, however the phone must go in a fixed position.

The approach more similar to our work is the one proposed by [12]. Although the final purpose is different from ours, since it is the distinction between driver and passenger, one of the intermediate steps they carry out is the identification of the driver. They use the sensors recorded in the user's smartphone, but they need both accelerometers and gyroscopes. In our case, we do not use gyroscopes, since many smartphones currently do not implement this sensor to save costs or space. In driver classification their best top-5 validation accuracy was 79%.

Table 1. State of the art in driver identification.

Ref.	Description	Signals	Method	Results
[8]	Identification of a driver out of a given set of drivers (10 cars and 64 total drivers), with sensor data from a single turn	Steering wheel angle, steering velocity, steering acceleration, vehicle velocity, vehicle heading, engine RPM, gas pedal and brake pedal positions, forward and acceleration, lateral acceleration, torque and throttle position	Random forest classifier	Accuracy for each of the 12 most frequently made turns: average of 76.9% for two-driver and 50.1% for five drivers
[12]	Drivers between passengers distinction	Accelerometer and gyroscope sensors: longitudinal acceleration, lateral acceleration and yaw	Neural network	In driver classification: best top-5 validation accuracy: 79%
[13]	Driver identification: 38 drivers for two months. Smartphone in a fixed position	GPS data measurements: speed, location, course and horizontal accuracy	Random forest	Drivers segregated in different groups and analyzed trips for each group. Accuracy of 82.3% for driver groups of 4–5
[14]	Trip driver identification using historical trip based data collected throughout one year. Participation of 217 families	Trip information, vehicle location and events of excessive maneuvers	Learning Vector Quantization (LVQ), Boosted C4.5, Random Forest and Support Vector Machines	~88% accuracy
[15]	Identification of 15 drivers. The same car, during the same time of day, in an isolated parking lot and in a predefined interurban loop spanning roughly 50 miles	CAN bus sensors of the vehicle: brake pedal, max engine torque, steering wheel, lateral acceleration, fuel, etc.	Machine learning algorithms: Support Vector Machine (SVM), Random Forest (RF), Naive Bayes and k-nearest neighbor (KNN)	~87.33% accuracy using brake pedal and 15 min of open-road. 100% using brake pedal and 1.5 h of journey. ~91.33% using all sensors in the parking lot and 8 min. 100% with all the sensors in the open-road
[16]	Driver identification and impostor detection. 11 drivers	CAN-bus signals, gas pedal sensor recordings, brake pedal sensor recordings, frontal laser scanner, accelerometers and measures of rotation rates	Feedforward Neural Network	80% for every group category and above 90% for groups of 2–3 drivers. Impostor detection rate above 80% when the car has a single genuine driver

(continued)

Table 1. (*continued*)

Ref.	Description	Signals	Method	Results
[17]	Recognition of driver behavior. 20 drivers in a simulator	Accelerator and steering wheel angle data	Hidden Markov Model (HMM)	Accuracy 85%
[18]	Driver identification of 10 drivers	Vehicle position, speed, steering wheel position, gas pedal position and brake pedal position	Support Vector Machine (SVM)	95% confidence with at most one false alert per day of driving
[19]	Real-time driver identification in 10 school bus drivers over 2 months. The smartphone was place rigidly and horizontally above driver console	Accelerometer and GPS data	Unsupervised Anomaly Detection and Feed-Forward Neural Network of three layers	81% accuracy, evaluated in 30 test examples for driver
[20]	Driver identification by means of acceleration variation	Acceleration from a fixed smartphone	Principal Component Analysis (PCA) and comparison with reference pattern components	They analyze the importance of selecting the right variables to correctly identify the driver. Optimal value between 5 and 6 variables
[21]	Real-time driver identification. The same route for all drivers	Brake pedal and gas pedal signals	Artificial Neural Networks and Cepstral Feature Extraction techniques	84.6% accuracy with 3 drivers. The identification rates decrease as the number of drivers to be identified increases
[22]	Vehicle's inertial sensors from the CAN bus to build a profile of the driver. Differentiation between two drivers	Lateral and longitudinal accelerations, and yaw angular velocity	Support Vector Machine (SVM) and k-mean clustering	Combining turning and braking events helps better differentiate between two similar drivers
[23]	Driver identification: 14 stable-health older drivers (70 years and older)	Vehicle location and speed	Supervised learning with multiclass linear discriminant analysis (LDA)	For 5 drivers. Events individually: 34% using accelerations, 30% decelerations. The most voted: 55.3% using accelerations, 49.1% decelerations, and 60.5% combining both

3 Procedure Description

Through the signals obtained from the accelerometers of the smartphone, we associate each recorded journey to each driver, within a defined set of drivers. To take advantage of the power of deeper and pre-trained networks, such as the ResNet50 model, it is necessary to convert the tri-axial accelerometer signals into images. To do this aim, every journey is divided into overlapping time windows and the accelerometer signals inside each window are transformed into an image, as it is illustrated in Fig. 1.

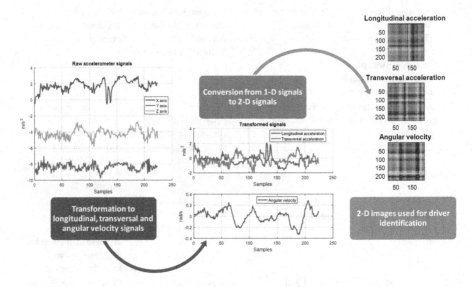

Fig. 1. Temporal signals transformations.

We must point out that the signals converted to images are not the original signals (recorded from the accelerometers of the smartphone). The signals transformed to images are the longitudinal and transversal forces associated with accelerations, and the angular velocity; which are obtained from the original signals from the tri-axial accelerometers. As it is shown in Fig. 1, once these three signals are obtained, they are transformed in order to obtain the three corresponding images that are used to train the driver identification network.

For the tests, a database of 25 drivers with more than 800 journeys per driver was used. The routes have been recorded through a smartphone application installed on both Android and iOS devices called Drivies (see page www.drivies.com, originally developed by company Telefónica R & D and recently became Telefónica spin-off PhoneDrive S.L.). The database is anonymous and the participating drivers have given their consent for the use of the routes. The dataset are not from professional drivers or dedicated to transportation services. The database is heterogeneous in terms of kind of trips and kind of terminals. For each driver there are present city trips, road trips and mixed trips. We have considered that city routes are developed mostly in urban areas and the maximum speed does not usually exceed 70 km/h, while road journeys are trips on highways or motorways with speeds greater than 80 km/h. Finally, the mixed routes do not have a dominant type (neither city nor road exceed 70% of the journey). Based on this, 50.35% of the total trips are mixed routes, 31.56% are city journeys, while 18.09% are road journeys. Although there are two types of strategies followed (to take the first 4 min of the journey or the maneuvering areas), it seems logical that the highest percentage of trips used are from city or mixed, since it is where maneuvers are more susceptible to detection. All drivers have routes from the three types both training and testing. Regarding the type of terminal, 40% of drivers have smartphones with iOS operating system, while 60% with Android operating system. A priori the car model used is unknown, each driver can present a different model or brand, as well as the experience of the drivers. During the capture no interaction between the driver and the application was required (automatic detection of driving) and the smartphone could be freely placed inside the vehicle, it did not need a fixed position. The accelerometer signals were saved from each recorded journey.

The general process is shown in the following figure (Fig. 2).

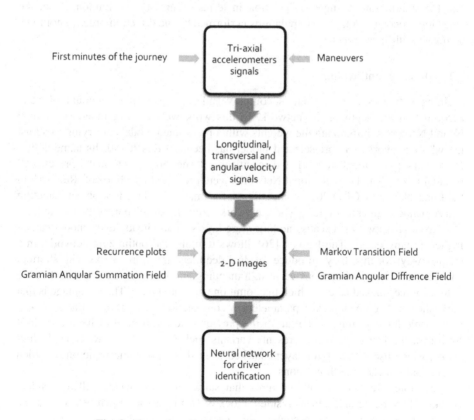

Fig. 2. General outline of the driver identification process.

3.1   Longitudinal, Transversal and Angular Velocity Signals

As we mentioned before, the temporal signals coming from the smartphone, the tri-axial accelerometer signals, are not used directly for the transformation to images. From these signals, we derive the longitudinal and transversal components associated with the car longitudinal and transversal accelerations and the angular velocity. Process for obtaining these signals from the raw signals of the accelerometers can be consulted in [24]. Summarizing this method, it is necessary to obtain the Vehicle Movement Direction (VMD), which is the "forward" movement direction vector in the smartphone reference system associated with driving along a trip. With the VMD, the longitudinal and transversal acceleration forces can be derived. An analysis of the main components is performed on the accelerations (filtered acceleration of the maneuvers and without gravity), giving signals and the directions associated with each component, which will transform the accelerometers to these directions. To obtain which component is associated with each of the forces, a trained Neural Network is used (for more details see the article mentioned [24]). The smartphone can be positioned freely, that is, it does not need to go in a fixed position. However, this process will be repeated every time that the smartphone changes its position in terms of gravity distribution. Also, the detection procedure detects manipulations performed by the driver, in order to not to be confused with maneuvers.

3.2   Driver Identification

Training networks from scratch is complex and requires a large amount of data, according to the depth of the network. In this work we have fine-tuned pre-trained Neural Networks, initializing the weights with a pre-trained model and trying to adjust the whole network. The pre-selected model has been the ResNet-50. Its name comes from the Deep Residual Network and specifically Resnet-50 refers to that it presents 50 residual layers. The Deep Residual Networks were presented by Microsoft Research in the ImageNet and COCO 2015 competitions, for image classification, object detection and semantic segmentation [25]. These networks come up to solve the difficulty to train Deep Neural Networks; because adding more layers to an already deep model causes higher training errors. The ResNet [26] allows training and optimizing networks in a simpler way; for this, they introduce residual deep learning frameworks with shortcut connections. These connections perform a mapping and their outputs are added to the outputs of the stacked layers, which may omit one or more layers. The advantage is that these shortcuts do not add extra parameters or computational complexity. Each block is responsible for adjusting the output of the previous blocks, instead of having to do it from scratch. The basic block presents variants such as the block "Residual Bottle-neck", which use sets of three layers stacked instead of two as the basic ones, which reduce and increase the dimensions.

Therefore, ResNets are architectures that stack building blocks, called Residual Blocks. The scheme of the basic residual block of two layers is shown below (Fig. 3). The block presents a shortcut connection that performs identity mapping, so at the output of the block the propagated information plus the output of the stacked layers is added, $F(x) + x$.

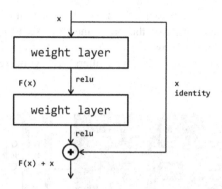

Fig. 3. Residual block (two-layer Residual Block).

In the ResNet50 model, the residual bottleneck of two layers (Fig. 3) is replaced with a three layer bottleneck block, with 1×1 convolutions and 3×3 convolutions. The general scheme of the Resnet-50 is shown in Fig. 4, where blocks of two-layer residual block have been colored in different colors, each color indicating a series of convolutions of the same dimension.

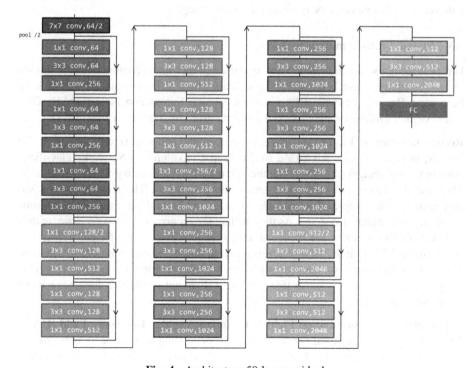

Fig. 4. Architecture 50-layer residual.

The general scheme of the network used for identification of drivers is as follows (Fig. 5). The first block consists of the pre-trained ResNet50 model, followed by a second block of the Recurrent Network. The Recurrent Networks we used are a Gated Recurrent Units (GRUs), because this type offers very good results for time series modeling.

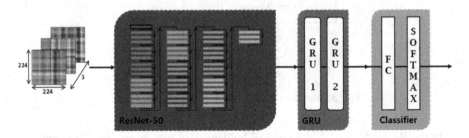

Fig. 5. Deep Neural Network architecture for driver identification.

The input to the ResNet-50 model must be an image, preferably of size 224 × 224 and of 3 channels, since the model is designed for RGB color images. In order to use this architecture, as commented before, we must to transform the 1-D longitudinal, transversal and angular velocity signals into 2-D images.

3.3   Signal Transformation to 2-D

Different methods that allow transforming the temporal signals coming from the mobile sensors to images have been tested. Specifically, four different transformations have been used: Recurrence Plots (RP), Gramian Angular Summation Field (GASF), Gramian Angular Difference Field (GADF) and the Markov Transition Field (MTF).

Recurrence Plots (RP). In [27], authors transform the temporal signals into 1D to 2D images, in order to take advantage of the deep Convolutional Networks classifiers. Among the advantages of CNN is that they do not require extracting features manually. They use the RPs to transform time series into 2D signals. The recurrence plot was described in [28], where they presented this graphical tool for measuring the time constancy of dynamical systems. As they define: *the recurrence plot is an array of dots in a NxN square, where a dot is placed at (i, j) whenever x (j) is sufficiently close to x (i)*. With i and j like times, so a recurrence plot describes time correlation information. These are usually very symmetric with respect to the diagonal $i = j$, since if x (i) is close to x (j) then x (j) is close to x (i); but it does not imply that they have to be completely symmetrical.

Gramian Angular Summation Field & Gramian Angular Difference Field (GASF &GADF). In [29], they used two frameworks for encoding time series like images, the Gramian Angular Summation/Difference Fields (GASF/GADF) and Markov Transition Fields (MTF). In the Gramian Angular Field, the temporal signals are represented in polar coordinates and then each element of the matrix is the cosine of the sum of the angles for the GASF, and the sine of the difference between each point for GADF.

Markov Transition Field (MTF). For Markov Transition Field, the transition probabilities of multiple intervals of the time series are coded, creating the matrix of Markov transitions. To do this, the data is divided into quantiles, and each element of the matrix indicates the probability of transition from one to another.

4 Results and Conclusions

Two different strategies have been applied to select the journey signal to be used for driver identification. As recommended in reference [12], the first strategy has been to use the first 4 min of the trip, in overlapping windows. The window size has been of 224 samples, 44.8 s, overlapped by 25%. The second strategy has been to use windows of the same size and overlap, but choosing only windows with maneuver information (either acceleration or braking or right-turn or left-turn or a mixture of all them).

As shown in Fig. 2, with the tri-axial signals of the accelerometers, we proceed to obtain the longitudinal, transversal and angular velocity signals with the method described in work [24]. Once these are obtained, the journeys are divided into overlapping windows following the strategies mentioned in the previous paragraph: 4 min without distinction of the information type or only maneuver windows. These windows are converted into images, following the 4 possible methods mentioned: RP, GASF, GADF and MTF.

Then, we perform fine-tuning in the architecture shown in Fig. 5, model Resnet50 adding two GRU stacked layers plus a Fully Connected layer and a Softmax layer. Although the training is at window level, the results obtained will be evaluated at journey level by driver. To do this, for each journey the probabilities obtained in each window are added, and the driver assigned will be who sums more probability. As mentioned before, two metrics have been considered to evaluate the results, the first one, *accuracy top-1*, refers to the percentage of success taking into account only the driver who has achieved the highest final probability at journey level, among the 25 drivers. The second one, *accuracy top-5*, refers to the success rate if at least the appropriate driver is among the 5 drivers with most probability at journey level, also given the total set of 25 drivers.

The training has been processed on a computer with a Nvidia Tesla K80 graphic card (dual GPU card, with 24 GB of GDDR5 memory, 480 GB/s of memory bandwidth and 4992 CUDA® cores). The open source software library Keras, with Python, was used to build the network.

The results obtained are shown in the following table (Table 2).

As expected, when we use the maneuvers as input signals, much better results are obtained than when we use blindly the first 4 min of the journey; regardless of the method used to transform the temporal signal into an image. Among the methods described in the literature to transform 1-D signals into 2-D signals, the GADF transformation offer the best result with a success rate of 69.92% at top-1 level and a rate of 90.31% at top-5 level.

The results have been compared using other network architectures, as well as bi-directional networks or directly using the 1-D signals, without transforming the signals into images and substituting the Resnet50 model for another CNN network. But the

Table 2. Result table for driver identification.

Input	Transformation	Accuracy top-1	Accuracy top-5
4 min	RP	28.83%	66.39%
Maneuvers	RP	54.70%	87.66%
4 min	GASF	12.5%	35.5%
Maneuvers	GASF	44.58%	78.61%
4 min	GADF	49.59%	82.45%
Maneuvers	GADF	69.92%	90.31%
4 min	MTF	25.39%	60.60%
Maneuvers	MTF	43.53%	81.08%

best results obtained have reached a rate of 53.57% of accuracy for the top-1 and 85.17% for the top-5. In order to prevent that the proposed Neural Network learns a certain type of route, for instance a habitual route of a driver, a heterogeneous group of journeys for each driver has been used (we have selected routes of the three classes for each one: mainly urban or city trips, mostly road and mixed). In addition, for all the drivers, maneuvers of different nature have been detected through a sliding window: both braking maneuvers (deceleration), accelerations, turns and combinations of these. As future work, we will try to expand the database of drivers as well as the number of journeys by driver. It will also be tested with other types of transformations.

Acknowledgments. We thank Drivies (PhoneDrive S.L.) for the support in the driving research and the access to the journeys database. This work has been partially funded by the Spanish Ministry of Economy and Competitiveness and the European Union (FEDER) within the framework of the project DSSL: "Deep & Subspace Speech Learning (TEC2015-68172-C2-2-P)".

References

1. European Commission: European Commission, 06 December 2017. https://ec.europa.eu/newsroom/just/item-detail.cfm?item_id=77704
2. Yan, S., Teng, Y., Smith, J.S.: Driver behavior recognition based on deep convolutional neural networks. In: International Conference on Natural Computation, Fuzzy Systems and Knowledge Discovery (ICNC-FSKD), Changsha, China (2016)
3. Guo, J., Liu, Y., Zhang, L., Wang, Y.: Driving behaviour style study with a hybrid deep learning framework based on GPS data. Sustainability **10**(7), 2351 (2018)
4. Ferreira Júnior, J., et al.: Driver behavior profiling: an investigation with different smartphone sensors and machine learning. PLoS ONE **12**(4), e0174959 (2017)
5. Ezzini, S., Berrada, I., Ghogho, M.: Who is behind the wheel? Driver identification and fingerprinting. J. Big Data **5**(1), 9 (2018)
6. Wallace, B., Goubran, R., Knoefel, F., Marshall, S., Porter, M., Smith, A.: Driver unique acceleration behaviours and stability over two years. In: International Congress on Big Data (BigData Congress), San Francisco, CA, USA (2016)

7. Nishiwaki, Y., Ozawa, K., Wakita, T., Miyajima, C., Itou, K., Takeda, K.: Driver identification based on spectral analysis of driving behavioral signals. In: Abut, H., Hansen, J.H.L., Takeda, K. (eds.) Advances for In-Vehicle and Mobile Systems, pp. 25–34. Springer, Boston (2007). https://doi.org/10.1007/978-0-387-45976-9_3

8. Hallac, D., et al.: Driver identification using automobile sensor data from a single turn. In: International Conference on Intelligent Transportation Systems (ITSC), Rio de Janeiro, Brazil (2016)

9. Horwitz, J.: Fake drivers and passengers are boosting Uber's growth in China, 9 June 2015. https://qz.com/423288/fake-drivers-and-passengers-are-boosting-ubers-growth-in-china/

10. Petrow, S.: I got taken for a ride by a fake Uber driver. Don't become the next victim, 12 October 2016. https://eu.usatoday.com/story/tech/columnist/stevenpetrow/2016/10/12/fake-uber-drivers-dont-become-next-victim/91903508/

11. Who'sDrivingYou? http://www.whosdrivingyou.org/rideshare-incidents

12. Fontaine, B.: Sentiance, 25 September 2017. https://www.sentiance.com/2017/09/25/deep-learning-on-passenger-and-driver-behavior-analysis-using-sensor-data/

13. Chowdhury, A., Chakravarty, T., Ghose, A., Banerjee, T., Balamuralidhar, P.: Investigations on Driver Unique Identification from Smartphone's GPS Data Alone. J. Adv. Transp. (2018)

14. Moreira-Matias, L., Farah, H.: On developing a driver identification methodology using in-vehicle data recorders. IEEE Trans. Intell. Transp. Syst. **18**, 2387–2396 (2017)

15. Enev, M., Takakuwa, A., Koscher, K., Kohno, T.: Automobile driver fingerprinting. Proc. Priv. Enhancing Technol. **2016**, 34–50 (2016)

16. Martínez, M., Echanobe, J., del Campo, I.: Driver identification and impostor detection based on driving behavior signals. In: 19th International Conference on Intelligent Transportation Systems (ITSC), Rio de Janeiro, Brazil (2016)

17. Zhang, X., Zhao, X., Rong, J.: A study of individual characteristics of driving behavior based on hidden markov model. Sens. Transducers **167**, 194–202 (2014)

18. Burton, A., et al.: Driver identification and authentication with active behavior modeling. In: 12th International Conference on Network and Service Management (CNSM), Montreal, QC, Canada (2016)

19. Tanprasert, T., Saiprasert, C., Thajchayapong, S.: Combining unsupervised anomaly detection and neural networks for driver identification. J. Adv. Transp. **2017**, 13 (2017). https://doi.org/10.1155/2017/6057830. Article ID 6057830

20. Phumphuang, P., Wuttidittachotti, P., Saiprasert, C.: Driver identification using variance of the acceleration data. In: International Computer Science and Engineering Conference (ICSEC), Chiang Mai, Thailand (2015)

21. del Campo, I., Finker, R., Martínez, M.V., Echanobe, J., Doctor, F.: A real-time driver identification system based on artificial neural networks and cepstral analysis. In: International Joint Conference on Neural Networks (IJCNN), Beijing, China (2014)

22. Van Ly, M., Martin, S., Trivedi, M.M.: Driver Classification and Driving Style Recognition using Inertial Sensors. In: IEEE Intelligent Vehicles Symposium (IV), Gold Coast, QLD, Australia (2013)

23. Fung, N.C., et al.: Driver identification using vehicle acceleration and deceleration events from naturalistic driving of older drivers. In: IEEE International Symposium on Medical Measurements and Applications (MeMeA), Rochester, MN, USA (2017)

24. Hernández Sánchez, S., Fernández Pozo, R., Hernández Gómez, L.A.: Estimating Vehicle Movement Direction from smartphone accelerometers using Deep Neural Networks. Sensors (2018)

25. Dietz, M.: Waya.ai, 2 May 2017. https://blog.waya.ai/deep-residual-learning-9610bb62c355

26. He, K., Zhang, X., Ren, S., Sun, J.: Deep residual learning for image recognition. In: IEEE Conference on Computer Vision and Pattern Recognition, Las Vegas, NV, USA (2016)

27. Hatami, N., Gavet, Y., Debayle, J.: Classification of time-series images using deep convolutional neural networks. In: Tenth International Conference on Machine Vision (ICMV 2017), Vienna, Austria (2018)
28. Eckmann, J.P., Oliffson Kamphorst, S., Ruelle, D.: Recurrence plots of dynamical systems. Europhys. Lett. **5**, 973–977 (1987)
29. Wang, Z., Oates, T.: Imaging Time-Series to Improve Classification and Imputation. In: IJCAI 2015 Proceedings of the 24th International Conference on Artificial Intelligence, Buenos Aires, Argentina (2015)

Dynamic Enterprise Architecture Capabilities: Conceptualization and Validation

Rogier van de Wetering[✉]

The Open University of the Netherlands, Valkenburgerweg 177,
6419 AT Heerlen, the Netherlands
rogier.vandewetering@ou.nl

Abstract. The notion of enterprise architecture (EA) and EA-based capabilities in IS literature has emerged as an important research domain. However, the conceptualizations of EA-based capabilities remain ambiguous, largely not validated and still lack a firm base in theory. This study, therefore, aims to rigorously conceptualize EA-based capabilities grounded in theory and puts forward the notion of dynamic enterprise architecture capabilities. These capabilities highlight the core areas in which organizations should infuse EA. The purpose of this study is to develop a reliable and valid measurement scale. This scale is validated using item-sorting analyses, expert reviews and an empirical study of 299 CIOs and enterprise architects. The outcomes support the validity and reliability of the scale. The dynamic enterprise architecture capabilities scale developed in this research contributes to theory development and the EA knowledge base. The scale may be used as an assessment or benchmarking tool in practice.

Keywords: Enterprise architecture · Enterprise architecture capabilities ·
Dynamic enterprise architecture capabilities · Dynamic capabilities view ·
Scale development and validation

1 Introduction

In today's dynamic business environment, there are various trends and market forces that drive the adoption of enterprise architecture (EA) within organizations. These include, e.g., growing regulatory pressure, the rising frequency and speed of business-driven and information technology (IT)-driven change opportunities, and an increased need for integration within and between business enterprises. Organizations are, therefore, adopting various forms of EA to facilitate the further integration of IT resources, assets and capabilities with business processes [1–3]. An EA can be considered a high-level representation of organizations' business processes and IT systems, their interrelationships [4]. While there is conceptual work that argues that EA allows organizations to add value across the organization [5, 6], much of the current literature still focus on EA artifacts, and their respective management [7–9]. Recently, the literature puts a greater emphasis on theory building and the EA-based capabilities that organize and deploy organization-specific resources to align strategic objectives with the particular use of technology [1, 10, 11].

© Springer Nature Switzerland AG 2019
W. Abramowicz and R. Corchuelo (Eds.): BIS 2019, LNBIP 354, pp. 221–232, 2019.
https://doi.org/10.1007/978-3-030-20482-2_18

Nonetheless, despite the recent growth in EA studies, substantial gaps remain in the literature. For instance, there is no conclusive evidence on how EA-based capabilities drive business transformation and deliver benefits [2, 4, 11–14]. Moreover, definitions and conceptualizations of EA-based capabilities remain ambiguous and large not validated [11, 15, 16]. Hence, despite the various scholarly contributions, the conceptualization of EA-based capabilities still lacks a firm base in theory. Following this assessment, there is a need to rigorously conceptualize EA-based capabilities grounded in theory and to develop a reliable and valid scale to measure its underlying dimensions. This study, therefore, aims to extend previous EA work and proposes a dynamic enterprise architecture capabilities scale that highlights the core areas in which organizations should infuse EA. Hence, we follow previous IT-enabled [17–19] and EA-based capability [1, 3, 20–22] scholarship that used dynamic capability-based approaches as a theoretical foundation. By doing so, this study embraces on a robust theoretical foundation that provides a rich vocabulary and empirically validated measures.

Given the above, this study addresses the following research question: *"how can a valid and reliable scale be developed to measure dynamic enterprise architecture capabilities and, thus, the core areas in which organizations should infuse EA?"*

This paper is structured as follows to answer this question. First, we briefly review the literature on dynamic and EA-based capabilities. Then, this work outlines the development process and methods. Sections four and five present the results, i.e., a validated conceptualization of dynamic enterprise architecture capabilities, the discussion of the results and the conclusions.

2   Theoretical Background

2.1   EA-Based Capabilities

The notion of differential organizational benefits that can be derived from EA-based capabilities has been subject of discussion in the past decade [4, 10, 11]. Recently, some researchers argue that managing and deploying an EA is, in fact, an organizational capability that can provide organizations with a competitive edge [1, 3]. In this particular context, a distinction can be made between EA capabilities and EA-based capabilities. EA capabilities include an organization's ability to create and maintain EA content, standards and guidelines. In essence, these capabilities focus on the development of EA artifacts, i.e., individual documents that describe various aspects of the EA [7, 23]. EA-based capabilities highlight the usage, deployment, and diffusion of EA in decision-making processes, and the organizational routines that drive IT and business capabilities [1, 3, 24]. Thus, these capabilities focus more on the development of unique competencies and capabilities that can leverage EA assets and resources. For instance, Shanks et al. [3] argue that EA capabilities are deemed necessary to provide advisory services to the organization. Likewise, Hazen et al. [1] provide foundational work that shows that EA-based capabilities can enhance organizational agility and indirectly enhance organizational performance. These outcomes are consistent with work by Foorthuis et al. [10] that demonstrate the importance of intermediate EA-enabled outcomes that contribute to the achievement of particular business goals and objectives. Hence, EA literature suggests that complementary EA capabilities enable firms to leverage their EA effectively [1, 4], contribute to IT efficiency and IT flexibility [25], and can drive alignment between business and IT [26].

This study concurs with this EA-based capability view and uses it to frame a new conceptualization of dynamic enterprise architecture capabilities systematically.

2.2 Dynamic Enterprise Architecture Capabilities

As organizations are investing in their EA, not all of them are successful and fail to deliver desired results and, therefore, they question the particular added value of EA [1]. This study claims that it is likely that the extent to which EAs are leveraged successfully within the organization depends on the dynamic capabilities that collectively use the EA to sense environmental threats and business opportunities, while simultaneously implementing new strategic directions. Dynamic capabilities can be considered an organizations' ability to integrate, build, and reconfigure internal and external competences to address the rapidly changing environment or when the opportunity or need arises [27]. Based on the dynamic capabilities view and recent EA-based capabilities work [1, 3, 11], this study defines dynamic enterprise architecture capabilities as *"an organization's ability to leverage its EA for asset sharing and recomposing and renewal of organizational resources, together with guidance to proactively address the rapidly changing internal and external business environment and achieve the organization's desirable state."*

Building on previous EA-based capability contributions and theoretically guided by the dynamic capabilities view, we synthesize the reach and range of EA-based capabilities through three related, but distinct capabilities, i.e., (I) EA sensing capability, (II) EA mobilizing capability, and (III) EA transformation capability. EA sensing capability refers to an organization's deliberate posture toward sensing and identifying new business opportunities or potential threats, and developing a greater reactive and proactive strength in the business domain using EA [3, 20]. EA mobilizing capability refers to organizations' capability to use EA to evaluate, prioritize and select potential solutions and mobilize resources in line with a potential solution or potential threats [3, 28, 29]. Finally, an EA transforming capability can be considered the ability to use the EA to successfully reconfigure business processes and the technology landscape, to engage in resource recombination and to adjust for and respond to unexpected changes [3, 17, 30, 31].

3 Construct Development Process and Methods

3.1 Construct Development and Specification

This research followed a staged approach to tackle a magnitude of challenges (e.g., selection indicators, reliability, and validity of constructs) that emerge during the development of new multi-item scales [32]. First, the principal investigator derived all the items to reflect the new construct from either previously cited in or implied by extant conceptual and empirical work. Hence, this study adapted validated measures from recognized empirical studies in information systems and sciences [3, 17, 29], management, organization, and decision sciences [18, 27, 31, 33]. Starting from the conceptualization of dynamic capabilities by [27], this study subsequently assigned measurement items to one of the three capabilities on the base of a review of primary scales present in the extant literature. The first pool of scale items was developed using a seven-point Likert-type scale, ranging from "strongly disagree" to "strongly agree." Then, two subsequent stages of scale development and purification followed based on previously outlined recommendations [32], i.e., (I) item-sorting analysis and expert review and (II) confirmatory analyses to assess the psychometric properties of the dynamic enterprise architecture capabilities scale.

Stage I: item-sorting analysis and expert reviews.

First, an item-to-construct sorting approach was employed to establish tentative item reliability and validity. Three Master students[1], doing their theses research, evaluated the initial item pool using a Q-sort approach during two three-hour intensive sessions. Through this iterative approach, the students were asked to sort the items according to the three underlying capabilities of the new construct. Hence, the inter-judge agreement was measured [34]. Next, the student reworded or deleted too ambiguous items as a result of the first stage, to improve the agreement between the judges [35]. These two steps enhanced the reliability and construct validity of questionnaire items at a pre-testing stage. This study omits the results of these intensive sessions for the sake of brevity.

To further enhance the content and face validity of the questionnaire times, the principal investigator asked ten experts with the appropriate competencies, familiarity with the research domain, and experience to evaluate all the scale items and offer improvement suggestions. These experts were enterprise architects (3), EA and MIS scholars (2), IT/business consultants and managers (5). The experts mainly looked at several criteria for testing the adequacy of questions including length, specificity and simplicity, and question order. Also, the experts were asked to reflect on any interpretation issues with the questions [36]. Outcomes of this stage offered many small iterations, improvements, and purifications to the questionnaire items and so formed a solid foundation for the final stage to assess the psychometric properties of the new scale. Table 1 shows the final items and the supporting literature for all the three capabilities.

Stage II: survey analyses.

This study applied confirmatory analyses to the dynamic enterprise architecture capabilities construct to assess the reliability and validity of the multi-item scales [32]. The conceptualization of dynamic enterprise architecture capabilities uses a formative higher-order construct that is composed of three underlying first-order dimensions [37]. As such, this second-order factor uses reflective first-order latent constructs. The manifest variables are, therefore, affected by the latent variables and are interchangeable [38, 39]. Thus, on the first-order level, the manifest variables reflect and depict the construct. The second-order factor (dynamic enterprise architecture capabilities), on the

---

[1] These students also governed the data collection process throughout this study.

Table 1. Final measurement items for dynamic enterprise architecture capabilities

Constructs and items	Supporting literature
(I) EA sensing capability	
S1. We use our EA to identify new business opportunities or potential threats.	[3, 17, 18]
S2. We review our EA services regularly to ensure that they are in line with key stakeholders wishes.	[3, 17, 18]
S3. We adequately evaluate the effect of changes in the baseline and target EA on the organization.	[3, 18]
S4. We devote sufficient time enhancing our EA to improve business processes.	[17, 18]
S5. We develop greater reactive and proactive strength in the business domain using our EA.	[17, 18, 28]
(II) EA mobilizing capability	
M1. We use our EA to draft potential solutions when we sense business opportunities or potential threats	[3, 28, 29]
M2. We use our EA to evaluate, prioritize and select potential solutions when we sense business opportunities or potential threats	[3, 28, 29]
M3. We use our EA to mobilize resources in line with a potential solution when we sense business opportunities or potential threats	[17, 42]
M4. We use our EA to draw up a detailed plan to carry out a potential solution when we sense business opportunities or potential threats	[28, 29]
M5. We use our EA to review and update our practices in line with renowned business and IT best practices when we sense business opportunities or potential threats	[33]
(III) EA transforming capability	
T1. Our EA enables us to successfully reconfigure business processes and the technology landscape to come up with new or more productive assets	[3, 17, 30, 31]
T2. We successfully use our EA to adjust our business processes and the technology landscape in response to competitive strategic moves or market opportunities	[3, 17, 43, 44]
T3. We successfully use our EA to engage in resource recombination to better match our product-market areas and our assets	[18]
T4. Our EA enables flexible adaptation of human resources, processes, or the technology landscape that leads to competitive advantage	[45]
T5. We successfully use our EA to create new or substantially changed ways of achieving our targets and objectives	[45]
T6. Our EA facilitates us to adjust for and respond to unexpected changes	[17, 27, 46]

other hand, is conceptualized through a formative mode. Such a model is called a reflective-formative type II model [40, 41]. Each of the three specified dimensions represents a unique trait of the higher-order construct. Removing a particular dimension would substantially alter the meaning and understanding of the overarching construct.

3.2 Data Collection Procedure

This study collected data as part of a Master course Enterprise Architecture of a Dutch University. Students[2] read recent academic articles on EA competences and capabilities, e.g., [1, 3, 4], and had to fill in a survey for their organization. Also, the participating students ($N = 235$) had to distribute this survey to two domain experts (professionals that are familiar with the material, e.g., CIOs, IT managers, and lead enterprise architects) following a snowball method. Thus, this research collected the data through respondent-driven sampling. Following [47] there is no reason to presume that the use of the respondent-driven sampling method resulted in an unacceptable (self-reported) bias that would jeopardize the outcomes of this research.

During the data collection, controls were built in, so that every organization completed the survey only once. Respondents were given an incentive to take part in the survey. They were offered a research report with the most important outcomes of this study. Following Podsakoff et al. [48] anonymity was guaranteed, and respondents could withdraw their scores if they wanted to.

The data collection phase started on the 17th of October 2018 and ended on the 16th of November 2018. In total 669 respondents from different organizations initially started the survey. Based on the final response, this study included a total of 299 usable questionnaires for the analyses. The majority of respondents operate in the private sector 57%, 36% from the public sector and only a small percentage (0.07) from other categories such as private-public partnerships (0.02%), and non-governmental organizations. The majority of responses were from large organizations with 3000 + employees (45%), 1001–3000 employees (14%), 301–1000 employees (13%), 101–300 employees (11%) and the remaining 16% had less than 1000 employees. 69% of the organizations are older than 25 years. Sub-group analyses for each dimension of the construct (using t-tests) showed no significant difference early (first two weeks) and later responses (final two weeks) to the survey. The data were obtained from a single source at one point in time. This study, therefore, controlled for common method variance (CMV) per suggestions of Podsakoff [48]. In doing so, Harman's single factor test was performed using IBM SPSS Statistics™ v24 on the primary study constructs. Hence, the construct variables were all loaded on to a single construct in an Exploratory Factor Analysis (EFA). Outcomes of this analysis showed that no single factor attributes to the majority of the variance; the sample is not affected by CMB [48].

4 Model Estimation and Validation

4.1 Model Estimation Procedure

For the model estimation, the present study ran parameter estimates for the measurement model. The analyses were done using SmartPLS version 3.2.7. [49], which is a Structural Equation Modeling (SEM) application using Partial Least Squares (PLS).

---

[2] Students that take part in this course are adults that have many years of working experience in either business or IT (management) functions.

This study uses PLS for theory development purposes and to validate the measurement model and examine the formative nature of our second-order factor model [41, 50, 51]. In the analyses, the factor weighing scheme within SmartPLS was applied. Also, a non-parametric bootstrapping procedure was employed to compute the level of the significance of the regression coefficients running from the first-order constructs to the second-order construct. In this process, 5000 replications were used to obtain stable results and to interpret their significance. Finally, the 299 organizations in the dataset far exceed all minimum requirements to run the SEM analyses [52, 53].

4.2 Confirmatory Analyses

The first-order constructs were subjected to internal consistency reliability, convergent validity, and discriminant validity tests to assess the psychometric properties following the suggestions of Ringle et al., [49] and MacKenzie et al. (2011). Hence, all Cronbach Alpha (CA) values were examined if they were above the threshold of 0.70 [53, 54]. Next, the measurement model is evaluated by its convergent and discriminant validity [50, 53]. The composite reliability (CR) values for each construct should typically between 0.60 and 0.90, as is the case in the present study (see Table 2). Also, the construct-to-item loadings were assessed, showing no violations. The average variance extracted (AVE) values were all above the lower limit of 0.50 [53, 55]. Discriminant validity was established through three different, but related tests. First, analyses showed that all cross-loadings (i.e., correlation) on other constructs were less than the outer loading on the associated construct [56]. Second, analyses showed that the square root of the AVEs, i.e., the Fornell-Larcker criterion, of all constructs was larger than the cross-correlation [50]. Third, and finally, a newly developed discriminant validity analysis was employed, i.e., the heterotrait-monotrait (HTMT) ratio of correlations approach by Henseler, Ringle, and Sarstedt [57]. All HTMT values showed acceptable outcomes well below the 0.90 upper bound. Table 2 shows the summary of the measurement model analyses that suggest that the first-order constructs are valid and reliable. As can be seen from Table 2, the included variance inflation factors (VIFs) values are well below a reported critical value of 5. These outcomes, in addition to the absence of non-significant relations between first-order capabilities and the second-order construct, indicate that no multicollinearity exists within our model [58].

The above outcomes confirm the three related, but unique EA capabilities that underly the formation of an organization's dynamic enterprise architecture capabilities. Figure 1 shows the respective significant path weights of the first-order constructs on the higher-order construct along with the construct-to-item loadings.

Table 2. Assessment of convergent and discriminant validity of the reflective constructs.

	CA	CR	AVE	VIF	(1)	(2)	(3)
(1) EA sensing	0.885	0.916	0.686	3.209	**0.826**		
(2) EA mobilizing	0.909	0.932	0.734	3.163	0.782	**0.857**	
(3) EA transforming	0.918	0.936	0.711	3.193	0.784	0.780	**0.843**

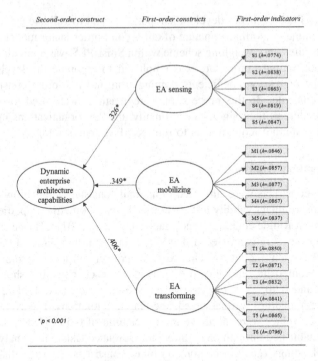

Fig. 1. Formative dynamic enterprise architecture capabilities measurement model

5 Discussion and Conclusion

5.1 Theoretical and Managerial Implications

This paper contributes to the extant literature on EA capabilities by constructing and validating a comprehensive capability and treating it as a dynamic capability. Second, by developing 16 measurement items across three dimensions, this study helps researchers conduct more systematic analyses on the organization's EA-based capabilities. Hence, this work enhances the theoretical underpinnings of empirical EA and dynamic capability research. Third, the dynamic enterprise architecture capabilities scale can guide IS scholars and practitioners in explaining IS-related phenomena. Specifically, it is well understood in the literature that EA-induced capabilities—following IT-enabled capabilities [59]—have an indirect effect on organizational benefits and competitive performance, by strengthening and renewing organizations' operational (ordinary) capabilities. Therefore, scholars can investigate how dynamic enterprise architecture capabilities influence organizational benefits following these theoretical suggestions. Also, the conception of dynamic enterprise architecture capabilities can also work as a mediating construct in a nomological value path in explaining how, e.g., EA resources, assets, and practices lead to enhanced business value (e.g., business-IT-alignment, innovation). Finally, research could conceive dynamic enterprise architecture capabilities as an outcome construct by studying its antecedents like EA competencies and approaches, principles that guide the design and evolution of EA, and architectural insights.

The dynamic enterprise architecture capabilities scale is a reliable and valid tool to measure the level of proficiency of the organization's deployment and usage of EA in the organization. IT and business managers can drive enterprise-wide transformational changes and provide an opportunity to build capabilities in parallel with implementing a new strategic direction using this scale. In doing so, they can use this scale for evaluation purposes. For instance, the scale can be used in a critical self-assessment of the organizations' EA strategy to unfold development opportunities, possible even within different departments, and layers within the organization. The scale encourages decision-makers to actively think about EA deployment in the organization and how they should allocate resources properly. Hence, they can look at dynamic enterprise architecture capabilities as a means to drive the organization's level of innovation and purposefully enhance its evolutionary fitness. Finally, the dynamic enterprise architecture capabilities scale provides a useful diagnostic and benchmarking tool with which they can assess and continuously monitor their organization's development needs and compare the results with other organizations that, e.g., operate in the same industry, market or segment.

5.2   Limitations and Concluding Remarks

This study has several limitations. A key limitation of this study is that all the data were gathered from Dutch-speaking organizations. So, comparing measurement scores across countries might well contribute to the generalizability of our findings. Second, the data included various demographic variables (e.g., type, size), but the present study did not consider possible differences among group segments and clusters. Notwithstanding, this study advances our understanding of EA-based capabilities by developing a reliable and valid scale that highlights the core areas in which EA should be infused to create value. Both scholars and practitioners can use the scale.

Acknowledgment. I want to thank Tom Hendrickx, Kevin Billen and Salo Langer for their contributions in the data collection and for sharing their perspectives in numerous discussions.

References

1. Hazen, B.T., et al.: Enterprise architecture: a competence-based approach to achieving agility and firm performance. Management **193**, 566–577 (2017)
2. Ross, J.W., Weill, P., Robertson, D.: Enterprise Architecture as Strategy: Creating a Foundation for Business Execution. Harvard Business Press, Boston (2006)
3. Shanks, G., et al.: Achieving benefits with enterprise architecture. J. Strateg. Inf. Syst. **27**(2), 139–156 (2018)
4. Tamm, T., et al.: How does enterprise architecture add value to organisations. Commun. Assoc. Inf. Syst. **28**(1), 141–168 (2011)
5. Bernard, S.A.: An Introduction to Enterprise Architecture, 3rd edn. AuthorHouse, Bloomington (2012)

6. Janssen, M.: Framing enterprise architecture: a metaframework for analyzing architectural efforts in organizations. In: Coherency Management: Architecting the Enterprise for Alignment, Agility and Assurance. Authorhouse (2009)
7. Kotusev, S.: Enterprise architecture and enterprise architecture artifacts: questioning the old concept in light of new findings. J. Inf. Technol. p. 0268396218816273 (2019)
8. Lange, M., Mendling, J., Recker, J.: An empirical analysis of the factors and measures of Enterprise Architecture Management success. Eur. J. Inf. Syst. **25**(5), 411–431 (2016)
9. Ahlemann, F., et al.: Strategic Enterprise Architecture Management: Challenges, Best Practices, and Future Developments. Springer, Heidelberg (2012). https://doi.org/10.1007/978-3-642-24223-6
10. Foorthuis, R., et al.: A theory building study of enterprise architecture practices and benefits. Inf. Syst. Front. **18**(3), 541–564 (2016)
11. Korhonen, J.J., Molnar, W.A.: Enterprise architecture as capability: strategic application of competencies to govern enterprise transformation. In: 2014 IEEE 16th Conference on Business Informatics (CBI). IEEE (2014)
12. Lankhorst, M.: Enterprise Architecture at Work: Modelling, Communication and Analysis. Springer, Heidelberg (2009)
13. van de Wetering, R., Bos, R.: A meta-framework for efficacious adaptive enterprise architectures. In: Abramowicz, W., Alt, R., Franczyk, B. (eds.) BIS 2016. LNBIP, vol. 263, pp. 273–288. Springer, Cham (2017). https://doi.org/10.1007/978-3-319-52464-1_25
14. Doucet, G., et al.: Coherency management: using enterprise architecture for alignment, agility, and assurance. J. Enterp. Architecture **4**(2), 1–12 (2009)
15. Greefhorst, D., Koning, H., Van Vliet, H.: The many faces of architectural descriptions. Inf. Syst. Front. **8**(2), 103–113 (2006)
16. Wilkinson, M.: Designing an 'adaptive'enterprise architecture. BT Technol. J. **24**(4), 81–92 (2006)
17. Mikalef, P., Pateli, A., van de Wetering, R.: IT flexibility and competitive performance: the mediating role of IT-enabled dynamic capabilities. In: 24th European Conference on Information Systems (ECIS) (2016)
18. Pavlou, P.A., El Sawy, O.A.: Understanding the elusive black box of dynamic capabilities. Decis. Sci. **42**(1), 239–273 (2011)
19. Wheeler, B.C.: NEBIC: a dynamic capabilities theory for assessing net-enablement. Inf. Syst. Res. **13**(2), 125–146 (2002)
20. Toppenberg, G., Henningsson, S., Shanks, G.: How Cisco Systems used enterprise architecture capability to sustain acquisition-based growth. MIS Q. Executive **14**(4), 151–168 (2015)
21. Abraham, R., Aier, S., Winter, R.: Two speeds of EAM—a dynamic capabilities perspective. In: Aier, S., Ekstedt, M., Matthes, F., Proper, E., Sanz, J.L. (eds.) PRET/TEAR -2012. LNBIP, vol. 131, pp. 111–128. Springer, Heidelberg (2012). https://doi.org/10.1007/978-3-642-34163-2_7
22. Labusch, N., Aier, S., Winter, R.: Beyond Enterprise Architecture Modeling-What are the Essentials to Support Enterprise Transformations? (2013)
23. Winter, R., Fischer, R.: Essential layers, artifacts, and dependencies of enterprise architecture. In: 10th IEEE International Enterprise Distributed Object Computing Conference Workshops. IEEE (2006)
24. Brosius, M., et al.: Enterprise Architecture Assimilation: An Institutional Perspective. Association for Information Systems (2018)
25. Schmidt, C., Buxmann, P.: Outcomes and success factors of enterprise IT architecture management: empirical insight from the international financial services industry. Eur. J. Inf. Syst. **20**(2), 168–185 (2011)

26. Hinkelmann, K., et al.: A new paradigm for the continuous alignment of business and IT: combining enterprise architecture modelling and enterprise ontology. Comput. Ind. **79**, 77–86 (2016)
27. Teece, D.J., Pisano, G., Shuen, A.: Dynamic capabilities and strategic management. Strateg. Manag. J. **18**(7), 509–533 (1997)
28. Overby, E., Bharadwaj, A., Sambamurthy, V.: Enterprise agility and the enabling role of information technology. Eur. J. Inf. Syst. **15**(2), 120–131 (2006)
29. Sambamurthy, V., Bharadwaj, A., Grover, V.: Shaping agility through digital options: reconceptualizing the role of information technology in contemporary firms. MIS Q. **27**(2), 237–263 (2003)
30. Pavlou, P.A., El Sawy, O.A.: From IT leveraging competence to competitive advantage in turbulent environments: the case of new product development. Inf. Syst. Res. **17**(3), 198–227 (2006)
31. Drnevich, P.L., Kriauciunas, A.P.: Clarifying the conditions and limits of the contributions of ordinary and dynamic capabilities to relative firm performance. Strateg. Manag. J. **32**(3), 254–279 (2011)
32. MacKenzie, S.B., Podsakoff, P.M., Podsakoff, N.P.: Construct measurement and validation procedures in MIS and behavioral research: integrating new and existing techniques. MIS Q. **35**(2), 293–334 (2011)
33. Wilden, R., et al.: Dynamic capabilities and performance: strategy, structure and environment. Long Range Plan. **46**(1–2), 72–96 (2013)
34. Nahm, A.Y., et al.: The Q-sort method: assessing reliability and construct validity of questionnaire items at a pre-testing stage. J. Mod. Appl. Stat. Meth. **1**(1), 15 (2002)
35. Moore, G.C., Benbasat, I.: Development of an instrument to measure the perceptions of adopting an information technology innovation. Inf. Syst. Res. **2**(3), 192–222 (1991)
36. Presser, S., et al.: Methods for testing and evaluating survey questions. Public Opin. Q. **68**(1), 109–130 (2004)
37. Coltman, T., et al.: Formative versus reflective measurement models: two applications of formative measurement. J. Bus. Res. **61**(12), 1250–1262 (2008)
38. Wetzels, M., Odekerken-Schröder, G., Van Oppen, C.: Using PLS path modeling for assessing hierarchical construct models: guidelines and empirical illustration. MIS Q. **33**(1), 177–195 (2009)
39. Petter, S., Straub, D., Rai, A.: Specifying formative constructs in information systems research. MIS Q. 623–656 (2007)
40. Becker, J.-M., Klein, K., Wetzels, M.: Hierarchical latent variable models in PLS-SEM: guidelines for using reflective-formative type models. Long Range Plan. **45**(5–6), 359–394 (2012)
41. Jarvis, C., MacKenzie, S., Podsakoff, P.: A critical review of construct indicators and measurement model misspecification in marketing and consumer research. J. Consum. Res. **30**(2), 199–218 (2003)
42. Teece, D., Peteraf, M., Leih, S.: Dynamic capabilities and organizational agility: risk, uncertainty, and strategy in the innovation economy. Calif. Manag. Rev. **58**(4), 13–35 (2016)
43. Kim, G., et al.: IT capabilities, process-oriented dynamic capabilities, and firm financial performance. J. Assoc. Inf. Syst. **12**(7), 487 (2011)
44. Fischer, T., et al.: Exploitation or exploration in service business development? Insights from a dynamic capabilities perspective. J. Serv. Manag. **21**(5), 591–624 (2010)
45. Protogerou, A., Caloghirou, Y., Lioukas, S.: Dynamic capabilities and their indirect impact on firm performance. Ind. Corp. Change **21**(3), 615–647 (2012)

46. van Oosterhout, M., Waarts, E., van Hillegersberg, J.: Change factors requiring agility and implications for IT. Eur. J. Inf. Syst. **15**(2), 132–145 (2006)
47. Warkentin, M., Johnston, A.C., Shropshire, J.: The influence of the informal social learning environment on information privacy policy compliance efficacy and intention. Eur. J. Inf. Syst. **20**(3), 267–284 (2011)
48. Podsakoff, P.M., et al.: Common method biases in behavioral research: a critical review of the literature and recommended remedies. J. Appl. Psychol. **88**(5), 879 (2003)
49. Ringle, C.M., Wende, S., Becker, J.-M.: SmartPLS 3. Boenningstedt: SmartPLS GmbH (2015). http://www.smartpls.com
50. Hair Jr., J.F., et al.: A Primer on Partial Least Squares Structural Equation Modeling (PLS-SEM). Sage Publications, Thousand Oaks (2016)
51. Ringle, C.M., Sarstedt, M., Straub, D.: A critical look at the use of PLS-SEM in MIS Quarterly. MIS Q. **36**(1) (March 2012)
52. Hair, J.F., Ringle, C.M., Sarstedt, M.: PLS-SEM: indeed a silver bullet. J. Mark. Theory Pract. **19**(2), 139–152 (2011)
53. Hair Jr., J.F., et al.: Advanced Issues in Partial Least Squares Structural Equation Modeling. SAGE Publications, Thousand Oaks (2017)
54. Nunnally, J., Bernstein, I.: Psychometric Theory. McGraw-Hill, New York (1994)
55. Fornell, C., Larcker, D.: Evaluating Structural Equation Models with Unobservable Variables and Measurement Error. J. Mark. Res. **18**(1), 39–50 (1981)
56. Farrell, A.M.: Insufficient discriminant validity: a comment on Bove, Pervan, Beatty, and Shiu (2009). J. Bus. Res. **63**(3), 324–327 (2010)
57. Henseler, J., Ringle, C.M., Sarstedt, M.: A new criterion for assessing discriminant validity in variance-based structural equation modeling. J. Acad. Mark. Sci. **43**(1), 115–135 (2015)
58. Kock, N., Lynn, G.: Lateral collinearity and misleading results in variance-based SEM: an illustration and recommendations (2012)
59. Rai, A., Tang, X.: Leveraging IT capabilities and competitive process capabilities for the management of interorganizational relationship portfolios. Inf. Syst. Res. **21**(3), 516–542 (2010)

Re-engineering Higher Education Learning and Teaching Business Processes for Big Data Analytics

Meena Jha[1]([⊠]), Sanjay Jha[1], and Liam O'Brien[2]

[1] Central Queensland University, Sydney, NSW 2000, Australia
{m.jha,s.jha}@cqu.edu.au
[2] Home Affairs, Canberra, ACT 2609, Australia
Liamob99@hotmail.com

Abstract. Big Data analytics need to be combined with higher education business processes to improve course structure and delivery to help students who have struggled to stay in the course by identifying their engagement and correlation with different variables such as access to documents; assignment submission etc. Online activity data can be used to keep students on track all the way to graduation and universities struggling to understand how to lower dropout rates and keep students on track during their study program. In this paper we discuss how Big Data analytics can be combined with higher education business processes using re-engineering for structured data, unstructured data, and external data. In order to achieve this objective, we investigate the core business processes of learning and teaching and define a re-engineered higher education business process model.

Keywords: Big Data analytics · Business process · Re-engineering ·
Higher education · Learning and teaching

1 Introduction

From a review of the literature in the Business Information System (BIS) research most of the research is concerned with the use of BIS in industry. The education sector is rarely addressed in the research. We believe there is an opportunity for the BIS community to address the changing requirements of educational business processes. One of the key business applications used in the education sector is Learning Management Systems (LMS). An LMS is a software application for the administration, documentation, tracking, reporting and delivery of educational courses, training programs, or learning and development programs. Learning Management Systems make up the largest segment of the higher education.

Data within LMSs and other sources within the University should be utilized through the introduction of business analytics to provide additional information to business processes within the academic institutions. The data within these systems individually would not be considered Big Data however combining the structured and

© Springer Nature Switzerland AG 2019
W. Abramowicz and R. Corchuelo (Eds.): BIS 2019, LNBIP 354, pp. 233–244, 2019.
https://doi.org/10.1007/978-3-030-20482-2_19

unstructured data from all of the various systems, as well as external data, turns it into a Big Data problem that needs to be addressed by a Big Data solution.

There has been growing interest in the higher education sector to take advantage of Big Data [2–4] to improve the learning performance of students, enhance working effectiveness of academic faculty and reduce administrative workload [5]. The overarching issue in institutions of higher education across the world is academic success and the retention of students [6, 7].

In this paper we propose and investigate the core business processes of learning and teaching and define a re-engineered education business process model to show how Big Data can be combined with existing higher education business processes. This business process model can shape the idea of correlating the student's requirements regarding the developed course structure and delivery used by higher education institutions. This business process model is addresses the importance of understanding student's requirements and their needs from course content and course delivery.

Big Data incorporates the emergent research field of learning analytics [11] which is already a growing area in education. However, research in learning analytics has largely been limited to examining indicators of individual student and class performance. The work on analytics within higher education is coming from interdisciplinary research, spanning the fields of Educational Technology, Statistics, Mathematics, Computer Science and Information Science. A core element of the current work on analytics in education is centred on data mining [2] and has not been addressed and focused on business processes.

A large shift towards Big Data from traditional approaches has been observed to handle different business processes and to develop better predictive models for the organization. Business intelligence and analytics is helping many companies to improve their efficiency in customer satisfaction. Business analytics is becoming standard to communicate data-driven business decision making. Big Data analytics have been one of the major reasons in drastically changing the products and services provided by companies in recent years [1]. There is an opportunity to apply the same within the Education sector.

Analytics goes beyond business intelligence, in that it is not simply more advanced reporting or visualization of existing data to gain better insights. Instead, analytics encompasses the notion of going behind the surface of the data to link a set of explanatory variables to a business response and outcome.

Producing meaningful, accessible, and timely management information has long been the holy grail of higher education administrative technology [8]. Marsh et al. [12] have observed that it is important for higher education institutions to use Big Data analytics in order to deliver the very best of learning environments for the good of society. Once the data is analysed it promises better student placement processes; more accurate enrolment forecasts, and the results of the analysis can be fed into an early warning system to provide early and better intervention that identify and assist students at-risk of failing or dropping out. In recent years studies suggest [1, 7, 10] leveraging Big Data technology and taking appropriate actions are able to enhance students' graduation rate, success rate and retention rate.

The widespread introduction of LMSs [29] such as Moodle and Blackboard resulted into increasingly large datasets. Each day, a LMS accumulate increasing amount of students' interaction data, personal data, systems information and academic information [9]. However, a LMS does not collect data from interaction among peers through blogs, social media, external review sites, students feedback surveys, tweets directed at higher education or on related hashtags, reviews on Facebook to name few external data sources. Higher education should not overlook what students are saying about them on social media. Students posting comments on Facebook wall is also a review feature where students can leave a detailed comment about higher education institutions and its offered courses. Current LMSs does not support external and unstructured data source.

To combine Big Data analytics with higher education business processes, we need to re-engineer existing higher education business processes by identifying input and output from the core process of learning and teaching and identifying how unstructured and external data sources can be helpful so that Big Data analytics can be applied and work effectively for decision making [19]. The obvious reason behind it is Higher Education business processes are not capable to handle different varieties of data. By definition of Big Data variety is one of the characteristics of Big Data. The technical and managerial issues resulting from the adoption and application of Big Data analytics is worth exploring [10]. Analysing and understanding online behaviour of students results in predictive capability important for intervention and is a cornerstone of effective personalised learning [2].

The biggest benefits of Big Data analytics in higher education institutions are: the ability to analyse, track and predict student performance; improve graduation and retention rate; and adjust teaching strategies for just-in-time intervention. Traditionally, it has been difficult for higher education institutions to provide critical and timely intervention simply because they did not know a student was struggling until it was too late. Organisational silos and a dearth of data specialists are the main obstacles to putting Big Data to work effectively for decision making [18]. Most of the legacy systems were developed without process models or data models which are now required to support data standardization. Higher education institutions need to improve processes and technological advances that can be integrated in the development of efficient processes through business process re-engineering and business process innovation [26]. A successful integration of Big Data analytics and business processes may create a "new class of higher education economic asset" and help higher education institutions redefine their business and outperform their competitors. By introducing Big Data to education institutions the Big Data becomes another BIS and the changed business processes that use Big Data make significant improvements in the way the academic processes operate.

The remainder of the paper is structured as follows. Section 2 describes Big Data analytics in higher education institutions. Section 3 describes Big Data analysis requirements in higher education institutions. Section 4 describes re-engineering business processes of higher education. Finally Sect. 5 concludes the paper and outlines future work.

2   Big Data Analytics in Higher Education

Big Data is described as a framework to predict future problems, future performance outcomes and prescribe solutions about academic programs, teaching, learning and research [20]. Siemens and Long [11] stated that Big Data could offer many benefits to the future of higher education. Big Data provides an opportunity to institutions to use their information technology resources strategically to improve educational quality and guide students to higher rates of completion, and to improve student persistence and outcomes. Higher education can use Big Data for many reasons. Some of the identified reasons are: improve students' performance; customise programs; create customised programs for individual students; prevent dropouts and learn from the results; reassess curriculum etc.

Data analytics can be applied at different levels, which includes national level, course level, curriculum level, and institutional level. In higher education Big Data tools and techniques and analytics can be used in various departments including, financial planning, student success systems, recruitment, donor tracking [22] marketing and institutional research. Big Data analytics concerned with improving learner's success at the learning and teaching level is referred to as learning analytics [14]. Learning analytics is concerned with the measurement, collection, and analysis and reporting of data about learners and their contexts, for purposes of understanding and optimising learning and the environments in which it occurs [11].

The Rio Salado Community College in Arizona developed a learning analytics application for tracking student progress in courses. The college enrolls more than 41,000 students in online courses. The application was developed to focus more on personalization – assisting non-traditional students achieving academic goals through personalized interventions [17]. Another learning analytics application is the Course Signals System popular known for its competence to identify student at risk and increase student success in the classroom. The Course Signals Systems is characterised by its ability to provide real time feedback, interventions starts early, and it also provides frequent and ongoing feedback [22].

Evans and Lindner [22] broadly grouped analytics into three categories; descriptive, predictive and prescriptive. These different forms of analytics are used to empower decision-makers in higher education to make evidence-based decisions. For example, the University of Nevada, Las Vegas used analytics to enhance the student experience, by collecting student learning and navigational patterns using learning devices and environments to analyse current and historical data. This was achieved with the implementation of a Big Data enterprise system that automatically predicted the population of at-risk students and prescribed real solutions that triggered alerts to students at risk [21].

Wagner and Ice [15], traced how technological advancements have triggered the development of analytics in higher education. Siemens, Dawson and Lynch [27] have reported a consolidated work on how higher education sector is improving the quality and productivity deploying learning analytics. Authors have reported ten case studies from different universities from USA, UK and Australia. There has been a lot of research [10–12, 16, 17] to suggest that Big Data analytics is used in higher education institutions, however none of them explicitly discusses how learning and teaching core processes can be re-engineered to combine Big Data analytics. Also, none of the research work focuses on how Big Data analytics can be used at each process of learning and teaching so that students benefit from the outcome of each process of learning and teaching. At each process of learning and teaching students are required to be monitored to gain the overall benefits of Big Data analytics. This generates the requirement for higher education stakeholders to establish how Big Data analytics can be used at each process of learning and teaching so that students benefit in their overall journey of student life. The prime focus of Big Data analytics is on developing a data-based architecture that enables the addition of a variety of sources so that various types of data can be analysed and put to use at each process of learning and teaching. The goal is to broaden the analytics process available and subsequently improve management of the institution and its resources by addressing the issues of students engagements related to the course content.

3   Selecting Core Business Processes of Learning and Teaching for Re-engineering: Big Data Analysis Requirements

Re-engineering is a systematic process of analysis, design and implementation. Hammer and Champy [13] have defined Business Process Re-engineering as *"the fundamental rethinking and radical redesign of business processes to achieve dramatic improvements in critical, contemporary measures of performance, such as cost, quality, service and speed"*. Hammer and Champy [13] also define a process as "a collection of activities that takes one or more kinds of input and creates an output that is of value to the customer". In Higher education the output from each process of learning and teaching should be used for Big Data analytics so that the output generated by each process is of value to students, educators and other stakeholders. Higher education institutions are finding it challenging to achieve their objectives and goals due to complicating factors arising from making decisions at the right time. Khalid et al. [16] have coined the term *"Educational Process Re-engineering"* based on the established concept of Business Process Re-engineering for process improvement of learning and teaching activities, academic administration and evaluation and assessment. However their work lacks combining Big Data analytics into the business processes. The dramatic transformation of society requires that the educational system adapt and change, or face obsolescence. The stimulus for re-engineering higher education business processes is the combination of four ongoing transitions in the society. They are: enhancing learning and implementation of learning activities; changing educational needs; requirements for alternative learning opportunities; and extensive availability of digital technology.

A re-engineered educational activity is a significant enhancement to traditional decision making process. Re-engineering higher education learning and teaching processes to combine Big Data analytics will exploit significant amounts of student level data.

Big Data analytics should be used to guide the re-engineering educational process to address the followings [15]:

- Educational needs are to be met by a proposed educational activity. This can range from the institutional to the individual lesson/module level.
- Redesign a curriculum to define the scope, general content, and structure of the proposed activity (program or area of study).
- For each area of study define the elements or units (class topics, modules, exercises, etc.) that are appropriate for concentrated study.
- Develop well defined learning objectives and expected outcomes. These are derived from the documented or perceived needs for the activity.
- Develop an analysis of the proposed learners (students) with respect to educational background, location (geographical distribution) and times available for learning activities.
- Apply established principles of learning to define the materials, media, and methods to design each learning unit.
- Develop or acquire the learning materials and media as required.
- Identify and make available additional references and resources to support the learning activities.
- Develop or acquire access to facilities with sufficient technology infrastructure to support the learning activities.
- Develop an appropriate management and administration system for the educational activities.
- Develop a faculty and collaborators with "real world" experience and the ability to guide and stimulate the learning activities.
- Provide faculty with learning and development opportunities to enhance their effectiveness as learning facilitators.
- Provide faculty with incentives to develop and use technology to enhance their teaching and academic activities.

Table 1 shows the identified core business processes of learning and teaching with its input and output. The core processes of learning and teaching to address the above points are identified as: prepare learning and teaching resources; implement learning and teaching resources; outcomes of learning and teaching resources; and review of learning and teaching resources. These business processes are required to be re-engineered to fit data driven outcomes in higher education environment. Each business process has input and output. Output from these business processes will help higher education to make near real time evidence based decisions if processes are combined with Big Data analytics. Input to these business processes should not be only data from student's success rate and graduation. Nowadays higher education generates a large amount of log data in their LMS. The benefits of analyzing this log data are to be found when processing them *en masse*. For instance, if a teacher is interested in tracking behaviour of a student over time, reconstructing user sessions, it is much more convenient to operate over all the logs. Logs are a good fit for Big Data architecture as MapReduce is a perfect fit to analyse logs. First, logs usually follow a certain pattern, but they are not entirely structured, so a RDBMS cannot be used to handle them. Secondly, logs represent student's behaviour use case where scalability not only matters, but is also a key to keeping the system sustainable. As services grows within higher education system, so does the amount of log and the need of getting something meaningful out of them.

Table 1. Core business processes of learning and teaching.

Core processes of learning and teaching	Prepare learning and teaching resources	Implement learning and teaching resources	Outcomes of learning and teaching resources	Review of learning and teaching resources
Input	Teaching strategies; Learning Materials	Develop course contents Coordinate courses	Graduates	Review course Review subjects Review industry feedback
Output	Attract students Select students	Graduation Retention rate Success rate	Scholarship of learning and teaching	Teachers and Student prospective

4 Re-engineering Learning and Teaching Business Processes of Higher Education

The model generated by us for re-engineering higher education business processes and combining it with Big Data analytics architecture is shown in Fig. 1. The identified inputs, to Big Data architecture are the output generated form the core business processes, which include: attract and select students; retention and success rate; scholarship of learning and teaching; and students, educators and external prospective. The data generated with these core business processes combined with educational side database and external database containing Google Reviews enabled us to find out correlations between different variables which helped us in identifying the cause and need of student's course requirements such as redesigning of course structure.

The key components of re-engineered higher education business process model to combine Big Data analytics in higher education are illustrated in Fig. 1. Data (external and internal) is acquired and organized as appropriate and then analyzed to make meaningful decisions. A variety of underlying platform provide a critical role. LMSs collect data from students that, properly analyzed, can provide educators and students with the necessary information to support and constantly improve the learning process [22, 23]. Unfortunately, these platforms do not provide specific tools to allow educators to thoroughly track and assess all students' learning process.

To test our e-business process model, we applied it to a course offered in mixed mode. We have used the re-engineered higher education business process model for analyzing LMS Moodle log file data, which collects set of data from students, educators and other stakeholders. We have also analysed Facebook reviews and University Google review data to see the satisfaction rate of course and content delivery. LMS course activity report, showing the number of views for each activity and resource (and

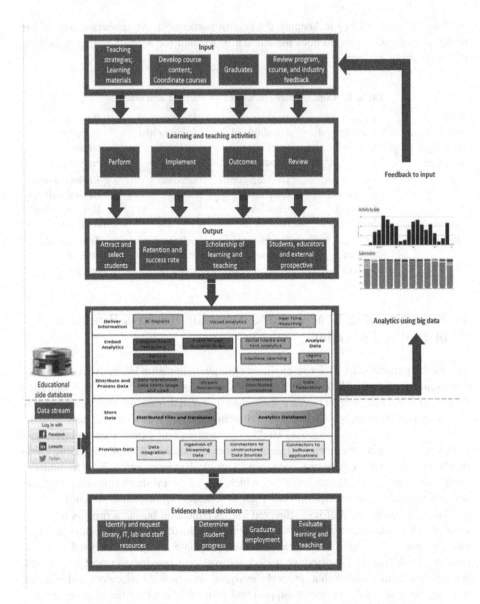

Fig. 1. Re-engineered higher education business process model.

any related blog entries), can be viewed by teachers using Moodle Activity Viewer (MAV).

MAV shows activity views as heatmap – colouring links lighter or darker according to how often records are accessed. Moodle data when analysed can serve as a tool for teachers to follow student's behaviour to identify critical situations. Typical data relates to the competency breakdown, logs, live logs, activity reports, activity completion, number of accesses, duration of accesses, paths traversed in the platform, tools used, resources used or downloaded, participations in the forum and other activities [24, 25]. Data was collected for two semester (6 months) for three courses. The population consisted of 620 participants. We have analysed data generated by MAV for students online activity to courses they have passed to see the correlation between these two variables. At the Deliver Information layer as shown in Fig. 1, we have used Apache Zeppelin web-based notebook which brings data ingestion, data exploration, visualization, sharing and collaboration features to Hadoop and Spark.

For the Embed Analytics layer and Analyse Data we have used Apache Spark. Apache Spark has as its architectural foundation the Resilient Distributed Dataset (RDD) and data items distributed over a cluster of machines. Apache Spark supports and can interface with Hadoop which we have used at layer Store Distribute and Process Data. For layer Provision Data, we have used Hive to import datastores into Hadoop. We have uploaded data into Hadoop Distributed File System (HDFS) Files View and created Hive queries to manipulate data. We created a table called *temp_students* to store data. We created another table *students*, so we can overwrite that table with extracted data from the *temp_students* table we created earlier. Then we did the same for *temp_online_activity* and *online_activity*. Finally, created queries to filter the data to have the result show the sum of online activities and courses passed by each student. Table 2 shows the data that were integrated into each of the two table.

Table 2. Contents of data used in our case study

Data sources	Indicator count
Online access to assignment 1	220914
Online access to assignment 2 marks	90722
News forum	7653
General discussion	6858
Question and answers	2645
Facebook review	556
University Google reviews	76

These data sources reflect student's engagement to learning activities and review on how the course is delivered and structured. The main objective in selecting these sources is to understand the correlation between students' academic success and student's engagement to learning activities. Facebook Reviews and University google reviews were used to improve on lecture delivery, course content and resources developed by educators.

For analytics we used Apache Spark with language R. For data sets that are not too big, calculating rules with Association rules (arules) in R is not a problem. Since association mining deals with students, the data had to be converted to one of class students, made available in R through the arules pkg. This is a necessary step because the apriori() function accepts students data of class students only. We included library (arules). We used several correlation scores calculated using access and participation data from Moodle activities and the grades obtained by the students. Student's online participation data in the unit and passing unit had a positive correlation. More online participation reflects higher grades in the unit. We used association based mining techniques which include Apriori. We selected Apriori algorithm. In R there is a package **arules** to calculate association rules, it makes use of the Apriori algorithm. We first connected to Apache Spark and uploaded the data to Spark. Association rule mining is a rule-based machine learning method for discovering interesting relationships between variables in large datasets. It is intended to identify strong rules discovered in data using some measures of interestingness [26]. The correlation between online activities and student's grades were analysed as shown in Table 3.

Table 3. Correlation scores calculated

Variables	Type of data	Correlation
Access to summative assessment documents/Final grades	Structured	Correlation final = 0.6593
Assignment submission	Structured	Correlation final = 0.55896
Participating in forums	Structured/Semi structured	Correlation final = 0.22589
Academic misconduct	Structured/Semi structured	Correlation final = 0.1052
Absence rate	Structured	Correlation final = 0.1238
Behavioral engagement data/Sentiment data	Text data unstructured	Correlation final = 0.1135

In our case study the correlation was used to gain the understanding of student's online activity to their academic success and how much these online activities are used by students. The regression algorithm shows the correlation between access to documents/final grades; assignment submission; participating in forums; academic misconduct; absence rate and behavioral engagement data with online activities. The result shows strong correlation between online activity and courses passed however, students are not very active on forums and not working towards tutorial exercises. We identified that the student's online activity decreased towards the end of the term. This resulted in a poorer exam outcome. Interestingly their grades remain good. The reason behind this was to pass the course it was not necessary to pass the exam. Students are active on summative tasks. This requires the course to be redesigned to address the issues of students not using all the required learning and teaching resources. This gives an input to re-design and re-develop course contents.

Students are posting comments on University Google reviews about the quality of the course contents and course delivery style. The decision tree algorithm is used to predict potential performance using online activities, courses attempted, and estimation of success factors. Our research findings are that student's academic success is not only affected by the extent of online activities, and experience with LMS, it is also affected by the design and structure of the course. Student's online activities increase and decrease during the term according to the overall contribution of formative and summative assessment to pass the course. With our results we were able to re-design the course and delivery style.

5 Conclusion and Future Work

Using the Big Data architecture for our case study, we found out that it was a positive first step in allowing our students and teachers to quickly gain an understanding of the integrated data and to visually extract interesting correlation. These correlations suggests the improvement on redesigning a curriculum to define the scope, general content, and structure of the proposed activity (program or area of study) based on observing student performance (which is outward and visible sign of student learning) and analysing the factors that improve student achievements, such as teaching practices when they are not engaged with online resources. Also Facebook reviews and University Google reviews have suggested improving course structure and content delivery. Future work is to further demonstrate and expand the scope of the use of proposed re-engineered higher education business process model and its use of Big Data analytics.

References

1. Baesens, B., Bapna, R., Marsden, J.R., Vanthien, J., Leon Zhao, J.: Transformational issues of Big Data and analytics in networked business. MIS Q. **40**(4), 807–818 (2016)
2. Daniel, B.: Big Data and analytics in higher education: opportunities and challenges. Br. J. Edu. Technol. **46**(5), 1–17 (2014)
3. Liebowitz, J.: Big Data and Business Analytics, 1st edn. Auerbach Publications, CRC Press, Taylor and Francis group (2013)
4. Chen, H., Chiang, R., Storey, V.: Business intelligence and analytics: from Big Data to big impact. MIS Q. **36**(4), 1165–1188 (2012)
5. King, I.: Big education in the era of Big Data. In: International Proceedings on Federated Conference on Computer Science and Information Systems, Warsaw, Poland (2014)
6. Crosling, G., Heagney, M., Thomas, L.: Improving student retention in higher education. Aust. Univ. Rev. **51**(2), 9 (2009)
7. Liang, J., Yang, J., Wu, Y., Li, C., Zheng, L.: Big Data application in education: dropout prediction in Edx MOOCs. In: Proceedings of IEEE Second International Conference on Multimedia Big Data, pp. 440–443. Taipei, Taiwan (2016)
8. Goldstein, P.J.: Academic analytics: the use of management information and technology in higher education. Educause Centre Appl. Res. 1–12 (2005)
9. Romero, C., Ventura, S., Garcia, E.: Data mining in course management systems: moodle case study and tutorial. Comput. Educ. **51**, 368–384 (2008)
10. Riffai, M.M.M.A., Edgar, D., Duncan, P., Al-Bulushi, A.H.: The potential for Big Data to enhance the higher education sector in Oman. In: Proceedings of 3rd MEC International Conference on Big Data and Smart City, Piscataway, NJ, Muscat, Oman, pp. 1–6 (2016)
11. Long, P., Siemens, G.: Penetrating the fog: analytics in learning and education. Educause Rev. **46**(5), 30–40 (2011)

12. Marsh, O., Maurovich-Horvat, L., Stevenson, O.: Big Data and education: what's the big idea. In: Big Data and Education Conference. UCL (2014)
13. Hammer, M., Champy, J.: Re-Engineering the Corporation: a Manifesto for Business Revolution. Nicholas Brealey Publishing Ltd., London (1993)
14. Jones, S.: Technology review: the possibilities of learning analytics to improve learner-centered decision-making. Commun. Coll. Enterp. 18(1), 89–92 (2012)
15. Wagner, E., Ice, P.: Data changes everything: delivering on the promise of learning analytics in higher education. Educause Rev. 47(4), 32 (2012)
16. Khalid, S. Md., Hossain, M.S., Rongbutsri, N.: Educational process re engineering and diffusion of innovations in formal learning environment. In: Proceedings of the 19th International Conference on Computers in education. Chiang Mai, Thailand (2011)
17. Picciano, P.: The evolution of Big Data and learning analytics in American higher education. J. Asynchronous Learn. Networks 16(3), 9–20 (2012)
18. Reid-Martinez, K., Mathews, M.: Harnessing data for better educational outcomes. A research report from The Center for Digital Education, Issue 3 (2015)
19. Capgemini: The deciding factor: Big Data and decision making. Economist Intelligence Unit, White Paper (2012)
20. Jha, S., Jha, M., O'Brien, L., Wells, M.: Integrating legacy system into Big Data solutions: time to make the change. In: Proceedings of the Asia Pacific Working Conference on CSE Plantation Island, Nadi, Fiji (2014)
21. Ndukwe, J.G., Daniel, B.K., Butson, R.J.: Data science approach for simulating educational data: towards the development of teaching outcome model (TOM). Big Data Cogn. Comput. 2(24), 1–18 (2018)
22. Evans, J.R., Lindner, C.H.: Business analytics: the next frontier for decision sciences. Decis. Line 43(28), 4–6 (2012)
23. García, E., Romero, C., Ventura, S., de Castro, C.: An architecture for making recommendations to courseware authors using association rule mining and collaborative filtering. User Model (2009)
24. Rice, W.H.: Moodle e-Learning Course Development a Complete Guide to Successful Learning Using Moodle. Packet Publishing, Birmingham (2006)
25. Arnold, K.E., Pistilli, M.D.: Course signals at Purdue: using learning analytics to increase student success. In: Proceedings of IEEE Frontiers in Education Conference (FIE), pp. 1–8 (2017)
26. Anand, A., Fosso Wamba, S., Gnanzou, D.: A literature review on business process management, business process reengineering, and business process innovation. In: Barjis, J., Gupta, A., Meshkat, A. (eds.) EOMAS 2013. LNBIP, vol. 153, pp. 1–23. Springer, Heidelberg (2013). https://doi.org/10.1007/978-3-642-41638-5_1
27. Siemens, G., Dawson, S., Lynch, G.: Improving the quality and productivity of the higher education sector. Policy and strategy for systems-level deployment of learning analytics. Office of Learning and Teaching. Australian Government (2013)
28. Tulasi, B.: Significance of Big Data and analytics in higher education. Int. J. Comput. Appl. 68(14), 23–25 (2013)
29. Paule-Ruiz, M.P., Riestra-Gonzalez, M., Sánchez-Santillan, M., Pérez-Pérez, J.R.: The procrastination related indicators in e-learning platforms. J. Comput. Sci. 21, 7–22 (2015)

Real-Time Age Detection Using a Convolutional Neural Network

Siphesihle Sithungu and Dustin Van der Haar[(✉)]

Academy of Computer Science and Software Engineering,
University of Johannesburg, Johannesburg, South Africa
{siphesihles,dvanderhaar}@uj.ac.za

Abstract. The problem of determining people's age is a recurring theme in areas such as law enforcement, education and sports because age is often used to determine eligibility. The aim of current work is to make use of a lightweight machine learning model for automating the task of detecting people's age. This paper presents a solution that makes use of a lightweight Convolutional Neural Network model, built according to a modification of the LeNet-5 architecture to perform age detection, for both males and females, in real-time. The UTK-Face Large Scale Face Dataset was used to train and test the performance of the model in terms of predicting age. To evaluate the model's performance in real-time, Haar Cascades were used to detect faces from video feeds. The detected faces were fed to the model for it to make age predictions. Experimental results showed that age-detection can be performed in real-time. Although, the prediction accuracy of the model requires improvement.

Keywords: Age detection · Convolutional Neural Network ·
Computer vision · Machine learning

1 Introduction

Large amounts of data are becoming increasingly available on the web. Software artifacts and computational models can be designed that use this data to improve the quality and understanding of human life. Human life in this context, also extends to how businesses and organizations, in general, can improve authentication and detection mechanisms in order to follow legal requirements or business processes.

Human beings are not perfect, and sometimes leaving the task of authentication to people presents several risks, especially when taking factors such as fatigue into account. In several cases, machine learning models have been trained to perform dedicated tasks until they reached a point where they surpassed human expert performance on the same tasks [1–3]. Therefore, the application of machine learning to specialized tasks can provide better accuracy and consistency.

The focus of current work is to investigate the possibility of automating the task of predicting a person's age in real-time using a lightweight Convolutional Neural Network (CNN). A lightweight CNN requires a small amount of storage and that is a huge advantage for applications that perform tasks in real-time. Moreover, a lightweight model is faster to train, while providing convenience in terms of reusability. Finally,

© Springer Nature Switzerland AG 2019
W. Abramowicz and R. Corchuelo (Eds.): BIS 2019, LNBIP 354, pp. 245–256, 2019.
https://doi.org/10.1007/978-3-030-20482-2_20

recognition tasks must be performed in a short space of time in some applications because the platform may be computationally limited [4].

The rest of the paper is arranged as follows: Sect. 2 offers the problem background and looks at similar works in literature. Section 3 explains the proposed solution as well as the experiment setup. Section 4 discusses experimental results, and Sect. 5 concludes the paper.

2 Problem Background

The aim of current work is to perform age detection using a modified CNN. The motivation behind the work is that age detection technology has important applications as is explored in the following subsection. Finally, there is not a considerably large amount of age detection research in literature [5].

2.1 Age Detection

The ability to detect the age of individuals is an important aspect of human life. In certain areas of business, age detection is an important aspect of day-to-day business processes. For example, it is important for liquor stores to only sell alcohol to human beings that are over a certain age in order to not violate the law.

Age detection is also important in the internet where content of any form is accessible from anywhere around the world. Internet businesses can also leverage age detection technology in order to provide more personalized search results, recommendations and advertisements [5].

In practical settings, computer vision-based age detection can be used, as explained in [10], by vending machines in order to determine if a user is eligible to purchase certain products. In the current age where day-to-day tasks are increasingly becoming automated, computer vision methods can be leveraged to enable the devices to determine a person's age.

Chen, Qian, Wang, You, Peng and Zhong conducted a study and used it to construct different types of Chinese language patterns [5]. They further used the constructed patterns to build a Support Vector Machine (SVM) that could classify Chinese bloggers into four different age classes with an accuracy of 88%. The researchers emphasized that this kind of research will provide insights into the Chinese language, and even potentially benefit areas of personalized web search, targeted advertising and recommendation systems [5]. Research work related to age-detection has also been conducted in the field of anomaly detection as shown in [6].

2.2 Convolutional Neural Networks

Recent work has seen Deep Neural Networks (DNN) perform well in classification, segmentation and compression tasks. Moreover, some DNN models have been built that outperformed human experts in specialized tasks [2]. Convolutional Neural Networks (CNNs) are a type of DNNs that were conceived with the aim of classifying images, but over the years their application has been extended to other tasks.

CNNs have led to several breakthroughs in recent years, especially in tasks related to computer vision [7]. One of the main limitations of regular neural networks that CNNs are built to alleviate, is the fact that regular neural networks are translation invariant. This means that, when a neural network learns to classify an object in an image, it may not be able to classify the same object when it appears at a different location in another image [8].

A CNN performs better because each layer of the network learns specific features that make up an object. Finally, CNNs have become a state-of-the-art model for object detection, image classification, feature segmentation, etc. [8]. Although a universal approach to improve the accuracy of CNNs is to increase the depth of the networks, this approach compromises the efficiency of the CNNs in terms of size and speed [4]. An example of a lightweight CNN architecture is the MobileNet architecture [4]. MobileNets were built mainly for use in mobile phones and embedded applications; they provide a trade-off between accuracy and efficiency while providing two hyper-parameters that can be finetuned to suit the desired application [4].

2.3 Similar Works

Fusion Network (FusionNet). The method proposed in [9] made use of a CNN architecture, known as Fusion Network (FusionNet), that extracted age-related facial attributes from face images in order to emphasize features that are specific to aging as part of its input.

Facial patch selection (performed with Bio-inspired Features and AdaBoost), a CNN and age regression were the three main parts of the solution pipeline [9]. The FusionNet was trained and evaluated on the MORPH II dataset. Experimental results showed that the FusionNet produced the lowest Mean Absolute Error (MAE) as compared to state-of-the-art models (i.e. OR-CNN, DEX and Ranking-CNN) on the same dataset.

Optimized CNN Architecture. In [10] Aydogdu and Demirci proposed an optimized CNN architecture consisting of 6 layers for age classification. The CNN's layers consisted of 4 convolutional layers and 2 fully connected layers. In order to discover this architecture, the authors performed multiple experiments to test 16 different architectures, ranging from 2 convolutional layers and 1 fully connected layer to 5 convolutional layers and 4 fully connected layers [10].

The performance of the architectures was compared using exact success, top-3 success and 1-off success criteria. Using this comparison criteria, the CNN with 4 convolutional layers and 2 fully connected layers reproduced the best performance with respect to the defined criteria and standard deviation values [10].

Matching Convolutional Neural Network. In [11] Cho, Jang and Park proposed the use of a matching CNN (consisting of an architecture similar to AlexNet) to perform age category estimation. The CNN took a facial image as input and output a comparison with a target image (*younger than*, *similar to* or *older than*) [11].

The input to the CNN was made by combining an input image and a discriminator image along the RGB channel axis. The CNN was then trained on the inputs for it to be able to differentiate between image pairs [11]. The difference between the CNN proposed by the authors and AlexNet was that the proposed CNN made use of batch normalization rather than local response normalization. Furthermore, 1024 was the chosen dimension for the last two fully connected (FC) layers [11].

Experimental results showed that the accuracy of the proposed CNN increased in proportion to the number of target images defined for each age category. When 16 target images were used, the accuracy of the proposed CNN reached an average of 94.5%, a score that outperformed existing methods [11].

3 Proposed Solution

Current work proposes the use of a modified LeNet-5 CNN to perform age detection in real-time. LeNet-5 (See Fig. 1) is a lightweight CNN that was introduced by LeCun, Bottou, Bengio and Haffner in [12] for document recognition. LeNet-5 has been improved and modified in several ways to solve computer vision problems [13–15].

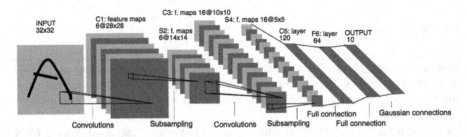

Fig. 1. The LeNet-5 CNN architecture.

The lightweight attribute of this architecture was the main motivation behind choosing it. Current work aims to achieve age detection in real-time. It is important, therefore, that the model being used to achieve this task provides age predictions in the shortest time possible. The solution's pipeline is shown in Fig. 2.

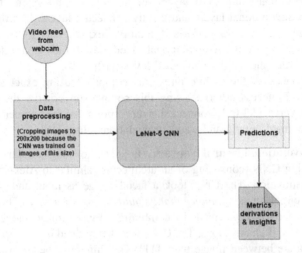

Fig. 2. A high-level pipeline of the proposed solution.

3.1   Experiment Setup

Hardware Components. The only hardware device that was used was a computer with a webcam.

Software Components. The software components composed of the following: (1) A dataset, (2) a set of libraries, and (3) a development framework.

Dataset. The dataset used was the UTKFace Large Scale Face dataset [16]. The original dataset consists of 23700 images. To achieve a balanced training scheme, 7700 images were used for the experiment. The reason behind using only a subset of the dataset is the limitation in terms of how much data could be loaded into memory for training the model.

An 80–20 split was used to divide the images to a training set and a test set. Only the ages of the subjects were taken into consideration when formulating labels for the data. Ages of individuals were separated into 7 classes. The dataset used had fewer images of subjects as their age increased. Therefore, images that represent subjects older than 70 years were excluded because including them would introduce an inconsistency with regards to the number of images per class. Table 1 depicts the age classes that were used.

Table 1. A distribution of age groups into categorical classes.

Age class	Age range/group (in years)
0	1–10
1	11–20
2	21–30
3	31–40
4	41–50
5	51–60
6	61–70

Development Framework. The application was developed using Python and Java. Figure 3 Shows a snapshot of the application during use. The two age classes (out of the 7 classes) for which the CNN has the highest certainty (i.e. the two output nodes that fired with the highest probabilities from a SoftMax activation point-of-view) are visualized on the pie chart. For example, the CNN had 19% certainty that the subject in the image is within the range 21–30 years (age class 2).

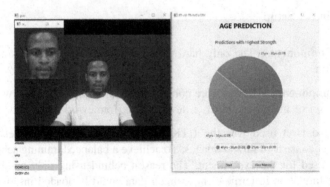

Fig. 3. A snapshot of the application during use.

3.2 Real-Time Image Extraction from Video Feed

The first step in the solution pipeline is to collect image data from a video feed in real-time. In this experiment, the video feed was obtained from a computer's webcam. Upon obtaining the video feed, the second step involves data preprocessing. For a network to be able to make a prediction on the data, that data must be fed to the network in the form of a 200 × 200 RGB image.

To accomplish this, a *MAX_FRAME_INTERVAL* variable is defined. For example, if the value of *MAX_FRAME_INTERVAL* is 100, the 100th frame will always be selected and fed to the CNN. In this way, the CNN can make predictions on an incoming feed in real-time and return the results without the feed having to stop. Therefore, after selecting the 100th frame to send to the CNN, Haar Cascades are used to detect the face of the subject in the image.

3.3 Convolutional Neural Network Model

Architecture. Figure 4. Depicts the architecture of the proposed modified LeNet-5 CNN. The second fully-connected (FC) layer was replaced with a dropout layer. The dropout layer was introduced to try and limit the model from overfitting the data. The replacement increased the training speed of the CNN significantly and produced a slight increase in accuracy.

The CNN takes 200 × 200 RGB images as input. Therefore, each image is converted to a NumPy array of shape (200, 200, 3). The first convolutional layer makes use of 6 7 × 7 filters. The second convolutional layer makes use of 16 3 × 3 filters to focus more on finer details of the convolved images. Max pooling is applied after each convolutional step to only focus on the most highlighted features of convolved images. Both the convolutional layers make use of Rectified Linear Unit (ReLU) activation. The first fully connected layer consists of 120 nodes that make use of ReLU activation and is followed by the dropout layer of the same number of nodes.

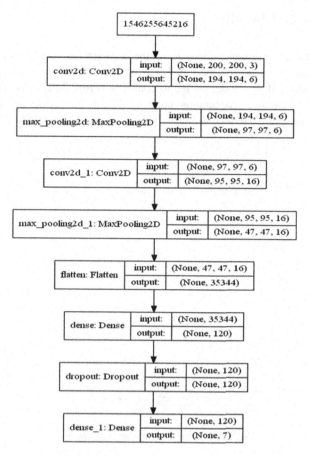

Fig. 4. The architecture of the CNN. The image was plotted using the Keras deep learning library.

Each node in the dropout layer has a 50% probability of firing and therefore does not make use of an activation function. The dropout layer was introduced as an attempt minimize overfitting of the training data by the CNN. The output layer consists of 7 nodes representative of the 7 different age classes. Seven classes were chosen (instead of three as in [11]) so that smaller age ranges are used. This makes the predictions more specific. The output layer makes use of SoftMax activation to show a probability distribution of the networks prediction as it provides more information as compared to discreet 1's and 0's.

Training. The network was trained on 6160 samples of the extracted dataset. The loss metric used was Categorical Cross Entropy and the network was evaluated on Accuracy. The network was trained for 100 epochs with a mini-batch size of 32.

4   Experimental Results

Accuracy and Loss. Figures 5 and 6 depict the accuracy and loss curves respectively. The overall accuracy score of the CNN was 45.3%. As can be seen in Fig. 5, the validation accuracy of the model started to converge before 20 epochs. The validation loss, however, showed a continuous increase. We believe that this phenomenon occurred because features related to facial aging were not focused on during the feature extraction phase.

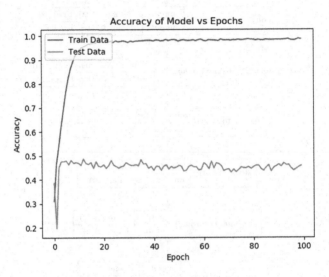

Fig. 5. Accuracy curve after 100 epochs.

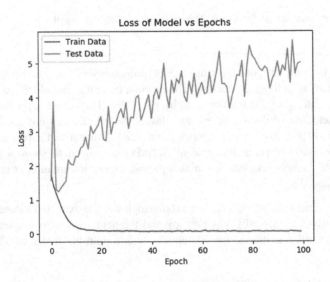

Fig. 6. Loss curve after 100 epochs.

Confusion Matrix. The CNN's confusion matrix is shown in Fig. 7. A confusion matrix that represents an ideal model should only have the diagonal blocks/indices shaded with dark blue. The rest of the diagonals must be approximately white in colour. The confusion matrix in Fig. 7 shows that the CNN made more correct than incorrect predictions for all the age classes. However, for age class 3 (31–40), for example, the CNN predicted the age class to be 4 (41–50), the immediate neighboring class, 28% of the time. This is a very high number of false positives for class 4.

It can also be seen that, for the majority of the time, the CNN made erroneous predictions that were only off the correct class by 1 (1-off). This is a promising observation as it shows that the CNN's predictions were close to the truth. Table 2 explains this more clearly by showing the CNN's performance if 1-off predictions were considered correct.

More Performance Metrics. Table 3 depicts the precision, recall and F1 scores for the different age classes. High precision and recall are ideal as they represent a model with both few false positives and false negatives. A high F1 score also indicates a good relationship between a model's precision and recall scores.

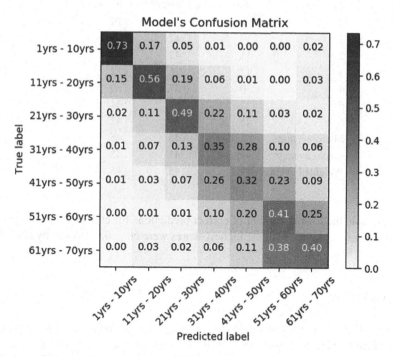

Fig. 7. A confusion matrix depicting the CNN's performance with respect to each age class. Darker shaded blocks represent higher percentages of correct predictions made for an age class. (Color figure online)

The lowest precision, recall and F1 scores for the CNN were for age class 4 (41–50). The scores correlate with the information represented in the confusion matrix since

Table 2. A summary of the model's performance when one-off predictions are also taken into consideration.

Age class	% of correct predictions (1-off excluded)	% of correct predictions (1-off included)
0	73	90
1	56	90
2	49	82
3	35	76
4	32	81
5	41	86
6	40	78
Average (%)	46.6	83

Table 3. A depiction of the CNN's precision, recall and F1 scores.

Age class	Precision (%)	Recall (%)	F1 score (%)
0	89	71	79
1	59	42	49
2	45	42	43
3	42	26	32
4	30	27	29
5	32	69	44
6	45	40	42
Micro average	45	45	45
Macro average	49	45	45
Weighted average	49	45	45

the CNN made the lowest number of correct predictions for class 4 when compared to other classes. Age classes with the highest scores were class 0, followed by class 1 and class 2.

5   Conclusion

A modified LeNet-5 CNN was trained to perform age detection across a wide variety of people of different gender and ethnicity in real-time. The real-time component of the solution worked efficiently. The CNN had an overall accuracy score of 45.3%. However, when considering 1-off predictions the estimated accuracy was 83%.

Future work should focus on applying techniques such as batch normalization and data augmentation - while keeping the size of the CNN relatively small – to help the CNN learn to generalize better. Architectures such as AlexNet or VGG16 should be deployed to compare results and to assess their performance in real-time settings, since both architectures are larger in size than the LeNet-5 architecture.

More improvements can be made by implementing an imbalanced training scheme so that the whole dataset can be used. Finally, more work should be done with regards to feature extraction in order to highlight more important facial landmarks related to aging.

References

1. Codella, N.C.F., Lin, C.-C., Halpern, A., Hind, M., Feris, R., Smith, J.R.: Collaborative Human-AI (CHAI): evidence-based interpretable melanoma classification in dermoscopic images. In: Stoyanov, D., et al. (eds.) MLCN/DLF/IMIMIC -2018. LNCS, vol. 11038, pp. 97–105. Springer, Cham (2018). https://doi.org/10.1007/978-3-030-02628-8_11
2. Dodge, S., Karam, L.: A Study and Comparison of Human and Deep Learning Recognition Performance Under Visual Distortions (2017)
3. Bianco, S.: Large Age-Gap face verification by feature injection in deep networks. Pattern Recogn. Lett. **90**, 36–42 (2016)
4. Howard, A.G., et al.: MobileNets: Efficient Convolutional Neural Networks for Mobile Vision Applications. Comput. Res. Repository 1704(04861) (2017)
5. Chen, L., Qian, T., Wang, F., You, Z., Peng, Q., Zhong, M.: Age detection for chinese users in Eeibo. In: Dong, X.L., Yu, X., Li, J., Sun, Y. (eds.) WAIM 2015. LNCS, vol. 9098, pp. 83–95. Springer, Cham (2015). https://doi.org/10.1007/978-3-319-21042-1_7
6. Bae, I.-H.: A rough set based anomaly detection scheme considering the age of user profiles. In: Shi, Y., van Albada, G.D., Dongarra, J., Sloot, P.M.A. (eds.) ICCS 2007. LNCS, vol. 4490, pp. 558–561. Springer, Heidelberg (2007). https://doi.org/10.1007/978-3-540-72590-9_78
7. Wu, Y., Li, J., Kong, Y., Fu, Y.: Deep convolutional neural network with independent softmax for large scale face recognition. In: Proceedings of the 24th ACM International Conference on Multimedia, New York (2016)
8. Ayyadevara, V.K.: Convolutional neural network. In: Pro Machine Learning Algorithms: A Hands-On Approach to Implementing Algorithms in Python and R, pp. 179–215. Apress, Berkely (2018)
9. Wang, H., Wei, X., Sanchez, V., Li, C.: Fusion network for face-based age estimation. In: 2018 25th IEEE International Conference on Image Processing (ICIP) (2018)
10. Aydogdu, M.F., Demirci, M.F.: Age Classification using an optimized CNN architecture. In: Proceedings of the International Conference on Compute and Data Analysis, Lakeland (2017)
11. Cho, J.H., Jang, D., Park, R.: Age category estimation using matching convolutional neural network. In: 2018 IEEE International Conference on Consumer Electronics (ICCE) (2018)
12. LeCun, Y., Bottou, L., Bengio, Y., Haffner, P.: Gradient-based learning applied to document recognition. Proc. IEEE **86**, 2278–2324 (2018)
13. Pham, V.H., Dinh, P.Q., Nguyen, V.H.: CNN-based character recognition for license plate recognition system. In: Nguyen, N.T., Hoang, D.H., Hong, T.-P., Pham, H., Trawiński, B. (eds.) ACIIDS 2018. LNCS (LNAI), vol. 10752, pp. 594–603. Springer, Cham (2018). https://doi.org/10.1007/978-3-319-75420-8_56
14. LeCun, Y., Haffner, P., Bottou, L., Bengio, Y.: Object recognition with gradient-based learning. Shape, Contour and Grouping in Computer Vision. LNCS, vol. 1681, pp. 319–345. Springer, Heidelberg (1999). https://doi.org/10.1007/3-540-46805-6_19

15. Ma, M., Gao, Z., Wu, J., Chen, Y., Zheng, X.: A smile detection method based on improved LeNet-5 and support vector machine. In: 2018 IEEE Smartworld, Ubiquitous Intelligence Computing, Advanced Trusted Computing, Scalable Computing Communications, Cloud Big Data Computing, Internet of People and Smart City Innovation (2018)
16. Zhang, Z., Song, Y., Qi, H.: Age Progression/Regression by Conditional Adversarial Autoencoder. In: IEEE Conference on Computer Vision and Pattern Recognition (CVPR) (2017)

An Inventory-Based Mobile Application for Warehouse Management to Digitize Very Small Enterprises

Daniel Staegemann[(⊠)], Matthias Volk, Christian Lucht,
Christian Klie, Michael Hintze, and Klaus Turowski

Otto-von-Guericke-University Magdeburg,
Universitaetsplatz 2, 39106 Magdeburg, Germany
{daniel.staegemann, matthias.volk, christian.lucht,
christian.klie, michael.hintze, klaus.turowski}@ovgu.de

Abstract. In today's world, digitization has reached an important role. While more and more enterprises are interested in a realization and aware of its importance, it rather lacks on the implementation. This applies most of all for those with a low budget for new investments, such as small and medium enterprises (SME), as well as even smaller forms (VSE). For this reason, a sectoral digitization approach based on a real-world VSE is pursued. By using the design science research methodology, a low cost and easy to use warehouse management system for the optimization of the inner logistics of VSEs is developed, evaluated and presented in this work.

Keywords: Mobile application · Warehouse management ·
Very small enterprises · Digitization · Use case

1 Introduction

The topic of digitization has drawn plenty of interest in the scientific community as can be seen in relevant journals, search engines or condensed in digests like [1, 2], as well as outside of it, like in politics and the economy [3]. The general term includes several facets and scopes, such as the digitization of countries [4, 5], the economy [5–7] or distinct methods and processes, for example to obtain customer feedback [8], getting digitized. Even though, apparently the general awareness exists, the practical implementation is in many cases still uncommon. While big enterprises often have room for improvements and investments, small and medium enterprises (SME) are in general positioned way worse in that regard [2, 5, 6, 9]. Especially in the area of logistics huge potentials can be expected [10]. The European Commission defines SMEs as enterprises, which have less than 250 employees and a yearly turnover of 50 million Euros at most. Enterprises with less than 50 employees and a yearly turnover of 10 million Euros or below form the subgroup of small enterprises [11]. As widely acknowledged, the aforementioned deficiency constitutes with a decreasing size of the company [9, 12–15]. Albayrak and Gadatsch [16] highlighted that this is accompanied with lack of maturity for digital transformation.

© Springer Nature Switzerland AG 2019
W. Abramowicz and R. Corchuelo (Eds.): BIS 2019, LNBIP 354, pp. 257–268, 2019.
https://doi.org/10.1007/978-3-030-20482-2_21

The partial negligence of small enterprises in the topic of digitization also shows in several surveys, which ignore very small enterprises (VSE) and only consider participants with for example at least ten employees or even more [17–19]. This circumstance is reinforced while observing the usually high number and importance of small and medium enterprises in relation to the whole economy that is being agreed on in various publications [7, 20–24]. In Germany for instance, those very small enterprises, with fewer than 10 employees and a yearly turnover of 2 million Euros or less [11], constitute over 80% of the total number of enterprises [21]. At the same time the general level of digitization is lower than for comparison in the USA [6] and is assessed as lackluster [2]. Motivated by the described situation and the claim of Albayrak and Gadatsch [16], that implies that digitization has to be seen as a holistic task instead of a matter of several small steps, we wanted to explore, if digitization of very small enterprises is possible and beneficial. In particular, the scope of this research focuses on the internal logistics process, as it has been observed as one challenging part in the area of digitalization [10]. In doing so, the following research questions shall be answered in the course of this work.

Q1: Is it possible to digitize very small enterprises, whose staff is technology averse, with negligible expenditures and burdens on the employees?

Q2: Can digitization benefit very small enterprises and convince formerly adverse employees?

The challenges, deterring small and medium enterprises in general and especially VSEs from digitizing themselves are diversified. They prominently include but are not limited to high innovation costs, lack of qualification amongst the employees and security concerns [14, 25]. Therefore in the majority of cases an attempt at digitization of an VSE needs to rebut those concerns of the decision makers by offering inexpensive, easy to realize and use solutions, that do not pose potential security threads for the business and offer clear and visible value, to convince them to overcome their old process flows and modernize their enterprise. The following sections will describe the methodology and the concrete use case. Afterwards the artifact itself is described and evaluated. In the end a conclusions a given, which also outlies possible future prospects.

2   Methodology

The contribution at hand follows the Design Science Research (DSR) paradigm and its seven guidelines formulated by Hevner et al. [26], since those have proven to be an valuable structure for the execution of formative endeavors in real world scenarios. It culminates in an inventory-based mobile application for warehouse management, which focusses on very small companies, as a usable artifact (Guideline 1). To conform to Guideline 2 (Problem Relevance) and Guideline 3 (Design Evaluation) the evaluation will be conducted according to the four stepped framework of Sonnenberg and vom Brocke [27], which is widely accepted and also focuses on the real world applicability of the regarded artifacts. Since the digitization of VSEs is not sufficiently advanced, the publication at hand contributes by illustrating a possible way to progress the corresponding development (Guideline 4). Research rigor is addressed by using the existing knowledgebase, e.g. known concerns of enterprises regarding digitization, as a starting-point for the requirements, applying modeling techniques for the conception of the artifact and utilizing a well-known and accepted framework for the evaluation (Guideline 5). While the existing knowledgebase provides a good starting-point, it is still necessary to embed the experiences and needs of the target audience. Therefore, the needs of the designated users of the artifact were included and formulated as requirements from the very beginning of the development, and are still being considered in the course of the ongoing evaluation (Guideline 6). This publication, as a form of communication of research, constitutes Guideline 7 of the DSR paradigm, feeding back into the knowledgebase. All guidelines, in relation with the described realizations, are summarized in Table 1. For the structure of this work, this means that in the following sections at first the use case is introduced to illustrate the challenges and needs that are incorporated in the requirements. Afterwards the artifact and its development are presented. In succession the evaluation process, that went along with the progression and is still ongoing, is described. Finally, a preliminary conclusion is drawn and possible future expansions are depicted.

Table 1. The conducted guidelines of the design science research according to [26].

Guideline	Realization
1. Design as an artifact	Creation of usable artifact
2. Problem relevance	Evaluation according to Sonnenberg and vom Brocke (cp. EVAL 1)
3. Design evaluation	Evaluation according to Sonnenberg and vom Brocke (cp. EVAL 2–4)
4. Research contribution	Advancing digitization of VSE
5. Research rigor	Usage of accepted proceedings and scientific knowledge
6. Design as a search process	Inclusion of insights gained during development and evaluation
7. Communication	Contribution at hand

3   The Use Case

The initial inspiration for this work stems from an incident in a pottery workshop with attached store in eastern Germany. A customer wanted to buy some specific products and was told that there are none in stock and that it would be necessary to craft them from scratch, which would take some time. As a result, the customer forewent the purchase and left the shop. Later on, the employees discovered that several boxes of the earlier requested products were in the warehouse. Thereupon, the owner of the pottery was being convinced to host a case study considering the prospects of digitizing the stock keeping to benefit the business.

The pottery has four employees, a yearly turnover in the high five-figure Euro range and is therefore categorized as a VSE. With the store, the rooms for the production and the warehouse there are several potential locations for storing pre-products and finished products. Additionally the warehouse contained several racks in which boxes with those items were randomly piled without any system or at least labeling. This resulted in a lack of knowledge regarding the inventory, leading to monetary disadvantages through wrong decisions and lackluster customer consulting that impeded potential sales. For example, it happened several times, that, after an inspection of the warehouse by an employee, potential customers were told the processing time of a bigger order is estimated with several weeks. This was due to the absence of any available pre-products, resulting in the need to produce everything from scratch, which takes a lot of time because of the nature of the pottery craft. This timeframe deterred the prospective buyers who decided to forego the order, for example because a set deadline would have been violated. Later on the employees found several boxes of the necessary pre-products, which would have allowed them to fulfill the order in less than a week, because only the (relatively time inexpensive) glazing and baking was necessary. But even smaller orders of only one or two objects, which are not uncommon for tourists, constituting a large portion of the potential buyers, take noticeably longer when they have to be made from scratch, therefore in some cases exceeding the available time frame, resulting in a lost order, despite the pre-product being available without the staff knowing. Therefore, by improving the warehousing, the overall situation of the enterprise could be enhanced, allowing for more effective customer consulting and better informed decisions regarding production plans and possibly also pricing or bundling depending on available supply of certain items. However, as in many SMEs [14, 25], there were several constraints regarding the preconditions, namely very low financial resources, respectively the apprehension of a negative outcome in financial regards and the absence of any technical affinity in the staff, which had to be considered regarding the design.

4 An Inventory-Based Mobile Application

To counteract those identified weaknesses in the warehousing of the regarded enterprise (as a representative of numerous comparable SMEs and VSEs), and to also correspond to the known obstacles of digitization, several requirements were formulated. Those can be categorized into functional and non-functional ones and are described in Table 2. While the functional requirements are predominantly focusing on issues of the currently existing warehouse management solution, the non-functional requirements target the uncomplicated and affordable implementation within the VSE. All requirements were developed in cooperation with the employees of the pottery workshop, increasing their involvement and avoiding to discourage and overburden them. However, this was on the expense of their quality by reducing the degree of formalism and precision. Derived from these requirements, the structure of the developed system is relatively simple, consisting of only three components. Those are the database, the client and an abstraction layer. As a result, no sophisticated hardware stacks and only commodity hardware elements are needed, which allows to use inexpensive standard solutions, keeping the costs for implementation and operation low. To fulfill the functional requirements, first, the database schema depicted as an Entity Relationship (ER) diagram in Fig. 1, was developed.

Table 2. The developed functional and non-functional requirements

No.	Type	Requirement
1.	Functional requirements	All products, pre-products, packages and shelves that are stored or storing vessels are represented in the system
2.		The system allows to locate the sought-after items by facility, shelf and storing vessel
3.		Items of the same kind can be stored in multiple packages and shelves, even across different facilities
4.		Not only the available type of items, but also their quantity shall be easily observable in the app
5.		The number of entries that can be administered in the system is sufficient to represent every item in enterprises of the targeted size
6.		For security reasons a possible attempt to overburden the maximum load of packages or shelves hast to be detected
7.	Non-functional requirements	Establishing the system is inexpensive. This includes hardware (servers, devices), training and initiation
8.		The operating costs of the system are low
9.		The artifact should offer a high usability. From a user's perspective it has to be time-saving and comfortable
10.		The system is uncritical regarding security concerns. Under no circumstance any serious harm for the enterprise is possible through its usage
11.		Using the system is intuitive, minimizing the required amount of training for the employees

Basically it consists out of five different entities, named *storage objects* (the stored items), *type*, *package*, *shape factor* and *shelf*, which also form the main tables of the database. Various relations were used for the connection of the entities, such as *has* and *stores*. The *stores* connections from *package* and *shelf* to the *storage object* also include a *quantity*. This allows to only scan one item of a kind and manually provide the number of deposits or withdrawals instead of scanning every single item, therefore increasing the usability. It also increases clarity, since users can comfortably see the available supply of each item. By using a self-designed ID as primary key, each entry of each table can be easily identified. Additionally to that, for specification purposes, different attributes were used for each element. This includes for instance the identification of the type of the package as *leak proof* or as a *cooler*, which were most of all integrated for later extensions and other application scenarios, increasing the generality and reducing the likeliness that customizations might be needed. Furthermore, direct measurements for each storage object can be saved.

This comprehensive scheme allows to store the properties of each item, as well as the packaging and the shelfs, enabling the system to automatically check, if items are too heavy for the specification of their designated repository. Since issues concerning the size would become apparent on eyesight, those do not require an automated check, even though the dimensions are stored, allowing for future extensions in that regard.

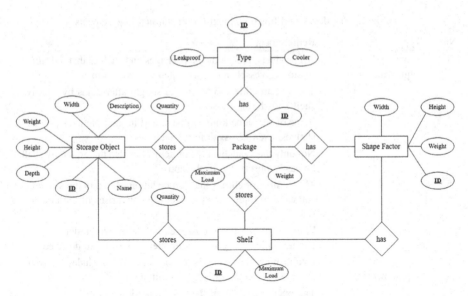

Fig. 1. The Entity Relationship (ER) diagram of the database scheme.

Therefore, theoretically, it would be possible to upgrade the system to automatically determine the optimal storage location for an item or packaging.

The ID, as the main identifier, consists out of two alphanumerical parts and is decomposed within Fig. 2. While the first part indicates the targeted type and includes a self-defined tag, the second is directly related to the product itself. The benefits provided by this structure, which is comprehensible for humans, are two-folded. On the one hand, the indication of the type of the database entry and the additional tag, which could for example indicate the facility a shelf is located in, helps to reduce the time a search requires. This is a direct consequence of the structure of the pottery workshop presented in the use case, preventing the necessity to remember in which facility a shelf is located. On the other hand skepticism towards the new technology and the barrier of acceptancy in VSEs, without obtaining own IT staff, can be diminished.

Due to the fact, that some categories of items (e.g. standardized plates) are stocked in huge numbers, it is possible to input the quantity of items of the same kind in a repository instead of registering every single one. This allows to quickly determine the number of certain items being available and also often reduces the number of steps necessary for storing or withdrawing items. Consequently the effort for using the system is lowered and therefore the usability increased. Even though the usage of the new warehousing tool can provide crucial advantages, no sensitive data are handled, therefore avoiding potential critical harm in case of security issues.

The design of the database is intended to be generic enough for an applicability in a plethora of enterprises, therefore avoiding the necessity of customizing it in most use cases. By now, there is also no option to deposit pictures of the listed items, since this would increase the complexity of the solution and also the size of the database. However, this might be a useful addition in the future.

Package/Shelf

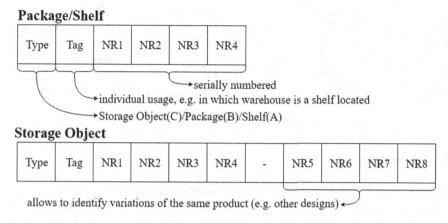

Type	Tag	NR1	NR2	NR3	NR4

→ serially numbered
→ individual usage, e.g. in which warehouse is a shelf located
→ Storage Object(C)/Package(B)/Shelf(A)

Storage Object

Type	Tag	NR1	NR2	NR3	NR4	-	NR5	NR6	NR7	NR8

allows to identify variations of the same product (e.g. other designs) ←

Fig. 2. Structure of the ID.

Utilizing the system is facilitated via an app developed for Android smartphones. Using smartphones allows a high mobility and usability while negating the costs for purchasing new equipment, since already existing private devices can be used. Furthermore the familiarity helps to alleviate possible reluctance and reduces the required efforts for training [28]. The decision regarding the operating system was based on the already existing technical endowment of the staff. In the app, users can create and delete entries. To increase usability, it is possible to create (and stick the printed code to the particular object) or scan QR-Codes instead of inserting the respective ID by hand. Users can also assign items to packages or shelves, respectively packages to shelves, search for entries and display properties and positions. To increase security, it is necessary to log in with a registered account before changing any data within the app. Once logged in, this status is maintained by a session token that expires after 25 min of inactivity. This token is utilized during usage, for the authentication while creating and changing data. The chosen timeframe is intended as a compromise between convenience, preventing the need of inserting the credentials all the time, and security. To allow customers to independently check the availability of items and as a preliminary stage for a potential web shop, read only access does not require an account.

The connection between the database and the app is constituted by the abstraction layer, which acts as an application programming interface (API). It receives HTTP requests by the user's smartphone, transforms the inquiry into the structured query language (SQL) format and forwards it to the database. The response of the database, received as raw data, will be handled afterwards and send in JSON format to the user's smartphone. This avoidance of direct communication serves as an additional security measure. Furthermore, by this approach features like the search function are less complicated to implement and more performant, since the well-established MySQL could be utilized instead of crafting a completely new solution. For other use cases, the smartphone and database component can be exchanged with alternative solutions. This could be for instance related to other operating systems, such as Apple's iOS, or alternative database solutions, like PostgreSQL. An overview of all components and their relations are depicted as the main artifact in Fig. 3.

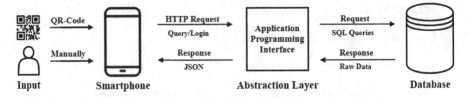

Fig. 3. An overview of the developed system

5   Evaluation

As mentioned beforehand, the evaluation, which corresponds to Guideline 2 and Guideline 3 by Hevner et al. [26], is conducted according to the framework from Sonnenberg and vom Brocke [27]. Therefore the evaluation consists of the four phases (EVAL1 to EVAL4) depicted in Table 3, covering different stages of the development.

Table 3. The conducted evaluation steps.

Evaluation step	Description	Status
EVAL1	Literature and demand of target audience	Completed
EVAL2	Logical reasoning	Completed
EVAL3	Prototype tested in typical use cases	Completed
EVAL4	Case study	Ongoing

EVAL1, which is also the foundation of the first chapter of the publication at hand, seeks to ensure that the problem that is being focused is actually meaningful in the context of design science research and that its solution provides a scientific benefit. The general interest in digitization and the forecasted benefits emphasize its importance. At the same time small and very small enterprises, which constitute an important part of the economy, lag behind in the according development and are, despite their importance, slightly disregarded by the scientific community and responsible organizations. For this reason there is a lack of studies, focusing on strategies for the digitization of those enterprises. Also from a practical point of view, there was explicit demand for a technical solution, like the proposed artifact, identified in the enterprise that is described in the chapter concerning the use case. Therefore, the publication at hand and the accompanying artifact are justified by their contribution to closing the gap in research and improving the working experience and business opportunities of real users.

EVAL2 aims to show, that the proposed design concept is a promising approach for handling the identified problem. As stated before, the most important barriers identified by (very) small enterprises related to their digitization are high innovation costs, a lack of qualification amongst their employees and security concerns [14, 25]. Those were incorporated from the beginning of the design process. The costs of establishing the system, as well as the operating costs are exemplarily depicted for a low-cost approach in Table 4. While the first model results in one-time costs for buying commodity hardware, such as the latest version of a Raspberry Pi 3 Model B+ [29] as a server and operating costs for acquiring the domain [30] and expenses [31] for the power usage [32], the second model has no initial investment costs, since it relies on renting the server [33]. As shown, in both cases the costs are marginal totaling to around 120€ for 2 years. At the same time, the employees gain new insight into their inventory, allowing them to quickly convey, if a certain product is available and to also determine the exact position. This can potentially, as described in the use case, allow for additional sales or prevent a loss of sales, due to insufficient or false customer consulting. Considering the price of a single plate starting at 15 Euros, even a small number of potential orders that is lost due to inadequate warehousing easily exceeds the expenses for the operation of the proposed system.

Table 4. Exemplary calculation of a low cost approach

Expense factor	Model 1 – buying the hardware		Model 2 – renting the hardware	
	Initial	Monthly	Initial	Monthly
The server	38€	0€	0€	5€
Power – 0,3€/kWh	0€	2,7€	0€	0€
Acquiring a domain	0€	0,8€	0€	0€
Total costs	38€	3,5€	0€	5€
Avg. monthly costs over 2 years	5,08€		5€	

Even though this is only a single example, a general impression of the situation in many SMEs can be derived. Therefore, one of the main concerns of the target audience can be avoided and, in addition to more comfortable working conditions, even directly related positive financial results are to be expected. The simplicity of the appliance also allows reducing the necessary briefing of employees to a minimum, making extensive trainings or service centers for user support unnecessary. Since no sensitive data are handled, there is also no risk of serious security issues. Even in a worst-case scenario, no critical damage can occur. Therefore, the proposed artifact complies with the identified needs of the target audience, depicted in Table 2.

EVAL3 is using a prototypical instance of the artifact to proof its applicability. For this purpose, an artificial setting is created, in which the artifact has to prove its feasibility for achieving the expected results. This was realized by simulating typical use cases that included creating and deleting items as well as sorting them into shelfs (registration of the position in the database) and searching their position aided by a prototypical implementation of the system. The used hardware was a rented virtual server in the described price range. Those tests were continually conducted during the development process to help refining the artifact and proving the sufficient capabilities of the chosen hardware. Since this simulation was successful and the last iteration met the expectations, EVAL3 can be deemed as successfully completed.

EVAL4, as the final stage of evaluation, examines the usefulness of the artifact by applying an instance in a real world situation. In this case, the artifact was implemented and integrated in the workflow, including the registration of every stored item along with an instruction of the employees, and used in daily business for the VSE that is described in the use case. While in theory the registration of the items constitutes an enormous expense factor, since it consumes a lot of paid working time, this did not apply in reality. Due to the nature of the enterprise, there are many situations where employees are not occupied, freeing them up to successively fulfill this task. In the first week, additionally a contact person was on site to assist the employees when needed, ask them for their evaluation as well as suggestions and to directly observe the usage as well as potential flaws. During the recording of the items it was detected that about 5000 products and pre-products with a combined value of approximately 120.000€ were stored in the facilities of the pottery workshop, retrospectively giving even more justification for the ongoing project. This applies even more in relation to the yearly turnover that is way below that sum. Since the objectives were reached by enhancing the warehousing while maintaining the other constraints, like costs and ease of use, and the users were, despite their initial skepticism, satisfied, the assessment is positive. However, to get a more extensive impression, it would be desirable to implement the proposed solution in a greater number of SMEs and VSEs. Therefore, the fourth stage of the evaluation is not concluded yet.

6 Conclusion

Even though additional case studies are planned, a first conclusion can already be drawn. However, the limited sample size and the focus on enterprises that are operating a warehouse have be considered. While the claim of Albayrak and Gadatsch [16] regarding a lack of maturity for digitization might be true in general, it was shown, that even very small enterprises with employees that are skeptical towards technical novelties can still benefit from minor but specific measures and implement them into their daily routine. For that reason, both research questions are positively answered in consideration of the limitations. This way of small steps also allows circumventing common barriers like high costs or a lack of qualification of the employees, which can also lead to a reluctance towards the usage of the new system. Additionally it does not force the enterprise to fully commit to something that is beyond their area of expertise. Therefore, this approach is facilitating the decision for a modernization and constitutes an entry-point for possible further projects in the future. While the showcased enterprise at this point only digitized its warehousing, many other modernizations, like an integration of cash register and warehousing, customer relations management or a web shop could be implemented in the future. For a more widespread usage it will be also necessary to expand the list of supported operating systems beyond the currently only option of using Android powered smartphones. In general it will be essential to focus on a broad usability, provided by a centralized and standardized development and distribution, instead of individualized solutions, because the low expenses can only be realized, when it is not necessary to factor in high costs for extensive expansions which might otherwise arise.

References

1. Hirsch-Kreinsen, H., ten Hompel, M.: Digitalisierung industrieller Arbeit: Entwicklungsperspektiven und Gestaltungsansätze. In: Vogel-Heuser, B., Bauernhansl, T., ten Hompel, M. (eds.) Handbuch Industrie 4.0 Bd.3, vol. 20, pp. 357–376. Springer, Heidelberg (2017). https://doi.org/10.1007/978-3-662-53251-5_21
2. Kollmann, T., Schmidt, H.: Deutschland 4.0. Springer Fachmedien Wiesbaden, Wiesbaden (2016). https://doi.org/10.1007/978-3-658-13145-6
3. Wolf, T., Strohschen, J.-H.: Digitalisierung: Definition und Reife. Informatik Spektrum **41**, 56–64 (2018)
4. Sabbagh, K., Friedrich, R., El-Darwiche, B., Singh, M., Ganediwalla, S.: Maximizing the impact of digitization. In: The Global Information Technology Report, pp. 121–133. World Economic Forum, Geneva (2012)
5. Bertschek, I., Briglauer, W., Fuest, C., Kesler, R., Ohnemus, J., Rammer, C.: Innovationspolitik in Deutschland: Maßnahmen für mehr Innovationen im Zeitalter der Digitalisierung. Mannheim (2016)
6. Böhm, M., Müller, S., Krcmar, H., Welpe, I.: Digitale transformation in ausgewählten Ländern im vergleich. In: Oswald, G., Krcmar, H. (eds.) Digitale Transformation, pp. 73–85. Springer Fachmedien Wiesbaden, Wiesbaden (2018)
7. Chaudhuri, P., Kumar, A.: Role of digitization and e-commerce in indian economic growth: an employment generation perspective. In: Proceedings of the 98th Annual Conference of Indian Economic Association (2015)
8. Dellarocas, C.: The digitization of word of mouth: promise and challenges of online feedback mechanisms. Manag. Sci. **49**, 1407–1424 (2003)
9. Bley, K., Leyh, C., Schäffer, T.: Digitization of German enterprises in the production sector – do they know how "digitized" they are? In: AMCIS 2016 Proceedings (2016)
10. Taschner, A.: Improving SME logistics performance through benchmarking. Benchmarking **23**, 1780–1797 (2016)
11. Empfehlung der Kommission vom 6. Mai 2003 betreffend die Definition der Kleinstunternehmen sowie der kleinen und mittleren Unternehmen. 2003/361/EG (2003)
12. Taiminen, H.M., Karjaluoto, H.: The usage of digital marketing channels in SMEs. J. Small Bus. Ente Dev. **22**, 633–651 (2015)
13. Saam, M., Viete, S., Schiel, S.: Digitalisierung im Mittelstand: Status Quo, aktuelle Entwicklungen und Herausforderungen. Mannheim (2016)
14. Demary, V., Engels, B., Röhl, K.-H., Rusche, C.: Digitalisierung und Mittelstand. Eine Metastudie. Institut der deutschen Wirtschaft Medien GmbH, Köln (2016)
15. Zentralverband des Deutschen Handwerks: Digitalisierung der Geschäftsprozesse im Handwerk. Ergebnisse einer Umfrage unter Handwerksbetrieben im ersten Quartal 2014 (2014)
16. Albayrak, C.A., Gadatsch, A.: Sind kleinere und mittlere Unternehmen (KMU) bereits auf die Digitale Transformation vorbereitet? (2018)
17. Bitkom: Der Weg zum digitalen Büro ist erst zur Hälfte geschafft. https://www.bitkom.org/Presse/Presseinformation/Der-Weg-zum-digitalen-Buero-ist-erst-zur-Haelfte-geschafft.html. Accessed 10 Dec 2018
18. Bundesamt, S.: Unternehmen und Arbeitsstätten. Nutzung von Informations- und Kommunikationstechnologien in Unternehmen, Wiesbaden (2017)

19. Homann-Vorderbrück, S., Sauer, J., Schröder, H.: Management von Digitalisierungspro-jekten – eine Bestandsaufnahme. In: Mikusz, M., Volland, A., Engstler, M., Fazal-Baqaie, M., Hanser, E., Linssen, O. (eds.) Projektmanagement und Vorgehensmodelle 2018, PVM 2018, pp. 15–23. Gesellschaft für Informatik e.V. (GI), Bonn (2018)

20. Gergin, Z., et al.: Industry 4.0 scorecard of Turkish SMEs. In: Durakbasa, N.M., Gencyilmaz, M.G. (eds.) ISPR 2018, pp. 426–437. Springer, Cham (2019). https://doi.org/10.1007/978-3-319-92267-6_37

21. Sommer, L.: Industrial revolution - industry 4.0: are German manufacturing SMEs the first victims of this revolution? JIEM **8**, 1512–1532 (2015)

22. Müller, J.M., Voigt, K.-I.: Sustainable industrial value creation in SMEs: a comparison between industry 4.0 and made in China 2025. Int. J. Precis. Eng. Manuf.-Green Tech. **5**, 659–670 (2018)

23. Mutula, S.M., van Brakel, P.: E-readiness of SMEs in the ICT sector in Botswana with respect to information access. Electron. Libr. **24**, 402–417 (2006)

24. Bouwman, H., Nikou, S., Molina-Castillo, F.J., de Reuver, M.: The impact of digitalization on business models. Digit. Policy Regul. Gov. **20**, 105–124 (2018)

25. Rische, M.-C., Schlitte, F., Vöpel, H.: Industrie 4.0 – Potenziale am Standort Hamburg. Studie im Auftrag der Handelskammer Hamburg (2015)

26. Von Alan, R.H., March, S.T., Park, J., Ram, S.: Design science in information systems research. Manag. Inf. Syst. Q. **28**, 75–105 (2004)

27. Sonnenberg, C., Vom Brocke, J.: Evaluations in the science of the artificial – reconsidering the build-evaluate pattern in design science research. In: Hutchison, D., et al. (eds.) Design science research in information systems, pp. 381–397. Springer, Heidelberg (2012)

28. Disterer, G., Kleiner, C.: BYOD bring your own device. Procedia Technol. **9**, 43–53 (2013)

29. Amazon: Raspberry 1373331 Pi 3 Modell B+ Mainboard, 1 GB. https://www.amazon.de/Raspberry-1373331-Pi-Modell-Mainboard/dp/B07BDR5PDW/ref=sr_1_3?ie=UTF8&qid=1544794679&sr=8-3&keywords=raspberry+pi. Accessed 14 Dec 2018

30. Strato AG: Domains. https://www.strato.de/domains/. Accessed 14 Dec 2018

31. Heidjann GmbH: Was kostet Strom? Strompreis in kWh. https://www.stromauskunft.de/strompreise/was-kostet-strom/. Accessed 14 Dec 2018

32. Raspberry Pi Foundation: Raspberry Pi 3 Model B+. https://www.raspberrypi.org/products/raspberry-pi-3-model-b-plus/. Accessed 14 Dec 2018

33. Strato AG: V-Server Linux V10. https://www.strato.de/server/linux-vserver/. Accessed 14 Dec 2018

Collaboration in Mixed Homecare – A Study of Care Actors' Acceptance Towards Supportive Groupware

Madeleine Renyi[1,2(✉)], Melanie Rosner[1], Frank Teuteberg[2],
and Christophe Kunze[1]

[1] Furtwangen University, Robert-Gerwig-Platz 1, 78120 Furtwangen, Germany
madeleine.renyi@hs-furtwangen.de
[2] School of Business Administration and Economics,
Department of Accounting and Information Systems, University Osnabrück,
Katharinenstr. 1, 49069 Osnabrück, Germany

Abstract. As more and more people reach high age the need for care, especially at home, rises. Caring involves the coordination of a wide variety of actors. Modern information and communication technologies (ICT) may improve care coordination and thus relieve all actors involved in outpatient care.

This paper presents the results of a study (n = 108), that aimed to find out about the attitude of care actors towards digital care coordination tools in Germany. The survey contained questions regarding the care situation, expectations, technology commitment, barriers and need for assistance.

The data were primarily evaluated according to the subgroups informal caregivers and professional actors. The study showed a lack of target group oriented provision and support of groupware. A mere provision of the technology does not lead to the desired acceptance of the offer because none of the actor groups sees the initiating role of technology use on their side. Personal instruction and support are in demand in both user groups, regardless of technology commitment. For the rather less technology-savvy informal caregivers, this can be explained through their rather tense care situations and the mostly rather high age and the associated restrictions. Professionals demand to learn the technology in order to integrate it as effectively as possible into their daily care routine.

Keywords: Technology acceptance · Groupware · Adoption of innovation · Mixed homecare · Outpatient care

1 Introduction

For many, the family is one of the most important supports in life. To stand up for and support all family members from birth to death has a high priority for them in return. In today's world, which is characterized by increasing employment [1] and great geographical distance, it is often difficult to cope with this challenge. The family loses its importance and the potential for provision till old age dwindles [2]. This poses great challenges for society. Particularly in view of the current demographic development

© Springer Nature Switzerland AG 2019
W. Abramowicz and R. Corchuelo (Eds.): BIS 2019, LNBIP 354, pp. 269–283, 2019.
https://doi.org/10.1007/978-3-030-20482-2_22

and the expected shortage of skilled workers in the nursing sector [3], it is neither possible to provide for the aging population through purely professional inpatient care structures nor is it desired by a large number of senior citizens [4, 5]. In future, home care will, therefore, depend to a large extent on a successful care mix, i.e. a combination of informal (family or neighbor) help and professional care [6, 7].

The rapid pace of technological development could open up new opportunities. However, while professional caregivers already use software tools for planning and documenting their work quite commonly, there are generally no formal communication and cooperation structures with informal caregivers. Recognizing the high potential groupware could provide to enhance the awareness of each other and the tasks conducted, several studies examined the needs and expectations of informal and professional caregivers towards computer supported collaborative work (CSCW) systems that could be used to coordinate care networks (c.f. [8–11]). Some studies resulted in the creation of research prototypes which were tested for several weeks in real life environments [12–15]. Common to all these studies is the emphasis on the importance and effort of preparing and conducting field studies in this environment to identify unforeseen processes and implementation hurdles. How difficult and varied these challenges can be the authors had to experience in their own studies. It is a long way from motivating individual test persons to the adoption of new technologies through whole networks in their everyday workflows. This insight provided the impetus for the study presented in this article, in which the acceptance of supporting groupware tools for care actors was examined in more detail. With this work the authors want to contribute to achieving technology adoption in care mix organization. The central research question of this contribution is:

What are the care actors' acceptance factors and how do they influence the implementation of groupware tools for care mix networks?

2 Background and Related Work

2.1 Collaboration in Care Mix

The terms "mixed care", "mixed homecare" or "care mix" describe the composition of various support services and activities for shaping everyday life on an individual level, combining and networking professional offers, civic involvement and help from relatives [6, 7, 16]. In addition to medical and nursing care, the social care of those in need of care is of great importance. The care mix therefore also includes, for example, transport services, meals on wheels or neighborly help [17]. Summarized, care occurs in a network of interactions among a lot of different actors, with different (work) attitudes, believes and from different generations. The challenging task of coordinating these diverse actors mostly lies within the responsibility of the care dependent person him-/herself or caring relatives, who, in most cases, are old themselves.

2.2   ICT Supported Collaboration in Care Mix

The role that technical support systems will play in care mix in the future has been investigated only in rudimentary form. Görres et al. [17] see "the potential of a division of labor (…) organized through the use of technology still far from being exhausted" and see, among other things, the coordination of the individual offers as well as the control and optimization of care processes as the objectives of the use of technology. As Pinelle and Gutwin [18] describe it, care mix structures are generally weakly structured and loosely coupled work processes in which actors have only limited access to information from other organizations. The low level of networking between the various actors also makes it difficult to introduce cross-organizational processes and systems for exchanging information. Typical barriers include role conflicts, unevenly distributed added values of use and the difficulty of reaching a "critical mass" of users, which would be necessary for a sensible use of the system.

The interdisciplinary research field CSCW deals with information and communication technologies to support the cooperation of individuals and working groups. The importance of CSCW research for the field of IT in the health care system as a whole is a result not least of the fact that the medical and nursing care of individuals is a highly cooperative task that comprehends diverse occupational groups and supply contexts. While research on medical information systems from a top-down perspective puts institutional goals (e.g. cost savings, increased efficiency) in the foreground, the CSCW research understands collaborative work processes as complex interactions between different actors in socio-technical systems [19]. Researchers contribute knowledge, often in the form of observational studies, aiming to improve working practices and opportunities through technical support. An example of this are studies that investigate the use of paper notes and -documentation, which are commonly used additionally to "official" electronic documentation systems to transfer information between occupational groups or to support an informal exchange of information [20–23].

Since the health care system differs considerably from other fields of application (e.g. regarding data protection requirements, trust, work cultures specific to occupational groups, etc.), CSCW research findings from other fields of application are not readily transferable to the health care system. Thus e.g., Robertson et al. [24] stated that the previous findings on computer-assisted telecooperation seem to be of little importance for multidisciplinary team meetings in health care.

In an early study on the topic of computer-aided cooperation and coordination in outpatient care, Mynatt et al. [25, 26] used the Family Portrait System to investigate the extent to which the provision of status information on the activities of persons in need of care can promote awareness of the situation and stimulate interaction. Christensen, Grönvall, Bossen et al. [12, 27] developed various approaches to information sharing, such as a digital pen and RFID reader to supplement paper documentation and digital voice notes in the home of the patient. An example of work in the field of care mix in outpatient care is the study by Pinelle and Gutwin [18, 28, 29], which examined the exchange of information and cross-organizational cooperation in home care. Based on a comprehensive qualitative investigation of the application context, they developed and tested an asynchronous groupware system called "Mohoc", which supported the exchange of case-related activities, explicit messages and group discussions.

The system was evaluated in two field studies for about 3 months each. Considerable implementation barriers were identified, which are among others attributed to the low networking of the different actors and insufficient user involvement. Another example is the study by Bossen et al. [12], where a tablet-based system called "CareCoor" was developed, which supports the coordination of tasks and appointments and the exchange of messages in mixed homecare networks. This was tested in 5 care networks for six weeks. All these developments did not overcome the prototype state. A summary of groupware functions that can be important for the caregivers is provided in an earlier publication by the authors [11].

2.3   Adoption of ICT for Care Mix

General studies about health-related ICT adoption show, that older adults – and therefore most caring relatives – are not yet ready to adopt health-related ICT [30]. This may be one of the reasons, why in spite of the high potential scientists see in the use of groupware to support outpatient care collaboration, the adoption and use of ICT is rather low amongst outpatient healthcare stakeholders. The adoption of a new technology stands at the end of a process, divided into six steps: (1) orientation, (2) insight and understanding, (3) acceptance, (4) change and (5) maintaining the change [31, 32]. Common to all presented projects is, that end-user participation is taken seriously in the development process of the (health-related) ICT. A sufficient orientation and insight and understanding of the technology, its benefits and needed training can, therefore, be assumed. Nevertheless, most projects for home-care groupware tools have only a little case number. A lack of acceptance to try out something new in mixed homecare therefore hinders the further adoption process.

An understanding of this lack of care actors' acceptance towards supportive groupware is needed. Known already is that acceptance is rather low when the new technology is understood as extra work and not as a facilitator. Insights and understanding of the technology are, therefore, essential for the acceptance and as a consequence the improvement of work and organization processes [33]. Also, facilitating conditions at organizational and system level are needed [34]. In the context of care mix, technology adoption has to be understood as a social process, even more than a technical matter [31].

Since the acceptance of technology depends on various factors, numerous models and theories for determining the acceptance of technology have been designed over the course of time. One of these frameworks to evaluate the acceptance and the implementation of new technologies is the Unified Theory of Acceptance and Use of Technology (UTAUT) [35]. This framework is currently wildly used to analyze mHealth applications and services (e.g. [36–38]) and also served as a basis for this study. The model postulates that four basic constructs (performance expectancy, effort expectancy, social influence, facilitating conditions) have a direct influence on behavioral and usage intention. Performance expectancy (PE) describes "the degree to which an individual believes that using the system will help him or her to attain gains in job performance" [35]. Effort expectancy (EE) is defined "as the degree of ease associated with the use of the system" [35]. Social Influence (SI) is defined "as the degree to which an individual perceives that important others believe he or she should use the new system" [35]. The "degree to which an individual believes that an organizational and technical infrastructure exists to support use of the system" is defined as facilitating conditions (FC) [35].

Summarized, in recent years, several studies have dealt with individual aspects of the coordination of home care using mobile services. Significant prior work in CSCW and related venues has sought to understand the nature of coordination among formal and informal caregivers (c.f. [9, 39, 40]), and has proposed and evaluated specific technologies to support this coordination (c.f. [12, 14, 41, 42]). However, none of the developed prototypes has been able to sustainably assert itself in everyday use in nursing care. The adoption angle on mobile health technologies has only just recently shifted into the focus of research attention (c.f. [37, 43]) and is therefore incomplete for the specific use case of care mix groupware where not the adoption of a single person but a complex and diverse group must be achieved.

3   Methods

3.1   Questionnaire Construction

To gain a deeper understanding of the current attitude towards digital support tools of the informal and professional actors in mixed homecare in Germany as well as their technology expectations and to derive prerequisites to enhance the acceptance of technology for care mix collaboration, a questionnaire[1] was designed and made available online and on paper in German language. Based on the four constructs of the UTAUT model, own questions were formulated and combined with an already existing validated questionnaire for technology commitment [44]. In the survey period from mid-May to the end of December 2018 108 usable questionnaires were generated, 94 online and 14 paper-based. The questionnaire contained the following sections:

1. **Role in Care**: To find out who is answering the survey, questions were asked about the participant's age, gender, role in care and the number of cared for persons.
2. **Introduction to the Topic**: Because it can be assumed that every participant has different previous experiences and assumptions regarding technology possibilities and usage in care mix a video[2] was shown early in the online survey. For the paper-based surveys, the video was transformed into a storyboard and was handed out with additional information to the participants. This material served to demonstrate the possibilities of groupware for care mix illustrated by use cases and ensured a basic understanding by everyone.
3. **Current Care Situation and Organization**: The adoption of a groupware for mixed homecare organization, not only depends on the acceptance of a single, easy to classify person, but the acceptance of a multifaceted group in an individual care situation. Knowledge about the current care situation and organization is therefore

---

[1] https://tinyurl.com/BIS2019Renyi.

[2] https://youtu.be/mDkuJrocxL4 (informal actors), https://youtu.be/ZLSYalm9bT0 (professional actors); English subtitle available.

essential to evaluate the transferability and comparability of the results. Questions regarding this topic were the duration of care, distance to the person in need of care, main activities, current use of technology, actors involved, contact with care actors, the current situation in the care organization and participation of important actors. Depending on the question lists with radio buttons, check boxes or 5-point Likert scales were used.

4. **Expectations and Requirements**: The needs and requirements for care mix groupware identified in previous studies [11] were listed and queried for evaluation and weighting.

5. **Technology Commitment (TB)**: The validated questionnaire by Neyer et al. [44][3] was used to identify the personal attitude towards new technical developments. The TB measures the technology commitment on three subscales: technology acceptance (e.g. "I am very curious about new technology"), technology competence (e.g. "When dealing with modern technology, I am often afraid of failing."), and technology control beliefs (e.g. "Whether I am successful in using modern technology depends essentially on me.").

6. **Barriers**: After rating their current care organization (4-point Likert scale), participants had the chance to indicate what circumstances would hinder them to use digital tools to help organize care (4-point Likert scale).

7. **Additional Need for Support**: The need for support like introduction videos, training courses, and supportive documents was evaluated.

8. **Further Interest**: The participants could indicate if they would like to test a groupware prototype, receive further information or do not wish to be contacted further.

3.2 Pretest

A pretest of the questionnaire was carried out with 12 participants in April 2018. The pretest checked the questionnaire for comprehensibility before use. It also gave information on the duration of the questions to be answered, the occurrence of answers and the appearance of incomprehension. Because it is important to carry out the pre-test under the most realistic conditions possible [45], the persons were individually chosen according to their profession and background (caring relatives, nurses, health care students and scientists with nursing background). From reactions, answers, and questions from the respondents, it was possible to draw conclusions about the quality of and make changes to the questionnaire. The data from the pretest was not included in the final dataset.

3.3 Sampling and Recruiting

Everyone involved in outpatient care was counted to the potential target group – former, current or future caring relatives, healthcare professionals or civil engaged. The survey was made available in two forms: online, using the tool LimeSurvey, and

---

[3] Free online access to the questionnaire: https://zis.gesis.org/skala/Neyer.

offline, printed out on paper together with additional information to replace the introduction video. To reach out to a broad majority of the target group the survey was advertised via e-mail, social media, flyers or personal contact. Multipliers like care service stations, self-help groups or associations throughout the whole of Germany were mostly contacted with the wish to distribute the survey. Two self-help groups for caring relatives were visited personally during group meetings, the video was shown to the members and the paper-survey handed out.

3.4 Analysis

All raw data were imported into the analysis tool SPSS Statistics and cleaned up. The data analysis was carried out as a descriptive evaluation as well as inferential statistics. The inferential statistics was conducted using the t-Test for independent sample mean values and cross tables in combination with the Chi^2-Test x^2. Results were accepted as significant for $p \leq 0.050$.

4 Results

Role in Care. The participants (n = 108) could choose between former (n = 14), current (n = 41) or future (n = 4) caring relative, civil engaged with (n = 1) or without (n = 0) professional structures, professional caregiver (n = 40), temporary care worker (n = 1) and other professional care actor (n = 7). These subgroups were necessary to make it easier for participants to choose a role. For the analysis only two subgroups were built: informal actors (former, current or future caring relatives; civil engaged without professional structures) (n = 59; 54.6%) and professionals (professional caregiver, temporary care worker, other professional care actor, civil engaged with professional structures) (n = 49; 45.4%). A detailed table of the descriptive statistic of the sample ordered by the two subgroups can be downloaded under: https://tinyurl.com/BIS2019RenyiResults.

Care Situation. Caring relatives mainly look after one needy person, whereas professionals mostly care for more than three. The duration of care is generally longer than twelve months. Over half of the participating informal caregivers live in the same quarter or household as the person in need of care. But at least also ten persons who live more than ten km away from the person in need of care took part in the survey. Furthermore, 14 caring relatives answered to be working simultaneously full-time and eight to be working part-time.

The network of caregivers is spun mostly between caring relatives, nurses, doctors and other healthcare providers. Civil engagement, friends or distant family members play a minor role in the questioned care mix cases.

Care Organization. The fields of activities of the two groups partially distinguish from each other. While the three main activities of caring relatives are domestic tasks, organization and basic care, professionals spent most of their time for basic care and needs, drug distribution and mobilization. The role of the care organization therefore mostly lies in the responsibility of the caring relatives.

Informal caregivers use significantly (p = 0.028) less technology than professionals to organize their care activities. While over the half of the professionals use technology, it is only less than a third of the informal caregivers. The 40 persons stating to use technology for care organization were then asked to explain for which tasks they use digital help. For professionals, the telephone is still the most preferred communication channel, whereas informal caregivers prefer instant messaging for general communication matters. Issues where instant feedback seems necessary nevertheless are preferably handled directly (via telephone).

Expectations and Requirements. The participants were asked, what they would hope to improve in their care arrangement by using a groupware tool. Informal caregivers hoped to enhance communication, optimize appointment coordination and to have a better overview over care actors. Professionals hope to improve collaboration, fasten information flows and enhance communication. All expectations of the informal were in mean lower than those of the professional caregivers. Significantly higher hopes do the professionals have for the improvement of the collaboration (p = 0.005) and the distribution of tasks (p = 0.011), as well as the fastening of the information flow (p = 0.003).

Additionally, the participants should answer what functionalities they expect from a groupware tool for care collaboration. For informal caregivers the three most important requirements are to be able to inform oneself, to be reachable for others and to have all contacts at hand quickly in case of emergency. The professionals top three are to have all contacts at hand quickly in case of emergency, to be able to inform oneself and to erase uncertainties. For four questioned requirements the values of the informal caregivers are significantly lower than of the professionals: avoid paper notes (p = 0.017), derive and structure tasks (p = 0.001), maintain contact to the person in need of care (p = 0.042) and receive case knowledge (p = 0.047).

According to the participants, the groupware must be used by caring relatives, nurses and physicians to achieve the most added value. For professionals, the participation of the caring relatives is significantly (p = 0.024) more important.

Technology Commitment. The analysis of the technology commitment showed, that the informal caregivers are in means less technology committed than professional actors, as well as the standardized controlled group of Neyer [44]. Additionally, the technology commitment of the sample decreases with increasing age of the participants (p = 0.06).

Barriers. Participants were asked to rate their current care organization to be able to identify barriers and optimization chances through groupware. On average, the current care arrangements are rated at 3.2 on a 4-point Likert scale. There are no significant differences between informal (M = 3.16) and professional (M = 3.24) actors. Although the care organization is overall rated as more or less good, informal caregivers seem to be more dissatisfied with the collaboration, appointment coordination and information flow than professionals.

The biggest barrier to use groupware for all participants of the survey is data privacy aspects (informal caregivers: M = 3.07; professionals: M = 2.82). At this point, it has to be mentioned that the survey was conducted at a time, when data privacy was strongly present in the media because of new European laws becoming effective in May 2018. For caring relatives, second important is the fear of losing personal contact through technology (M = 2.65), followed by fears about additional costs and extra work with the use of technology. Concerns that the system might not live up to their expectations seem to worry some professional actors (M = 2.49). It is also noticeable that the fear of overburdening is a significantly (p = 0.032) higher burden for informal caregivers.

Introduction to the Topic and Additional Support. 39 informal and 33 professional caregivers watched the introduction video. The video was rated as a helpful introduction to the topic (84.85%), interest arousing (70.31%) and sufficient (62.26%) without significant differences in the two subgroups.

A desire for face-to-face (F2F) events like training courses was expressed by most participants (informal caregivers: 82.50%; professionals: 93.75%). Although the main reason for technology training in both groups is learning the new technology, this is significantly more important (p = 0.047) for professionals (M = 0.83[4]) than for informal caregivers (M = 0.61 (see footnote 4)). Getting to know other people motivates for participation the least.

Further Interest. Additionally, the participants were asked, if they would be interested in testing a groupware prototype developed by the authors. Of the 66 participants answering this question, 26 informal caregivers and 14 professionals had no further interest. However, twelve informal caregivers and nine professionals wanted to be further informed and one informal caregiver and four professionals even stated to be willing to directly try out the groupware. A comparison of the 66 participants regarding further interest and current technology usage for care organization showed, that 40% already used technology, regardless of their further interest.

A comparison of means between those with and without further interest and their technology commitment showed an average TB of 3.8 (further interest) and 3.47 (no interest). It therefore seems reasonable to assume that an increased technology commitment is conducive to an interest in groupware testing. With the previously confirmed correlation between age and TB, it could thus be concluded that younger people with a high willingness to use technology are more open to try out something new. This once again emphasizes the adoption problem, manifesting itself in only small numbers of test persons in groupware tests as described at the beginning, since the target group of informal caregivers are commonly older people with little affinity for technology.

5 Discussion and Conclusion

5.1 Discussion of Further Correlations and Acceptance Factors

Summarizing the results, it got obvious, that technology that simplifies the establishment of contact is desired by caring relatives as well as professionals. A place to store common information, provide information in an emergency and eliminate ambiguities

---

[4] Participants were asked to tick the relevant reasons. 0 means "not relevant" and 1 "relevant".

would support the care mix for everyone and satisfy the technology expectancies. To answer the main research question the four determinants of the UTAUT model serve as a structure for the discussion.

Performance Expectancy. According to Klein [46], there is in general a difference in the acceptance of technology between those affected and nursing staff. As the data showed this might be traceable to the differences in the expectations and the requirements for technology in the two groups. The different roles of the actors lead to different main tasks, therefore, different information interest and thus different technology requirements. Generally, more than 84.71% feel that the care organization is currently running rather well or well. The high potential to attain gains through groupware usage cannot be recognized and the added value through the usage of technology seems to be considered low in comparison to the effort especially by the informal caregivers. In line with Eggert et al. [47] we can say that the current care situation is accepted as a matter of fact and change is rather not aimed at. The satisfaction with the cooperation in care networks open for further information is significantly ($p = 0.048$) better than among the uninterested. Deviating from Eggert et al. [47], consequently, we were able to determine that the better the cooperation with others already runs, the more willingness to try out something new with these actors exists.

Effort Expectancy. In this study, not the ease associated with the use of one specific system was tested, but the multifaceted attitude towards groupware for care mix collaboration. "Whilst care actors cherish the access to information that may support them to enhance their caregiving skills and the social contact that those solutions allow for, they recurrently notice that these systems are not the simplest to use or the most intuitive to interact with" [14]. To help them find an easy access to such systems understanding is needed regarding barriers associated with the use of a care mix groupware. The biggest barrier mentioned – privacy aspects –, unfortunately, is nothing easy to overcome, especially for research prototypes, who aim to find out more about user behavior and intentions. Explainable through "the lack of [support and] guidance on how informal caregivers should coordinate care" [48], the fear of being overburdened ($p = 0.003$) by and missing support ($p = 0.029$) for technology usage significantly stronger burdens participants with a low TB than the ones with a high TB. People with an affinity for technology, therefore, see a clear chance of using technology instead of problems, whereas respondents with lower technical competence fear of problems when using technology. The effort expectancy to learn and use new technology, therefore, depends on the one hand on the technology commitment of the individual person and on the other hand on the given setting (facilitators and barriers).

Social Influence. Peek et al. [5, 49] have shown especially for seniors, that the organizational structures of the environment could constitute an obstacle to the use of technology and can thus affect the views of potential users. In accordance with the literature we, therefore, showed that important multipliers, like (family) doctors [50], need to be gained to promote the usage of new technology. Unfortunately, many professional caregivers don't see the necessity to communicate with informal caregivers on a frequent basis [51], thus have no motivation to promote technology usage and see the caring relatives responsible for the initiation.

Facilitating Conditions. In the literature, it is assumed that facilitating conditions are helpful to enhance the acceptance as well as the adoption of technology [35, 52, 53]. In this study, the fact if, F2F events like training courses could constitute such a facilitator for the implementation of groupware for care mix collaboration was examined. 87.50% expressed the wish for a F2F event. This, therefore, has practical consequences if one considers that software for private individuals does not normally provide F2F training, whereas software for professional users in the healthcare context does. One conclusion could be that the task of promoting and initiating groupware tools needs to be assigned to professional caregivers (to stabilize care arrangements). Importantly these training offers should be family-oriented including support and education for informal and professional caregivers likewise [9].

5.2 Limitations of the Study

In the light of demographic change, family ties are becoming less important. Future care mix will, hence, stronger depend on the involvement of civil engaged persons for example organized in neighborhood associations. In this study, only one suchlike semi-professional actor could be reached and in the questioned care mix cases civil engagement only plays a minor role.

Around ten percent of the people starting the survey dropped out at the page showing the introduction video, which therefore represented a barrier for the participation. The video, however, was needed to ensure an equal understanding of all participants. Additionally, it has to be admitted that the given information and video may have influenced the participants which might have shaped their answers considerably.

Despite a high effort of reaching potential participants, the reached sample size of 108 persons might not be representative for the overall target group of outpatient care actors (caring relatives 4.7 mio [54] + professionals 0.35 mio [55]). And even though not all findings of this study may be innovative and transferable in the research context, we emphasized the problems associated with the adoption of groupware for care mix collaboration and cleared the path for future work, that could after an optimization of the questionnaire regarding the quality criteria objectivity, reliability, and validity extend this study for example to examine amongst others organizational and legal aspects in more detail [34]. The attitude of nursing service organizations towards data privacy may also be of importance for the usage of the individuals.

5.3 Conclusion

Direct communication dominates current care mix organization. But especially in the informal sphere, asynchronous communication channels such as instant messengers are gaining importance for non-time-critical activities. Fear of being overburdened by and missing support for the usage of a groupware collaboration tool hinders a faster spreading of the technology adoption. Amongst other things, this study revealed a problem of provision and support for such software. Providers of purely private software usually do not offer any training for their systems and make their products available for self-download in the standard app stores, whereas software for professional users in the healthcare context comes along with training and support.

Most respondents wish for support such as face-to-face training. A solution would be the provision and support of the software by a care service operator. This survey, as well as personal experiences, show, however, that professionals see caring relatives in the responsible and initiating role. Like in related fields of application [56] the need for a neutral caretaker gets apparent by this. Acceptance seems only reachable if enough motivation and support are given, especially in the initial phase to minimize the fears of extra costs and work, as well as overburdening.

Network effects may bring forward the usage of groupware tools on an individual and even more on an organizational level as the more care arrangements use such tools the higher the additional value, especially for the organizations. The higher the added value for the organizations the higher their motivation to promote such tools. Reaching a critical mass, social network analytics could bring forward new strategies to manage customer relationships and optimize work processes. Another added value could be financial compensation for the coordination of care mix through health insurances. Negotiations in this direction should be fostered in the future.

Acknowledgments. We want to thank our participants for contributing their time. This work was supported by the Ministry of Social Affairs and Integration Baden-Württemberg.

References

1. Neubert, L., König, H.H., Brettschneider, C.: Seeking the balance between caregiving in dementia, family and employment: study protocol for a mixed methods study in Northern Germany. BMJ Open **8**, 1–9 (2018)
2. Bianchi, S.M.: A demographic perspective on family change. J. Fam. Theory Rev. **6**, 35–44 (2014)
3. Vannieuwenborg, F., Van der Auwermeulen, T., Van Ooteghem, J., Jacobs, A., Verbrugge, S., Colle, D.: Evaluating the economic impact of smart care platforms: qualitative and quantitative results of a case study. JMIR Med. Informatics **4**, e33 (2016)
4. Piau, A., Campo, E., Rumeau, P., Vellas, B., Nourhashemi, F.: Aging society and gerontechnology: a solution for an independent living? J. Nutr. Heal. Aging **18**, 97–112 (2014)
5. Peek, S.T.M., Wouters, E.J.M., van Hoof, J., Luijkx, K.G., Boeije, H.R., Vrijhoef, H.J.M.: Factors influencing acceptance of technology for aging in place: a systematic review. Int. J. Med. Inform. **83**, 235–248 (2014)
6. Jacobs, M., Van Tilburg, T., Groenewegen, P., Broese Van Groenou, M.: Linkages between informal and formal care-givers in home-care networks of frail older adults. Ageing Soc. **36**, 1604–1624 (2016)
7. Bäuerle, D., Scherzer, U.: Zukunft Quartier – Lebensräume zum Älterwerden. Themenheft 1: Hilfe-Mix – Ältere Menschen in Balance zwischen Selbsthilfe und (professioneller) Unterstützung, Netzwerk Soziales Neu Gestalten (SONG) (2009). http://www.netzwerk-song.de/fileadmin/user_upload/Themenheft1.pdf
8. Camarinha-Matos, L.M., Afsarmanesh, H.: Design of a virtual community infrastructure for elderly care. In: Camarinha-Matos, L.M. (ed.) PRO-VE 2002. ITIFIP, vol. 85, pp. 439–450. Springer, Boston, MA (2002). https://doi.org/10.1007/978-0-387-35585-6_47

9. Bratteteig, T., Wagner, I.: Moving healthcare to the home: the work to make homecare work. In: Bertelsen, O., Ciolfi, L., Grasso, M., Papadopoulos, G. (eds.) ECSCW 2013: Proceedings of the 13th European Conference on Computer Supported Cooperative Work, 21–25 September 2013, Paphos, Cyprus, pp. 143–162. Springer, London (2013). https://doi.org/10. 1007/978-1-4471-5346-7_8

10. Bosch, L.B.J., Kanis, M.: Design opportunities for supporting informal caregivers. In: Proceedings of the 2016 CHI Conference Extended Abstracts on Human Factors in Computing Systems - CHI EA 2016, pp. 2790–2797. ACM Press, New York (2016)

11. Renyi, M., Teuteberg, F., Kunze, C.: ICT-based support for the collaboration of formal and informal caregivers – a user-centered design study. In: Abramowicz, W., Paschke, A. (eds.) BIS 2018. LNBIP, vol. 320, pp. 400–411. Springer, Cham (2018). https://doi.org/10.1007/ 978-3-319-93931-5_29

12. Bossen, C., Christensen, L.R., Grönvall, E., Vestergaard, L.S.: CareCoor: augmenting the coordination of cooperative home care work. Int. J. Med. Inform. **82**, e189–e199 (2013)

13. Span, M., et al.: An interactive web tool for facilitating shared decision-making in dementia-care networks: a field study. Front. Aging Neurosci. **7**, 1–12 (2015)

14. Breskovic, I., de Carvalho, A.F.P., Schinkinger, S., Tellioglu, H.: Social awareness support for meeting informal carers' needs: early development in TOPIC. In: Bertelsen, O.W., Ciolfi, L., Grasso., M.A., Papadopoulos, G.A. (eds.) Proceedings of the ECSCW 2013, Paphos, vol. 2, pp. 3–8 (2013)

15. Boessen, A.B.C.G., Verwey, R., Duymelinck, S., van Rossum, E.: An online platform to support the network of caregivers of people with dementia. J. Aging Res. (2017)

16. Buboltz-Lutz, E., Kricheldorff, C.: Freiwilliges Engagement im Pflegemix - Neue Impulse. Lambertus-Verlag, Freiburg (2006)

17. Görres, S., Seibert, K., Stiefler, S.: Perspektiven zum pflegerischen Versorgungsmix. In: Pflege-Report 2016, pp. 3–14. Schattauer GmbH, Stuttgart (2016)

18. Pinelle, D., Gutwin, C.: A groupware design framework for loosely coupled workgroups. In: Gellersen, H., Schmidt, K., Beaudouin-Lafon, M., Mackay, W. (eds.) ECSCW 2005, pp. 65–82. Springer, Dordrecht (2005). https://doi.org/10.1007/1-4020-4023-7_4

19. Schmidt, K., Bannon, L.: Taking CSCW seriously. Comput. Support. Coop. Work **1**, 7–40 (1992)

20. Tang, C., Carpendale, S.: An observational study on information flow during nurses' shift change. In: Proceedings of the CHI 2007, p. 219. ACM Press, New York (2007)

21. Zhou, X., Ackerman, M., Zheng, K.: Doctors and psychosocial information: records and reuse in inpatient care. In: Proceedings of the CHI 2010, pp. 1767–1776 (2010)

22. Petrakou, A.: Integrated care in the daily work: coordination beyond organisational boundaries. Int. J. Integr. Care. **9**, e87 (2009)

23. Abou Amsha, K., Lewkowicz, M.: Shifting patterns in home care work: supporting collaboration among self-employed care actors. In: De Angeli, A., Bannon, L., Marti, P., Bordin, S. (eds.) COOP 2016: Proceedings of the 12th International Conference on the Design of Cooperative Systems, 23–27 May 2016, Trento, Italy, pp. 139–154. Springer, Cham (2016). https://doi.org/10.1007/978-3-319-33464-6_9

24. Robertson, T., Li, J., O'Hara, K., Hansen, S.: Collaboration within different settings: a study of co-located and distributed multidisciplinary medical team meetings. Comput. Support. Coop. Work **19**, 483–513 (2010)

25. Mynatt, E.D., Rowan, J., Craighill, S., Jacobs, A.: Digital family portraits: supporting peace of mind for extended family members. In: Proceedings of the CHI 2001, pp. 333–340. ACM Press, New York (2001)

26. Rowan, J., Mynatt, E.D.: Digital family portrait field trial: support for aging in place. In: Proceedings of the CHI 2005, p. 521. ACM Press, New York (2005)

27. Christensen, L.R., Grönvall, E.: Challenges and opportunities for collaborative technologies for home care work. In: Bødker, S., Bouvin, N.O., Wulf, V., Ciolfi, L., Lutters, W. (eds.) ECSCW 2011: Proceedings of the 12th European Conference on Computer Supported Cooperative Work, 24–28 September 2011, Aarhus Denmark, pp. 61–80. Springer, London (2011). https://doi.org/10.1007/978-0-85729-913-0_4

28. Pinelle, D., Gutwin, C.: Supporting collaboration in multidisciplinary home care teams. In: Proceedings of the AMIA 2002, pp. 617–621 (2002)

29. Pinelle, D., Gutwin, C., Greenberg, S.: Task analysis for groupware usability evaluation: Modeling shared-workspace tasks with the mechanics of collaboration. ACM Trans. Comput.-Hum. Interact. **10**(4), 281–311 (2003)

30. Heart, T., Kalderon, E.: Older adults: are they ready to adopt health-related ICT? Int. J. Med. Inform. **82**, e209–e231 (2013)

31. Rogers, E.M.: Diffusion of Innovations, 5th edn. Free Press, New York (2003)

32. Grol, R., Wensing, M.: What drives change? Barriers to and incentives for achieving evidence-based practice. Med. J. Aust. **180**, S57–S60 (2004)

33. Bleses, P., et al.: Zwischenbericht des Verbundprojekts KoLeGe. Verbundprojekt KoLeGe, Bremen (2017)

34. Greenhalgh, T., et al.: Beyond adoption: a new framework for theorizing and evaluating nonadoption, abandonment, and challenges to the scale-up, spread, and sustainability of health and care technologies. J. Med. Internet Res. **19**, e367 (2017)

35. Venkatesh, V., Morris, M.G., Davis, G.B., Davis, F.D.: User acceptance of information technology: toward a unified view. MIS Q. **27**(3), 425–478 (2003)

36. Khatun, F., Palas, J.U., Ray, P.K.: Using the Unified Theory of Acceptance and Use of Technology model to analyze cloud-based mHealth service for primary care. Digit. Med. **3**, 69–75 (2017)

37. Hoque, R., Sorwar, G.: Understanding factors influencing the adoption of mHealth by the elderly: an extension of the UTAUT model. Int. J. Med. Inform. **101**, 75–84 (2017)

38. Cimperman, M., Makovec Brenčič, M., Trkman, P.: Analyzing older users' home telehealth services acceptance behavior-applying an Extended UTAUT model. Int. J. Med. Inform. **90**, 22–31 (2016)

39. Grönvall, E., Lundberg, S.: On challenges designing the home as a place for care. In: Holzinger, A., Ziefle, M., Röcker, C. (eds.) Pervasive Health. HIS, pp. 19–45. Springer, London (2014). https://doi.org/10.1007/978-1-4471-6413-5_2

40. Schorch, M., Wan, L., Randall, D.W., Wulf, V.: Designing for those who are overlooked - insider perspectives on care practices and cooperative work of elderly informal caregivers. In: Proceedings of the CSCW 2016, pp. 785–797. ACM Press, New York (2016)

41. Eschler, J., et al.: Shared calendars for home health management. In: Proceedings of the CSCW 2015, pp. 1277–1288 (2015)

42. Bødker, S., Grönvall, E.: Calendars: time coordination and overview in families and beyond. In: Bertelsen, O., Ciolfi, L., Grasso, M., Papadopoulos, G. (eds.) ECSCW 2013: Proceedings of the 13th European Conference on Computer Supported Cooperative Work, 21–25 September 2013, Paphos, Cyprus, pp. 63–81. Springer, London (2013). https://doi.org/10.1007/978-1-4471-5346-7_4

43. Dwivedi, Y.K., Shareef, M., Simintiras, A., Lal, B., Weerakkody, V.J.: A generalised adoption model for services: a cross-country comparison of mobile health (m-health). Gov. Inf. Q. **33**, 174–187 (2018)

44. Neyer, F.J., Felber, J., Gebhardt, C.: Entwicklung und Validierung einer Kurzskala zur Erfassung von Technikbereitschaft. Diagnostica **58**, 87–99 (2012)

45. Porst, R.: Pretests zur Evaluation des Fragebogen(entwurf)s. In: Fragebogen. SS, pp. 189–205. Springer, Wiesbaden (2014). https://doi.org/10.1007/978-3-658-02118-4_15

46. Klein, B.: Neue Technologien und soziale Innovationen im Sozial- und Gesundheitswesen. In: Soziale Innovation, pp. 271–296. VS Verlag für Sozialwissenschaften, Wiesbaden (2010)
47. Eggert, S., Sulmann, D., Teubner, D.C.: ZQP-Befragung: Einstellung der Bevölkerung zu digitaler Unterstützung in der Pflege. ZQP Stiftung, Berlin (2018)
48. Ponnala, S., Weiler, D.T., Gilmore-Bykovskiy, A., Block, L., Kind, A.J., Werner, N.E.: Towards an understanding of informal care networks of persons with dementia: perceptions of primary caregivers. Proc. Hum. Factors Ergon. Soc. Annu. Meet. **62**(1), 561–562 (2018)
49. Peek, S.T.M., et al.: Older adults' reasons for using technology while aging in place. Gerontology **62**, 226–237 (2016)
50. Fachinger, U.: Technikeinsatz bei Pflegebedürftigkeit. In: Jacobs, K., Kuhlmey, A., Greß, S., Schwinger, A., Klauber, J. (eds.) Pflege-Report 2017: Schwerpunkt: Die Versorgung der Pflegebedürftigen, pp. 83–88. Schattauer, Stuttgart (2017)
51. van Wieringen, M., Broese van Groenou, M.I., Groenewegen, P.: Impact of home care management on the involvement of informal caregivers by formal caregivers. Home Health Care Serv. Q. **34**, 67–84 (2015)
52. Jeon, E., Park, H.: Factors affecting acceptance of smartphone application for management of obesity. Heal. Informatics Res. **21**, 74–82 (2015)
53. Silow-Caroll, S., Smith, B.: Clinical management apps: creating partnerships between providers and patients. Commonw. Fund. **30**, 1–10 (2013)
54. Wetzstein, M., Rommel, A., Lange, C.: Pflegende Angehörige – Deutschlands größter Pflegedienst. GBE kompakt. 6 (2015)
55. Statistisches Bundesamt: Pflegestatistik 2015 - Pflege im Rahmen der Pflegeversicherung Deutschlandergebnisse., Wiesbaden (2017)
56. Willard, S., Cremers, G., Man, Y.P., Van Rossum, E., Spreeuwenberg, M., De Witte, L.: Development and testing of an online community care platform for frail older adults in the Netherlands: a user-centred design. BMC Geriatr. **18**, 1–9 (2018)

Stress-Sensitive IT-Systems at Work: Insights from an Empirical Investigation

Michael Fellmann[1] ⓘ, Fabienne Lambusch[1(✉)] ⓘ, and Anne Waller[2]

[1] University of Rostock, Albert-Einstein-Str. 22, 18059 Rostock, Germany
{michael.fellmann, fabienne.lambusch}@uni-rostock.de
[2] University of Osnabrück, Katharinenstr. 3, 49069 Osnabrück, Germany

Abstract. High workload, complex and knowledge-intense tasks as well as increased expectations in respect to flexibility and timeliness give rise to work intensification. This can lead to permanent stress causing serious health problems. Thus, it is a major concern to take measures against stress in order to maintain workers' health and productivity. While information technology provides great potential to mitigate work-induced stress, preferences of workers regarding IT-based assistance are largely unknown. Against this research gap, we conducted a quantitative study on the acceptance and feasibility of implementation options for stress-sensitive systems. Our results are intended to inform future research in the design and development of such systems.

Keywords: Work stress · Information systems · Stress interventions · Quantitative research · User acceptance

1 Introduction

Work stress is a major challenge especially in developed countries. Globalization and the increasing use of information and communication technologies have a far-reaching impact on developments in the working world [1]. One of the aspects is the increasing competitive pressure due to globalization. In addition, the working population is more often expected to be highly flexible and mobile [2]. With the extensive use of information and communication technology, multitasking and interruptions of the workflow are part of everyday life in numerous companies [3]. Due to the fast transfer of tasks via email, users are faced with the difficulty of processing tasks faster and continuously viewing and prioritizing their email inbox. Projects in the enterprise can be carried out faster by the fact that spatial and temporal borders are overcome. Such developments pose new challenges and increase the stress for the responsible individual [4]. Studies prove a high work intensity. In the EU, for example, 33% of workers report to work at high speed about three-quarters of their work time and 10% of workers even report to 'never' or 'rarely' have time to do their job [5]. Such working situations can cause stress. The term *stress* describes strain that can affect the organism [6]. Stress is triggered by so-called stressors that activate the organism by means of challenge, threat, or harm. For stressors that are considered dangerous, the available resources to deal with the situation are assessed. If the resources are appraised inadequate, stress occurs [7]. A high exposure to stress can lead to overstrain, which presents a significant hazard

© Springer Nature Switzerland AG 2019
W. Abramowicz and R. Corchuelo (Eds.): BIS 2019, LNBIP 354, pp. 284–298, 2019.
https://doi.org/10.1007/978-3-030-20482-2_23

for health and may be e.g. resulting in burnout [8]. Therefore, it is becoming increasingly important for companies to take preventive measures against stress at the workplace in order to maintain workers' health, reducing sick leave, and avoid loss of overall productivity of the company. With the ongoing development in sensor technology and the IT-systems in companies there are various potential approaches to augment IT-based systems with stress-sensitive features. However, for stress-sensitive systems to be successful and useful in everyday working life, it is of utmost importance to take into account the preferences of workers regarding stress measuring methods, purposes of data processing, system feedback, and interaction with the system. Empirical research that systematically analyses user preferences for all the above mentioned questions is still missing. Hence in order to inform the design and development of stress-sensitive systems, we contribute to the existing research with a quantitative study answering the following research questions:

- RQ1: What is the acceptance of different stress-related measuring methods?
- RQ2: For which purposes is the processing of personal data accepted?
- RQ3: What is the preferred system feedback?
- RQ4: What is the preferred interaction mode with a stress-sensitive system?

The remainder of this article is structured as follows. In Sect. 2, we describe related work. Section 3 deals with the design space of stress-sensitive IT-systems because this motivates our empirical investigation. In Sect. 4, we then present the design and results of our empirical investigation. The results of the questionnaire used are described in five subsections according to the respective contents. In the last section, the results are discussed and conclusions are drawn.

2 Related Work

First of all, the topic of IT-supported personal stress management deals with the concept of stress. Therefore, *established theories about stress* are related. Established theories (e.g. [9–11]) deal with important aspects of the phenomenon and describe the cognitive processes associated to stress as well as the result of stress such as strains. Further, a review of organizational stress theories is conducted by Sonnentag and Frese [12]. While such works form a valuable underpinning, they predominantly focus on the phenomenon stress itself rather than on how to promote stress management or coping.

Managing stress in IT-contexts or IT-induced stress is an active research field. Coping with such stress has been investigated in early works (e.g. [13, 14]). More recently, approaches for managing IT-induced stress have emerged and are broadly referred to as research regarding *technostress* (e.g. [15–17]). Moreover, a blueprint for the technostress-aware design of information systems has been developed [18]. While the research community around technostress is mainly concerned with the management of stress *caused by IT* and the *design of information systems*, our main objective is to shed light on the *preferences* of users *to manage stress with the help of IT*. Hence, technostress as a research area is complementary to our research goal and our results can inform the technostress research community. Regarding more holistic *assistance systems for IT-supported personal stress management and their acceptance*, only a few works exist so far. Most notably, tools such as a workload monitor and an e-coach for activity recommendation (NiceWork eCoach) have been developed in the context of the SWELL project [19]. While sophisticated prototypes have been developed and empirically tested for their effectiveness, there is still not much research available in regard to an empirical analysis about the willingness of participants to provide data and to answer questions which is a necessary precondition for IT-supported stress management. In this direction, a first study on the needs for mobile coaching has already been conducted [20] as well as a comparison of specific sampling methods for stress [21]. However, none of these studies focuses in detail on the willingness of users to utilize the automatic stress measurement techniques proposed in the literature and on the preferences for certain system feedback and interaction types.

3 Questions on How to Design Stress-Sensitive IT-Systems from an End-User Perspective

Fundamentally, an information system acquires data (input), processes it and ultimately delivers results (output) to the user or other systems. In regard to stress-sensitive IT-systems, data acquisition means to capture relevant data about the situation, e.g. the stress-level of the user and other context data. Data processing means to analyze and interpret this data. Delivery of results entails to either directly provide information to the user or to leverage the results in order to provide support e.g. in the form of system adaptations [18] or process adaptations [22]. An example would be a messaging system that increasingly shields the user from new messages of specified types such as newsletters or advertisements depending on the current stress level of the user. These messages could be delivered later on if the user is in a more relaxed state.

In the context of data acquisition (input), different information channels exist that reveal data useful to infer the current user situation. Among these channels are physiological parameters that can be recorded through methods like eye tracking, skin conductance measurement, or electrocardiogram [23]. Furthermore, communication data such as received and sent email messages, phone calls or other communication data such as instant messaging can be analyzed. In addition, document and calendar data might serve as a proxy to determine the workload and engagement of a user at the workplace. A prominent device for data collection, particularly in a mobile context, is the smartphone. This type of device can monitor various stress-related data ranging from steps through phone calls to appointments [24]. The widespread use of smartphones makes them a valuable part of data collection. While the aforementioned channels refer to (semi-) automatic data collection, another approach is to ask the user about the level of stress via self-assessment questionnaires, e.g. by the Perceived Stress Scale [25]. All the different data collection options have in common that their feasibility depends on the characteristics of the workplace and on the acceptance of the user. We provide empirical insights on both of these aspects in Sect. 4. In the context of data processing and analysis, two modes of inquiry can be distinguished: First, *descriptive* statements about what currently is the case could be derived. Such statements can be used for personal insights as well as to inform colleagues about the current situation (e.g. that planning for a meeting is not feasible due to workload, even if the calendar has free time slots), or even to provide feedback for supervisors. Second, *prescriptive* statements focusing on what should be done can be derived that are grounded in knowledge about the current state. An example for this would be recommendations for improving the individual style of working (e.g. blocking distractions during the personal biological "prime time" in order to preserve this time for important tasks) or for improving the work organization for entire teams (e.g. distributing tasks depending on the employees' workload). Regardless of the statement type, a decision has to be made in regard to privacy. It has to be determined which types of analyses are allowed (also from a legal perspective) for personal use only, implying a private mode of analysis; and which analyses are allowed for groups of employees implying a public mode of analysis requiring effective anonymization mechanisms. In order to learn about user preferences, we deliberately did not consider legal aspects, but asked in our study about the acceptance and valuation of a wide range of purposes of descriptive and prescriptive analytics.

In the context of results delivery (output), a system can provide results either upon request thus being *reactive*, or it can provide results without an explicit request thus being *proactive*. Since results may be used for interventions, proactive delivery of results can be divided further into information only and autonomous execution. In the former case, the user is notified about possible interventions which then have to be executed manually. In the latter case, the system directly adapts itself according to the data processing results. An example for this has already been introduced in the form of a stress-sensitive message filtering and delay mechanism.

4 Empirical Investigation

The aim of the study was to gain insight into the field of research using a larger sample under standardized conditions. For this purpose, a quantitative research approach was chosen with an online survey. The online questionnaire consisted largely of closed questions. In addition, semi-open questions with an additional free text field for "Other" was used and the last question in the questionnaire was modeled as an open question for final comments. In order to gain initial experiences systematically, the representativeness, such as the number and composition of the sample, played a subordinate role in the survey [26]. The questionnaire was distributed in German language through several channels. For the study, the complete case analysis (CC) was used, which means that only complete data sets were used for the evaluation [27]. In total, 103 complete records were received and analyzed. The following sections describe the results of the survey. For a better overview, the questionnaire was divided into five different parts. At the beginning of the questionnaire, an introductory text motivated the topic of stress-sensitive information systems at work and described that such systems shall recognize and counteract stress in real time during the work time. The first questions were used to collect data on gender, age and current employment. The second part of the survey refers to the general work situation of the respondents. Questions were asked about job status, working hours and technology usage, among other things. Furthermore, participants should indicate their situation related to sources of stress, e.g. on workload and time pressure. The other parts of the survey relate to possible components of a stress-sensitive system. Thus, the third section covers the acceptance and feasibility of several measurement methods in the workplace. In the fourth section of the survey, the respondents should state for whom and for what purpose they would allow the processing of personal data. Additionally, the participants could choose the desired type of system feedback and express concerns about a stress-sensitive system. The last section is meant to determine the desired interaction mode of the system, or rather whether a stress-sensitive system should provide support automatically, semi-automatically, or only manually. At the end of the questionnaire, participants had the opportunity to comment on the survey.

4.1   Characteristics of Survey Participants

At the beginning of the questionnaire, data were collected on gender, age and current employment. Among the 103 participants, 44.7% were female and 55.3% male. The age varied between 19 and 67 years. Nearly half of the respondents were between 25 and 31 years old. The question "Do you work?" had the options of "yes", "no", and "student with part-time job/internship". 92 persons (89.3%) answered yes, two persons no, and nine respondents stated they were students with a part-time job or an internship. As the next part of the survey focused the persons' everyday working life, the use of a skip function led participants who did not work to the third part of the survey.

4.2   General Work Situation

The first question of this part concerned the professional status of the working respondents (n = 101). Employees accounted for the largest proportion at around 60%. The proportion of freelancers, officials, managers and students/interns ranges between 10.9% and 8.9%, whereas the proportion of apprentices is only 0.99% (1 person). Furthermore, it was asked for the company's sector. The most commonly represented sectors were "services" (26.7%), "education" (22.8%), and "banks/insurance" (15.8%).

The participants should also indicate their average working hours and the average working hours at a computer workstation. Most participants worked seven hours or more a day (94%). The mean value is 8.5 working hours a day. It is noticeable that 60.4% of the participants, who worked on average 7–8 h a day spent the same amount of time on a computer workstation on average. The median of the average working hours spent at a computer workstation is between 6 and 7 h and thus, many participants spent a large part of their average working day at a computer workstation. This may be beneficial for the imagination and assessment of IT-based, stress-sensitive systems. Another question concerned the use of a smartphone at the workplace for professional purposes. This is important for various measurement methods that can be carried out via a smartphone, but also in connection to the permanent accessibility of the worker. Occupational smartphone use shows an almost balanced ratio, with 52.5% of respondents confirming a use. In a next question, 42.6% of the respondents stated that they are also expected to be available in their free time.

At the end of this survey part, a table should be completed with seven questions as rows and answer options as columns (results in Table 1). The questions were intended to give an indication of the working atmosphere perceived by the participants. The answer options were subject to a ranking or ordinal scale [28] in a range of five levels.

Table 1. Perceived work situation on a five level rating scale

n=101	Strongly agree	Agree	Partly/partly	Dis-agree	Strongly disagree	Mean
1. I have enough time to do my work.	10.9%	25.7%	46.5%	14.9%	2.0%	2.71
2. My everyday work is stressful.	13.9%	33.7%	48.5%	4.0%	0%	2.43
3. My daily workload is high.	20.8%	41.6%	31.7%	5.0%	1.0%	2.24
4. I can use my knowledge at work.	40.6%	49.5%	8.9%	0%	1.0%	1.71
5. I consider interruptions by emails or phone calls annoying.	15.8%	28.7%	33.7%	18.8%	3.0%	2.64
6. The working atmosphere at my work is good.	30.7%	47.5%	17.8%	2.0%	2.0%	1.97
7. Supervisors inform employees adequately about developments & decisions.	11.9%	34.7%	26.7%	23.8%	3.0%	2.71

There is a slight tendency of the workers to agree in having a stressful working day (mean: 2.43), a high daily workload (mean: 2.24), and to be disturbed by emails/phone calls (mean: 2.64). These statements tend to be in line with the basic idea of this research, namely that everyday working life is becoming more stressful and the workload ever higher, so that the participants might relate well to the fictitious scenarios of the survey. In comparison, the result of the statement "I have enough time to do my work" shows a mean value of 2.71, which tends to be slightly opposite to the answers for the second and third statement. A possible explanation could be that the stressor is due to factors outside the employee's work responsibilities. On average, knowledge can be used and the working atmosphere is good. The results of the statement for sufficient information by supervisors show only a slight tendency for approval.

4.3   Acceptance of Measuring Methods for Stress Detection

The focus of the third section was on the assessment of measurement methods and thus on answering **RQ1**. An introductory text pointed out that the following methods for stress recognition shall contribute to a targeted stress reduction and the extent of data processing is intended to be self-determined. The non-employees returned to the survey in this part, so that all 103 participants had the opportunity to answer the questions. Several methods of state detection were presented with explanatory examples and grouped according to the produced data in order to increase clarity. The same scheme was used to query for the groups whether the participants consider a certain method feasible at their workplace and whether they would agree with such a measurement. Table 2 below shows the assessments by the participants. For better visualization, the votes for each method have been colored in different shades of red and green (yes = green, no = red). The color gradient depends on the strength of the vote which in turn shows at a glance the strength of approval or rejection of the respective methods.

Table 2. Acceptance and appraised feasibility of methods for stress detection

n=103		Is this data collection method feasible at your workplace?		Would you agree to the implementation of this method?	
		Yes	No	Yes	No
Neurophysio-logical data	1. Eye tracking	35.9 %	64.1 %	47.6 %	52.4 %
	2. Skin conductance measurement	32.0 %	68.0 %	34.0 %	66.0 %
	3. ECG (electrocardiogram)	31.1 %	68.9 %	41.7 %	58.3 %
Communi-cation data	4. Email / Instant messaging	43.7 %	56.3 %	60.2 %	39.8 %
	5. Phone data	43.7 %	56.3 %	54.4 %	45.6 %
Documents & Calendar	6. Appointment calendar / Work plan	50.5 %	49.5 %	74.8 %	25.2 %
	7. Edited documents	43.7 %	56.3 %	63.1 %	36.9 %
	8. Self-assessment questionnaire	57.3 %	42.7 %	81.6 %	18.4 %

With regard to the question of whether the participants would agree with the measuring methods presented, a slight increase in acceptance from the group of neurophysiological measurement methods (1–3) to self-assessment by questionnaire (8) can be observed. Neurophysiological measurement methods would be the least accepted ones, especially skin conductance measurement is rejected by many participants (66%). The collection of communication data receives higher support and an even greater acceptance can be seen for monitoring edited documents (63.1%) and calendar or work schedule entries (74.8%). The greatest popularity is achieved by the option of self-assessment through questionnaires with 81.6%. A similar course of consent can be found in answering the question of how to carry out such measurements at the workplace. However, a clear trend can only be seen in the case of neurophysiological measurement methods, where more than 64% of respondents deny any possibility of measurement. Furthermore, participants could indicate whether they would agree to the collection of certain smartphone data, even if they did not use a smartphone for their work. In total 76 participants answered the smartphone related questions. As can be seen in Fig. 1 below, many participants would agree to collect information about the data connection. This allows to determine whether a person is within the company network or works in the field. Relatively few participants agree to the recording of running apps or the GPS position.

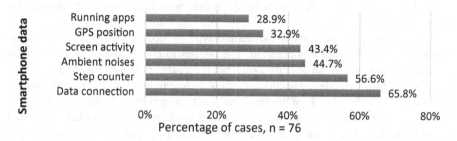

Fig. 1. Acceptance of measuring methods utilising smartphone data

4.4 Data Analysis and System Feedback

The fourth section of the survey was designed to determine the purposes for which a person would consent to the processing of personal data (**RQ2**) and what the feedback from the system should look like (**RQ3**). First, it should be clarified for what purpose the participants would allow the processing of personal data (see Fig. 2). This determines the extent to which the system should pass on information about a user's situation. All 103 participants answered this question, but since multiple choices were possible, a total of 211 answers were collected.

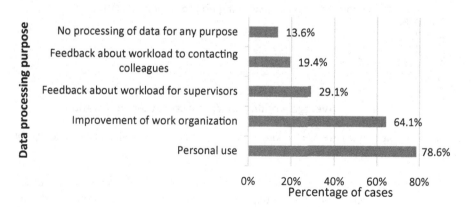

Fig. 2. Acceptance of data processing purposes

It can be seen from Fig. 2 that the information should mainly be used for personal purposes (78.6%), but there also seem to be a desire for an improvement of the work organization (64.1%), e.g. with a better distribution of work tasks in stressful situations. Still 29.1% of the participants would provide their supervisors information on their workload, but only 19.4% would allow their colleagues to access this data. Overall, the results show that data collection and analysis for a stress-sensitive system would be allowed by most participants, but stress-related information should just be passed to a limited group of people and for selected purposes. However, still 13.6% of the participants would not agree with any form of data processing. Due to a skip function in the questionnaire, these participants were forwarded to the next part of the questionnaire related to the user interaction. Subsequent questions in this part were answered by a total of 89 participants.

The next question was about the desired type of feedback (cf. Fig. 3). Multiple answers were allowed for this question. The options to choose were either a warning message on the screen or an audible signal, but it was also possible to write down other ideas in a textbox provided. The answers show a clear trend to give a warning on the screen (82%). Only 16.9% of respondents want an acoustic signal. Searching for correlations, a connection to the indicated screen working hours of participants could be found. In the group of people whose work time at a computer workstation is up to four hours, the desire for audible feedback from the system stands out in comparison to groups of people spending more time on the screen. The proportion in this category is 36.8%, while with longer working hours the demand for acoustic feedback in each case is 16% and far less. Furthermore, it can be seen that with a daily work time of five or more hours at a computer workstation an increase of the choices for "Other" is recorded. It can be deduced from these responses that, for example, feedback in the form of an email would also be desirable, so that the employee would not be disturbed by direct feedback in the workflow and can decide when to read it. Another suggestion was the use of a smartwatch for feedback. This could also be a good approach to combine measurement (e.g. heart rate data) and feedback in one device.

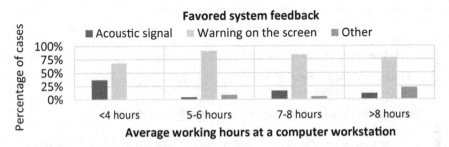

Fig. 3. Relation between average working hours at a computer workstation and desired type of feedback

Next, the participants could state whether the system should refer to the current state or to a predicted future state (e.g. if the system recognizes through task and appointment entries that the workload for the next day is too high). Primarily, feedback on the current status is desired (88.8%), while just over half of respondents (55.1%) would also like to receive feedback about a predicted state. Finally, there was a possibility that respondents raised concerns about data collection. Due to possible multiple choices, 106 replies were submitted in total. Privacy concerns are indicated by 40.4% of the participants and 37.1% agree with the thought that it could be impracticable. Almost 1/3 of the participants (31.5%) stated they had no concerns about data collection. Nine persons (10.1%) used the textbox for "Other". The entered free texts also refer to a difficult or complex implementation or to the point of data protection.

4.5   Interaction Mode and Final Remarks

The last part of the questionnaire addressed how to interact with a stress-sensitive system (**RQ4**) and the participants had the opportunity to comment on the survey and its topic. Figure 4 shows the results regarding the interaction with the system.

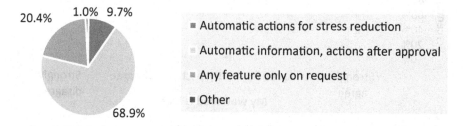

Fig. 4. Favored interaction with the system

Full automation, in which the system takes action without being asked, is desired by only 9.7% of participants (n = 103). Although 68.9% of participants would like the system to automatically inform them about measured stress data, they want stress-reducing measures to be taken only after approval. 20.4% of participants would like the system to even become active only on request. A slight trend in the response behavior to the desired interaction can be seen in connection with the responses to the earlier statements of a stressful working day and high workload (see Fig. 5). It is noticeable that the desire for a system that autonomously reduces sources of stress in the background is only present in the first three categories for the statements of a stressful working day and a high workload. This means that anyone wishing to have a fully automated stress reduction system has described the daily work as at least partially stressful and the workload also as at least partially high. The strongest form of automation of a stress-sensitive system seems therefore to be related to the level of stress in everyday working life. In addition, the first graph in Fig. 5 shows a predominantly increasing percentage from "Strongly agree" to "Disagree" in the desire for the system to become active only on request. Of those who strongly agreed to a stressful working day, 14.3% would want system activity only on request, while of those disagreed, it is even 50%. In at least half of the cases in each category of agreement, the semi-automatic version of the system is preferred.

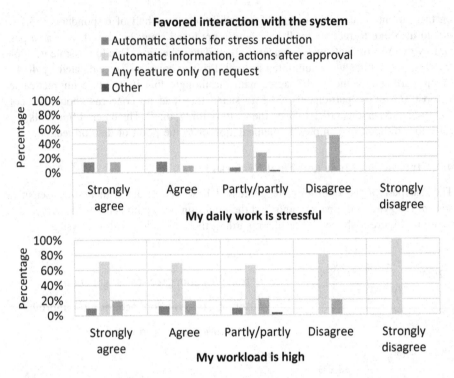

Fig. 5. Relation between perceived level of stress/workload and desired user interaction

Another interesting point is how people from different age groups responded to the question about interaction. This connection is shown in the following Fig. 6.

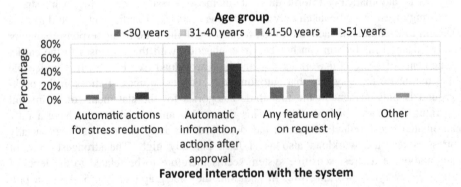

Fig. 6. Relation between age groups and user interaction

While just under 77% in the under-30 age group favor partial automation and only 16% want to request the system to become active, the response behavior of older age groups is a bit different. 50% of participants over 51 years prefer a partial automation and even 40% a system activity only on request. However, in both age groups the proportion of people who prefer autonomous actions of a system (response 1) is less than 10%. In contrast to these groups, the proportion of 31–40 year-olds stands out, of which 22.7% wish that the system tries to reduce sources of stress in the background. At the end of the survey, participants had the opportunity to comment on the questionnaire and its topic. Six participants took this opportunity. In the following, brief remarks will be made on tendencies that can be seen in the comments. There is a clear concern that information on the stress level could be detrimental to colleagues or supervisors. An interpretation of weaknesses that could endanger one's employment is a concern of several people. The privacy aspect also plays a significant role in the comments. It was mentioned that anonymizing the data should help to ensure that any use of data that is not in the user's interest would become unusable and that the data of the individual would thus be protected from misuse. Another aspect mentioned are deadlines. In some situations, where stress is unavoidable, e.g. as a result of a deadline, remarks of a stress-sensitive system could be inappropriate and even obstructive.

5 Discussion and Conclusions

With advancing developments like an intensification of work, people are often faced with high work pressure and stress. Permanent stress can lead to overstrain, which poses a significant health risk for the individual. Thus, reducing and managing work stress is one of today's major challenges. Advances in sensor technology offer the potential to augment IT-systems with stress-sensitive features, but for these features to be effective it is necessary to take into account the needs and preferences of intended users. This article focuses on the acceptance and feasibility of implementation options for IT-based assistance against individual stress in an organizational context. To this end, a survey was conducted among a convenience sample of 103 people. We have to acknowledge as a limitation of our research that our results cannot be generalized beyond this sample. However, in line with [26], in order to gain *initial* experiences on a subject in a systematic way, representativeness in terms of the number and composition of the sample plays a subordinate role. The merit of our investigation is nonetheless that we provide *preliminary insights and tendencies* on the acceptance and feasibility of implementation options for stress-sensitive systems that can be tested through further research. Since the participants answered the questions on a theoretical basis, an interesting point for future research would also be to test the acceptance in practice when providing a prototype to participants. In the following, we summarize and reflect on our most important findings in regard to the defined research questions RQ1–4.

Regarding the **acceptance of measuring methods (RQ1)**, the *most popular variant is the self-assessment by questionnaire* (81.6%). This could be linked to the fact that people do not like to leave it to measurements to determine their stress levels and instead prefer to validate these measurements by their own assessments. However, a sole assessment by questionnaire is not a reliable method of stress recognition. For example, people with pronounced ambition or perfectionism may be too influenced by their values to be able to set limits. Such attitudes can lead to overstrain, which often may be hardly perceived due to underlying desires and fears [8]. Also, the *utilization of calendar, document, and communication data is quite accepted.* The recording of an employee's calendar and work plan entries (74.8%) is quite accepted. The acceptance for the analysis of processed documents and email/messaging data with above 60% is a bit lower, and for an analysis of communication data still 54.4%. The slight difference in the acceptance could be explained by a possible higher degree of discomfort with the analysis of telephone conversations, especially in qualitative form (e.g. by word recognition), than with the analysis of written text. Perhaps people are also most interested in the analysis of work plan and calendar data, because these methods are more directly targeted to time management and are not perceived that private. *Neurophysiological measuring methods are least accepted.* Possibly this is due to the fact that these methods such as ECG, eye-tracking, or skin conductance measurement put the most noticeable restrictions on the subjects, e.g. in regard to the required use of additional devices (e.g. wearables; camera) and the freedom to move (e.g. through wearing; limited camera area). Also, data from these sources may be perceived as too private and hardly controllable. In summary, our results imply that *integrating a user's own assessment of workload and stress into the measurement methods is greatly appreciated and could be complemented in particular by calendar, document, and quantitative communication data.*

Regarding the **purposes for which processing of personal data is accepted (RQ2)**, *data processing is accepted predominantly for personal use* (78.6%), e.g. to provide personal feedback on working habits, but there also seem to be a desire *for an improvement of the work organization* (64.1%), e.g. with a better distribution of work tasks in stressful situations. The emphasis on personal use might be due to the fear of an abusive use of personal stress data. Also possible competitive behavior between colleagues in the company may be a reason for fewer desire to provide supervisors or colleagues information on the stress level, because identified high loads could be interpreted as weakness. Accordingly, when developing a stress-sensitive IT system, transparency and the option to choose a private mode is important. The privacy of the user must be taken very seriously and legal conditions such as those imposed by the General Data Protection Regulation (GDPR) must be adhered to (e.g. by processing data only locally on behalf of one single user on his or her personal device).

Regarding the **system feedback (RQ3)**, it depends a lot on which working environment an employee is in. Since *information and recommendations are predominantly desired for private purposes*, messages in this direction that are displayed on a screen should be designed in a decent way to reflect the private nature of information. In addition, acoustic signals are also heard by colleagues or other persons and these could in turn draw conclusions about the personal degree of workload and stress, hence they should be optional. The system should therefore recognize the current situation of the user in real time and then silently provide information on the stress status, possibly also in connection with direct countermeasures for stress reduction such as micro-breaks or relaxation. The potential of newer devices like smartwatches to provide users information in a more decent way could be tested in a work context. *Much to our surprise, the survey participants showed only moderate interest in a prediction of future stress states* (recognisable e.g. through closely staggered appointment calendar entries).

Further investigations are necessary to determine the reasons. Perhaps people consider such predictions as inaccurate or they worry about getting more stressed by information about stress in advance.

Regarding the **preferred interaction mode (RQ4)**, it turns out that many of the survey participants want a possibility to influence the measures of the system and do not want decisions by the system that they cannot influence. A completely automated system is probably also still unimaginable for many due to its novelty. The strongest form of automation of a stress-sensitive system seems also to be related to the level of stress in everyday working life, as a higher perceived stress level in everyday working life seem to lead to a higher desire for automation of a stress-sensitive system. However, it is apparent that a semi-automated system, in which the user first has to agree to counter-measures, is the most desired version All in all, there is a *clear interest in predominantly semi-automated systems, which inform the user about current workload and stress, but only carry out countermeasures to reduce stress after approval.* This means that there is a general interest in assistance for stress reduction, but that the user should remain responsible for decisions to initiate any measures. An aspect mentioned in the comments are deadlines. In some situations, where stress is unavoidable, e.g. as a result of a deadline, remarks of a stress-sensitive system could be inappropriate and even obstructive. For these cases, additional settings for such a system should be considered. It would be conceivable to deactivate the interactivity of the system so that a user is not interrupted by warnings or restricted by stress reduction measures adopted by the system.

All in all, high workload and work pressure as well as the negative consequences they provoke such as burnout are among the most urgent problems of today's working world. Therefore, incorporating stress sensitivity into IT-systems is an important aspect that not only has the potential to maintain employee health, but also to promote long-term productivity and well-being. We hope that our empirical insights inspire more research on this important topic and help to delineate the design space for stress-sensitive IT-systems at work.

References

1. Walter, N., et al.: Die Zukunft der Arbeitswelt. Auf dem Weg ins Jahr 2030. Bericht der Kommission "Zukunft der Arbeitswelt" der Robert Bosch Stiftung (2013)
2. Siegrist, J.: Burn-out und Arbeitswelt. Psychotherapeut **58**(2), 110–116 (2013)
3. Freude, G., Ullsperger, P.: Unterbrechungen bei der Arbeit und multitasking in der modernen Arbeitswelt - Konzepte, Auswirkungen und Implikationen für Arbeitsgestaltung und Forschung. Zbl Arbeitsmed **60**(4), 120–128 (2010)
4. Kielholz, A.: Online-Kommunikation: Die Psychologie der neuen Medien für die Berufspraxis. Springer, Heidelberg (2008). https://doi.org/10.1007/978-3-540-78393-0
5. Eurofound: Sixth European Working Conditions Survey – Overview report (2017 update). Publications Office of the European Union, Luxembourg (2017)
6. Lohmer, M., Sprenger, B., von Wahlert, J.V.: Gesundes Führen: Life-Balance versus Burnout im Unternehmen. Schattauer, Stuttgart (2017)
7. Lazarus, R.S., Folkman, S.: Stress, Appraisal, and Coping. Springer, New York (1984)

8. Scharnhorst, J.: Burnout. Präventionsstrategien und Handlungsoptionen für Unternehmen. Haufe Lexware (2012)
9. Folkman, S., Lazarus, R.S.: If it changes it must be a process: study of emotion and coping during three stages of a college examination. J. Pers. Soc. Psychol. **48**, 150–170 (1985)
10. Folkman, S., Lazarus, R.S., Dunkel-Schetter, C., DeLongis, A., Gruen, R.J.: Dynamics of a stressful encounter: cognitive appraisal, coping, and encounter outcomes. J. Pers. Soc. Psychol. **50**, 992–1003 (1986)
11. Selye, H.: The Stress of Life. McGraw-Hill, New York (1956)
12. Sonnentag, S., Frese, M.: Stress in organizations. In: Handbook of Psychology, vol. 12, no. 2, pp. 560–592. Wiley, Hoboken (2013)
13. Benamati, J., Lederer, A.L.: Coping with rapid changes in IT. Commun. ACM **44**, 83–88 (2001)
14. Al-Fudail, M., Mellar, H.: Investigating teacher stress when using technology. Comput. Educ. **51**, 1103–1110 (2008)
15. Ayyagari, R., Grover, V., Purvis, R.: Technostress: technological antecedents and implications. MIS Q. Manage. Inf. Syst. **35**, 831–858 (2011)
16. Maier, C., Laumer, S., Weinert, C., Weitzel, T.: The effects of technostress and switching stress on discontinued use of social networking services: a study of Facebook use. Inf. Syst. J. **25**, 275–308 (2015)
17. Fischer, T., Riedl, R.: The status quo of neurophysiology in organizational technostress research: a review of studies published from 1978 to 2015. In: Davis, F.D., Riedl, R., vom Brocke, J., Léger, P.-M., Randolph, A.B. (eds.) Information Systems and Neuroscience, vol. 10, pp. 9–17. Springer, Cham (2015). https://doi.org/10.1007/978-3-319-18702-0_2
18. Adam, M.T.P., Gimpel, H., Maedche, A., Riedl, R.: Design blueprint for stress-sensitive adaptive enterprise systems. Bus. Inf. Syst. Eng. **59**(4), 277–291 (2017)
19. Koldijk, S.J.: Context-Aware Support for Stress Self-Management: From Theory to Practice (2016)
20. Harjumaa, M., Halttu, K., Koistinen, K., Oinas-Kukkonen, H.: User experience of mobile coaching for stress-management to tackle prevalent health complaints. In: Oinas-Kukkonen, H., Iivari, N., Kuutti, K., Öörni, A., Rajanen, M. (eds.) SCIS 2015. LNBIP, vol. 223, pp. 152–164. Springer, Cham (2015). https://doi.org/10.1007/978-3-319-21783-3_11
21. Atz, U.: Evaluating experience sampling of stress in a single-subject research design. Pers. Ubiquit. Comput. **17**(4), 639–652 (2013)
22. Fellmann, M., Lambusch, F., Waller, A., Pieper, L., Hellweg, T.: Auf dem Weg zum stresssensitiven Prozessmanagement. In: Eibl, M., Gaedke, M. (eds.) INFORMATIK 2017, 863-869. Gesellschaft für Informatik, Bonn (2017)
23. Dimoka, A., et al.: On the use of Neurophysiological tools in IS research: developing a research agenda for NeuroIS. Manage. Inform. Syst. Q. **36**(3), 679–702 (2012)
24. Gimpel, H., Regal, C., Schmidt, M.: myStress: unobtrusive smartphone-based stress detection. In: ECIS 2015 Research-in-Progress Papers (2015)
25. Cohen, S., Kamarck, T., Mermelstein, R.: A global measure of perceived stress. J. Health Soc. Behav. **24**(4), 385 (1983)
26. Bortz, J., Döring, N.: Forschungsmethoden und Evaluation für Human- und Sozialwissenschaftler. 4. überarb. Aufl. Springer, Heidelberg (2006)
27. Göthlich, S.E.: Zum Umgang mit fehlenden Daten in großzahligen empirischen Erhebungen. In: Methodik der empirischen Forschung. 3. Auflage, pp. 119–135. Gabler Verlag, Wiesbaden (2009)
28. Porst, R.: Fragebogen, 4th edn. Springer Fachmedien, Wiesbaden (2014). https://doi.org/10.1007/978-3-658-02118-4

Data Quality Management Framework for Smart Grid Systems

Mouzhi Ge[⊠], Stanislav Chren, Bruno Rossi, and Tomas Pitner

Faculty of Informatics, Masaryk University, 602 00 Brno, Czech Republic
{mouzhi.ge,chren,brossi,tomp}@muni.cz

Abstract. New devices in smart grid such as smart meters and sensors have emerged to become a massive and complex network, where a large volume of data is flowing to the smart grid systems. Those data can be real-time, fast-moving, and originated from a vast variety of terminal devices. However, the big smart grid data also bring various data quality problems, which may cause the delayed, inaccurate analysis of results, even fatal errors in the smart grid system. This paper, therefore, identifies a comprehensive taxonomy of typical data quality problems in the smart grid. Based on the adaptation of established data quality research and frameworks, this paper proposes a new data quality management framework that classifies the typical data quality problems into related data quality dimensions, contexts, as well as countermeasures. Based on this framework, this paper not only provides a systematic overview of data quality in the smart grid domain, but also offers practical guidance to improve data quality in smart grids such as which data quality dimensions are critical and which data quality problems can be addressed in which context.

Keywords: Smart grid · Data quality · Data quality problem · Smart meter

1 Introduction

Smart grids are developed to optimize the generation, consumption, and management of energy via intelligent information and communication technology. Its research involves smart meters, user-end smart appliances, renewable energy resources, digitalization in electricity supply networks, as well as new technologies to detect and react to the changes in electricity supply networks. As [1] stated, a smart grid reflects a combination between Information and Communication Technologies (ICT) and Internet of Things (IoT), whereby data services such as aggregation of sensor data and analysis of voltage consumption from smart meters [28] offer a foundation for the concept of "smartness". Since the quality of data can directly affect the output of data services, the security, quality, reliability, and availability of an electric power supply depends on the quality of data in the power system [19]. Thus, data quality has been considered as a

© Springer Nature Switzerland AG 2019
W. Abramowicz and R. Corchuelo (Eds.): BIS 2019, LNBIP 354, pp. 299–310, 2019.
https://doi.org/10.1007/978-3-030-20482-2_24

prominent issue in smart grids [8]. In a broader context, data quality has become a critical concern to the success of organizations [15]. Numerous business initiatives have been delayed or even canceled, citing poor-quality data as the main reason [16]. Therefore, data quality management can be regarded as an indispensable component in smart grid applications.

The current data quality problems in smart grid are addressed still in an ad-hoc style. For example, Chen et al. [8] focused on the outlier detection of electricity consumption data. Their solution tackles a specific quality aspect of electricity consumption data. However, this will obstruct practitioners to foresee the other data quality problems and delay the reaction on time for potential data quality problems. Thus, it is valuable to obtain a big picture of the different data quality problems in the smart grid network. Also, some of the data quality problems in smart grids may be interconnected. One data quality problem may be caused by another data quality problem. For example, an outlier in the electricity consumption data may be caused by missing data items or data attacks. Therefore, focusing on specific quality aspects can mislead the root causes of the data quality problems. Based on our review, there is a lack of a systematic framework for managing data quality in smart grids. Also, data quality is critical in the smart grid domain, as invoices of end users depend for example on the collected power consumption data.

In this paper, we propose a systematic data quality management framework for smart grids. It can not only profile a variety of data quality problems in the smart grid context, but also show how to categorize and organize the data quality problems based on data quality dimensions. In this framework, different data quality problems are identified and assigned to the related dimensions. It can therefore indicate which data quality dimensions are critical in the smart grid data quality improvement. Furthermore, the data quality problems assigned in the same dimension may need to considered together.

The remainder of the paper is organized as follows. Section 2 reviews the general data quality management and the state-of-the-art data quality research in Smart Grid. Section 3 identifies and summarizes a comprehensive set of the possible data quality problems in smart grid. Based on the identified data quality problems, Sect. 4 proposes a framework to categorize the data quality problems into established data quality dimensions. Finally, Sect. 5 concludes the paper and outlines the future research.

2   Data Quality and Smart Grid Research

Since the data quality problems are usually domain-specific, the importance of data quality dimensions may vary in different application domains. For example, Ge et al. [17] conducted a study to rank the overall importance of different data quality dimensions used in a variety of data quality studies. They further emphasized that prioritizing the importance of data quality can determine the focus of data quality improvement and management. Therefore, to find out which dimensions are important in the smart grid domain, assigning the data quality problems to dimensions can be used to facilitate the data quality measurement process.

There exists some research work that tends to classify the data quality problems in smart grids. for example, Chen et al. [8] proposed that the data quality issues in electricity consumption data can be divided into noise data, incomplete data and outlier data. Noise data refer to the data with logical errors or the data violating certain rules or specifications. These data that can in turn affect data analysis results. Incomplete data mean the missing values in the data sources, and outlier data are the data that deviate from standard data variation ranges. However, the scope of data in smart grids is broader than electricity consumption data. For example, [24] specified various types in smart grids such as sensor data, battery status data, and device downtime data. Therefore, the classification of the data quality problems in smart grids can be extended to a larger scale.

The smart grid domain encounters the Big Data Quality problems. Due to the massive number of smart meters, various sensors and other customer facilities, the smart grid network has been generating Big Data [11]. Zhang et al. [35] further described the big data characteristics in smart grids such as large amount of meter and sensor data (volume), real-time data exchange (velocity), and extensive data sources in a smart grid (variety). They further stated that data quality is a critical issue in processing the big smart grid data, where the data quality management is usually positioned in the data preprocessing phase. Zhang et al. [35] classified the big data quality problems by using three countermeasures, which are data integration, data cleansing, and data transformation. While data integration deals with the entity resolutions and data redundancy, data cleansing can be used to alleviate missing and abnormal data. Finally, data transformation serves to provide high-quality data formats for data analytics such as correcting data distribution and constructing new data attributes. This classification is especially designed for smart grid data analytics. In this paper, we will outline big and normal data quality problems in smart grids.

3   Data Quality in Smart Grid

When we discuss the term "data" in the context of smart grids, we cannot ignore the overall complexity of the infrastructure and the communication needs [10]. Due to the complexity, data in smart grids comes from a variety of sources, and can be structured, unstructured, but very often a mixture of both, making the analysis more complex [35].

3.1   Smart Grid Infrastructure

The smart grid infrastructure comprises several parts, each of them with different responsibilities regarding the energy and data transfer. The smart grid architecture depends on the standards used. According to the NIST standard, the smart grid has a hierarchical structure that includes the following domains: Wide Area Network (WAN) is responsible for communication between power generation plants, substations and transformer equipment. Neighbourhood Area

Network (NAN) serves as a bridge between customer premises and the sub-stations. This level focuses on the collection of data from the smart meters, which are aggregated by the data concentrator and further transferred to the data centers [10]. Furthermore, the customer premises network (CPN) consists of networks at the customer location. Depending on the type of customer, we can distinguish between Home Area Networks (HAN), Industrial Area Network (IAN) and Business Area Network (BAN). This layer enables communication between the smart meters, intelligent appliances and their connectivity to NAN. An overview of communication technologies in smart grids is shown in Fig. 1.

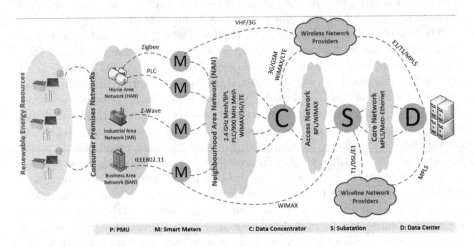

Fig. 1. Communication infrastructure in smart grids (from Al-Omar et al. [2])

3.2   Data in Smart Grid

Generally, smart grid data can be classified in three categories: measurement data (e.g., smart meters data), business data (e.g., customer data), and external data (e.g., weather data) [35]. In this context, we focus our analysis on *measurement data*, that is the type of data that can, more than other types, characterize SG data analysis needs. We focus in particular on data derived from two devices: Smart Meters [10] and PMUs [5,25].

In the smart grid infrastructure, there are two main components which produce the measurement data essential for the grid operation: smart meters and phasor measurement units (PMUs). *Smart meters* are devices which serve as replacements of traditional power meters installed at customer premises (e.g. households, industrial buildings, etc.). They record data about customer's power consumption (and possibly production if the customer utilizes renewable power sources). Smart meters enable two-way communication and a power distributor is also able use them to remotely control appliances such as water heaters. This becomes useful in various load management programs to balance the power flow in the grid. Besides the measurement data, smart meters are also able to report various events, for example meter failures, unexpected manipulation with the device or occurrence of over/under-voltage states [31]. On the other hand, *PMUs* are devices which measure phasor information in the power distribution, such as voltage and current. The PMUs collect the measurement data from many points in the power grid at very high frequency (up to 120 samples per second). The data are time synchronized based on the GPS radio clock. Measurement data are transmitted to various monitoring systems using them to analyze the current state of the power grid to discover potential stability issues.

There is a number of systems in smart grids that ensure reliability of the power supply and the availability of critical services and which rely on high quality data collected from smart meters or PMUs [9]: (1) Blackout Prevention Systems protect the grid from instabilities and failures. They cover the whole power grid, using the data from PMUs to obtain relevant information from the grid. (2) Supervisory Control and Data Acquisition Systems (SCADA) are one of the core systems of a Smart Grid that provide monitoring and support to operation activities and functions in transmission automation, dispatch centers and control rooms. In a SCADA system, a remote terminal unit collects data from smart meters or devices in a substation and delivers the data to a central Energy Management System. (3) Flexible Alternating Current Transmission Systems are responsible for reliable and secure transmission of power. They allow dynamic voltage control, increased transmission capability and capacity, and support fast restore of the grid after failure. (4) Feeder Automation Systems are responsible for the operation of medium-voltage networks including fault detection.

3.3   Data Quality Problems

Issues in data collected by smart grid devices are usually referred by literature as either "bad data" [29], "corrupted" [6], or "missing data" [6]. However, such definitions do not capture the diversified facets of smart grid data issues that are more refined in terms of specific issues. For example, Shishido and Solutions [32] discuss issues in smart meter data quality during the consumption data collection process. The main issues reported are duplicate items from the meter readings, zero record periods, and large spikes over periods of time. There are some issues in Smart Meters/PMU data that are peculiar of the smart grid context: non-trustful data points, data aggregation issues due to privacy concerns, timing issues with skewed timestamps of recorded events. We summarize the main Data Quality Problems in Table 1, as a series of issues that are derived from literature on smart meters and PMU data analysis.

Duplicate records from multiple devices (DQP1) mean that the same record is stored multiple times in the same way or with different values, causing duplication in the data [32]. A suggested strategy for the identification is cross-linking records across different devices to look for possible duplicated values, as well to search for repeating sequences [32].

Missing/incomplete data (DQP2) represents the case in which data recordings are missing for some periods of time, making this a problem of data imputation research [22]. Strategies in such cases go in the direction of linear

Table 1. Data quality problems in the smart grid context.

DQ problem	Description	Context	Countermeasures
DQP1. Duplicate data	Duplicate records from Smart Meter reading, can be caused by upgrading of Smart Meters (e.g., same reading from the old and new SM) [32]	SM	Cross-linking data from multiple devices and examining repeating sequences [32]
DQP2. Missing/ incomplete data	Some data can be expected to be available (e.g. regular smart meter reading) but due to some reasons (e.g. technical failure) they are not [32]	PMU/SM	Linear interpolation (short periods), creation of daily load profiles for historical patterns recreation (longer periods) [22,29]
DQP3. Zero Records Semantics	Detecting differences between data that was not transmitted/recorded by sensors and stand-by periods. All lead to difficulties in interpreting zero-valued ranges [29,32]	SM	Creation of daily load profiles [22,29], reasonability tests for allowed ranges and comparison of values from other devices [34]
DQP4. Data Outliers (out-of-range)	Large bursts (spikes), or low values compared to the average over a period of time [20,32]	PMU/SM	Reasonability tests for allowed ranges [34], application of anomaly detection algorithms, context-, collective-based [20,31]
DQP5. Measurement Errors	Datapoints that represent measurement errors due to hardware failures, signal interference, etc. [29]	PMU/SM	Reasonability tests for allowed ranges and comparison of values from other devices [34], signal analysis of smart meters for outliers detection [30]
DQP6. Non-trustful datapoints	Datapoints that were manipulated intentionally (e.g. data injection attacks: alter the measurements of SMs to manipulate the operations of the smart grid [7,23,27])	PMU/SM	Using historical data, statistical-based detection [23]
DQP7. Data anonymization	Aggregation of attributes/features for privacy preservation/anonymization can lead to issues for data analysis [12]	SM	Preserving data integrity for smart grid data aggregation, e.g. by hashing/signature checking against data tampering [26], Smart Metering data de-pseudonymization [21]
DQP8. Timing issues	Timing in which an event is recorded by PMUs/Smart Meters is not precise, causing difficulties in the integration of data, or in case of PMUs, wrong computations [13,34]	PMU/SM	Comparing values recorded by different systems, e.g. PMU and SCADA [34]

interpolation for short periods of missing records, or the creation of daily load profiles for historical patterns recreation in case of longer periods [22,29].

Zero record periods (DQP3) constitute a distinct case from the aforementioned missing/incomplete data scenario [32]. In this case, data is present but with zero recordings, making difficult to understand if such records were not recorded/transmitted, or missing values were due to some stand-by period [29].

There are different strategies that can be applied to understand the semantic of zero-record periods of time, like the creation of daily load profiles from smart meter data [22,29], or reasonability tests for allowed ranges and comparison of values from other devices for PMUs [34].

Data outliers or out-of-range values (DQP4) represent large bursts of data spikes or low values compared to the average values over periods of time [32]. Detection of these value ranges is part of the anomaly detection area, determining outliers based on context-, or collective-based algorithms [20,31]. For PMUs, reasonability tests for allowed ranges are important [34].

Measurement errors (DQP5) can represent a relevant issue for both smart meters and PMUs [29]. There can be many sources of such issues in smart grids data. According to Chen et al. [6], measurement errors can derive from smart meter problems, communication failures, equipment outages, lost data, interruption/shut-down in electricity use, but also components degradation and operational issues [3]. To address measurement errors, reasonability tests for allowed ranges and comparison of values from other devices can be used in the context of PMUs [34], while signal analysis of smart meters for outliers detection can be applied to smart meter data [30].

Non-trustful data points (DQP6) derive from potential cyber-attacks to the smart grid infrastructure. Such attacks do not only involve authentication issues, but also false data injection attacks to provide fake data-points as they were real recorded events [7,23]. The non-trustful data-points injected/modified by means of cyber-physical attacks, are meant to manipulate the overall operations in the SG by leading operators into false beliefs about the current state of the infrastructure [27].

Data aggregation issues due to privacy concerns (DQP7) come from the needs to preserve the privacy of data collected from smart meter readings. Some features collected by the different types of devices might be obscured or aggregated into other features, making the analysis process more difficult. Over the last years, many techniques have been developed to preserve the statistical properties of aggregated features [12], preserve data integrity from tampering [26], and algorithms that attempt at data de-pseudonymization [21]. While the removal of some features might be seen as a way to anonymize data from smart meters, this is however ineffective, as customers can be re-identified by other features [4].

Timing issues with skewed timestamps of recorded events (DQP8) can be an issue in both smart meters and PMUs. While in smart meters such errors might just involve issues in later data attribution between different devices [13], for PMUs such issues can involve subsequent wrong computations [34]. Comparing data from different devices can be a strategy to detect and correct timing issues, such as comparing timestamps from PMUs and SCADA systems [34].

Furthermore, there are two main peculiarities in smart grid data cleansing activities: (1) data are mostly generated from sensors, hardware-devices: root causes can be found in hardware failures, communication related problems [14]; (2) data cleansing in the data mining domain usually assume structural data, while in the smart grid context, mostly time-series approaches are needed for the identification of patterns/anomalies.

4   Data Quality Management Framework in Smart Grid

To propose a data quality management framework for smart grids, we have adopted the general data quality framework from Wang and Strong [33]. Thus, intrinsic, contextual, representational and accessibility are adopted to categorize the data quality concept. Further, the relations between data quality categories and data quality dimensions are also adopted from Wang and Strong [33]. However, since not all the data quality dimensions are important for smart grids, we have used the identified data quality problems to select the data quality dimensions in our framework. Therefore, our framework is intended to be domain-specific for smart grids. Based on the typical data quality problems that we revisited in the smart grid context, we derive the data quality framework in smart grid that classifies these data quality problems into dimensions, categories, as well as into contexts. This data quality framework is divided into five layers. The first layer from the top is the overall data quality in smart grid. This layer is usually used as one step in the whole big data analytics process e.g. before or after the data integration. The second layer divides the smart grid data quality into four quality aspects. Under each data quality aspect, it is the dimension layer that indicates which data quality dimensions are related to which smart grid data quality problems. Therefore, the third layer and fourth layer are data quality dimensions and specific problems. As data quality problems are derived from different contexts, the context in smart grids is the final layer.

It can be seen that there are seven data quality dimensions that are particularly important for the smart grid domain. These seven data quality dimensions are accuracy, consistency, timeliness, completeness, believability, accessibility, and interpretability. Accuracy is mainly defined as the data points falling into a normal range or interval. Thus, data outliers in smart grids belong to this dimension and detected data outliers can be considered as inaccuracy. The consistency dimension is used when there are different sources of smart grid data. In the smart grid domain, data can be generated from different devices. On one hand, this creates the data redundancy, on the other hand, the cross-reference approach can be used to validate the data consistency. Since time series analysis is usually used in smart grids, timing issues like wrong timestamps may cause problems to construct the time series data. Therefore, the timeliness dimension is to control if the data are recorded in a precise time or time interval. Completeness dimension involves the problems of missing and incomplete smart grid data. As there is a large number of devices in smart grids, the data completeness issue can be regularly caused by device malfunction. Although the believability dimension is not well discussed in other domains, trustful data are important in smart grids because the data manipulation in smart grid can be directly related to economical benefits. The accessibility dimension is related to the hardware and infrastructures in smart grid. Therefore, accessibility can be used to measure if the data can be accessed. Finally, interpretability is defined as how well

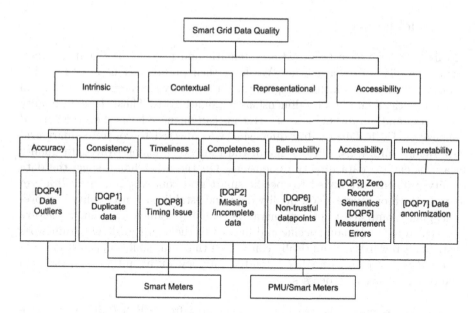

Fig. 2. Data quality framework in smart grid

the data can be interpreted. This can be balanced between the data privacy and the analytic details. Overall, not all the data quality dimensions from previous research are critical for the smart grids domain (Fig. 2).

Our framework is proposed in an operational and measurable level. For example, each data quality dimension can be measured by its related data quality problems. Likewise, the data quality categories such intrinsic or contextual data quality can be further measured by aggregating the related dimensions. Since most of the general data quality management frameworks are not domain-specific, their model granularity is refined only to the dimension level: it is therefore difficult to apply other data quality management frameworks and measure data quality in a domain. Our framework tackles this problem and relates the data quality measurement to specific quality problems.

Our framework can be further integrated into other data quality management frameworks. Since most of the existing data quality models or frameworks are based on the data quality dimensions [18], our framework is centralized by data quality dimensions and can be easily integrated into other frameworks or models by replacing the dimensions from this framework. Furthermore, our proposed framework can locate the root cause of low data quality dimensions by concrete quality problems in smart grids. After the assessment, the contexts and countermeasures can then be used for data quality improvement. For example, to determine which data quality problems occurred in which context and to determine the countermeasures to improve some data quality dimensions.

5 Conclusions

In this paper, we have proposed a systematic and practical taxonomy of data quality problems in smart grids. We have then proposed a new data quality management framework that adapted the data quality aspects and refined them to seven critical data quality dimensions for smart grids. Thus, the data quality assessment and improvement in smart grids can be more focused on the derived dimensions. Each data quality dimension is connected to concrete smart grid data quality problems. On one hand, the framework enables the data quality measurement for data quality dimensions. On the other hand, since the data quality problems are linked to specific smart grid contexts, it can facilitate to identify the root causes of low quality data and establish a data quality improvement plan. Compared to other general data quality frameworks, our framework is designed to be domain-specific and limited to the smart grid. The framework contributes towards automatically controlling the data quality in smart grid. As future work, we plan to further extend the framework by automating the data quality measurement processes.

Acknowledgements. The research was supported from ERDF/ESF "CyberSecurity, CyberCrime and Critical Information Infrastructures Center of Excellence" (No. CZ.02.1.01/0.0/0.0/16_019/0000822).

References

1. Abo, R., Even, A.: Managing the quality of smart grid data research in progress. In: IEEE International Conference on Emerging Technologies and Innovative Business Practices for the Transformation of Societies, pp. 5–8 (2016)
2. Al-Omar, B., Al-Ali, A., Ahmed, R., Landolsi, T.: Role of information and communication technologies in the smart grid. J. Emerg. Trends Comput. Inf. Sci. **3**(5), 707–716 (2012)
3. Alahakoon, D., Yu, X.: Smart electricity meter data intelligence for future energy systems: a survey. IEEE Trans. Industr. Inf. **12**(1), 425–436 (2016)
4. Buchmann, E., Böhm, K., Burghardt, T., Kessler, S.: Re-identification of smart meter data. Pers. Ubiquit. Comput. **17**(4), 653–662 (2013)
5. Burnett, R.O., Butts, M.M., Sterlina, P.S.: Power system applications for phasor measurement units. IEEE Comput. Appl. Power **7**(1), 8–13 (1994)
6. Chen, J., Li, W., Lau, A., Cao, J., Wang, K.: Automated load curve data cleansing in power systems. IEEE Trans. Smart Grid **1**(2), 213–221 (2010)
7. Chen, P.Y., Cheng, S.M., Chen, K.C.: Smart attacks in smart grid communication networks. IEEE Commun. Mag. **50**(8), 24–29 (2012)
8. Chen, W., Zhou, K., Yang, S., Wu, C.: Data quality of electricity consumption data in a smart grid environment. Renew. Sustain. Energy Rev. **75**, 98–105 (2017)
9. Chren, S., Rossi, B., Buhnova, B., Pitner, T.: Reliability data for smart grids: where the real data can be found. In: 2018 Smart City Symposium Prague, pp. 1–6 (2018)
10. Chren, S., Rossi, B., Pitner, T.: Smart grids deployments within EU projects: the role of smart meters. In: 2016 Smart Cities Symposium Prague, pp. 1–5. IEEE (2016)

11. Daki, H., El Hannani, A., Aqqal, A., Haidine, A., Dahbi, A.: Big data management in smart grid: concepts, requirements and implementation. J. Big Data **4**(1), 13 (2017)

12. Efthymiou, C., Kalogridis, G.: Smart grid privacy via anonymization of smart metering data. In: First IEEE International Conference on Smart Grid Communications, pp. 238–243. IEEE (2010)

13. Eichinger, F., Pathmaperuma, D., Vogt, H., Müller, E.: Data analysis challenges in the future energy domain. In: Computational Intelligent Data Analysis for Sustainable Development, pp. 181–242 (2013)

14. Gao, J., Xiao, Y., Liu, J., Liang, W., Chen, C.P.: A survey of communication/networking in smart grids. Future Gener. Comput. Syst. **28**(2), 391–404 (2012)

15. Ge, M., Helfert, M.: A framework to assess decision quality using information quality dimensions. In: Proceedings of the 11th International Conference on Information Quality, pp. 455–466. MIT, USA (2006)

16. Ge, M., Helfert, M.: Effects of information quality on inventory management. Int. J. Inf. Qual. **2**(2), 177–191 (2008)

17. Ge, M., Helfert, M., Jannach, D.: Information quality assessment: validating measurement dimensions and processes. In: 19th European Conference on Information Systems, Helsinki, Finland, p. 75 (2011)

18. Ge, M., O'Brien, T., Helfert, M.: Predicting data quality success - the Bullwhip effect in data quality. In: Johansson, B., Møller, C., Chaudhuri, A., Sudzina, F. (eds.) BIR 2017. LNBIP, vol. 295, pp. 157–165. Springer, Cham (2017). https://doi.org/10.1007/978-3-319-64930-6_12

19. Gellings, C.W., Samotyj, M., Howe, B.: The future's smart delivery system. IEEE Power Energ. Mag. **2**(5), 40–48 (2004)

20. Jakkula, V., Cook, D.: Outlier detection in smart environment structured power datasets. In: Sixth International Conference on Intelligent Environments, pp. 29–33. IEEE (2010)

21. Jawurek, M., Johns, M., Rieck, K.: Smart metering de-pseudonymization. In: 27th Annual Computer Security Applications Conference, pp. 227–236 (2011)

22. Kim, M., Park, S., Lee, J., Joo, Y., Choi, J.K.: Learning-based adaptive imputation method with kNN algorithm for missing power data. Energies **10**(10), 1668 (2017)

23. Kosut, O., Jia, L., Thomas, R.J., Tong, L.: Malicious data attacks on the smart grid. IEEE Trans. Smart Grid **2**(4), 645–658 (2011)

24. Leonardi, A., Ziekow, H., Strohbach, M., Kikiras, P.: Dealing with data quality in smart home environments lessons learned from a smart grid pilot. J. Sens. Actuator Netw. **5**(1), 5 (2016)

25. Li, F., et al.: Smart transmission grid: vision and framework. IEEE Trans. Smart Grid **1**(2), 168–177 (2010)

26. Li, F., Luo, B.: Preserving data integrity for smart grid data aggregation. In: 2012 IEEE Third International Conference on Smart Grid Communications, pp. 366–371. IEEE (2012)

27. Liu, Y., Ning, P., Reiter, M.K.: False data injection attacks against state estimation in electric power grids. ACM Trans. Inf. Syst. Secur. **14**(1), 13 (2011)

28. Matta, N., Rahim-Amoud, R., Merghem-Boulahia, L., Jrad, A.: Putting sensor data to the service of the smart grid: from the substation to the AMI. J. Netw. Syst. Manage. **26**(1), 108–126 (2018)

29. Peppanen, J., Zhang, X., Grijalva, S., Reno, M.J.: Handling bad or missing smart meter data through advanced data imputation. In: IEEE Innovative Smart Grid Technologies Conference, pp. 1–5. IEEE (2016)

30. Rao, R., Akella, S., Guley, G.: Power line carrier (PLC) signal analysis of smart meters for outlier detection. In: IEEE International Conference on Smart Grid Communications, pp. 291–296. IEEE (2011)
31. Rossi, B., Chren, S., Buhnova, B., Pitner, T.: Anomaly detection in smart grid data: an experience report. In: IEEE International Conference on Systems, Man, and Cybernetics, pp. 2313–2318. IEEE (2016)
32. Shishido, J., Solutions, E.U.: Smart meter data quality insights. In: ACEEE Summer Study on Energy Efficiency in Buildings (2012)
33. Wang, R.Y., Strong, D.M.: Beyond accuracy: what data quality means to data consumers. J. Manag. Inf. Syst. **12**(4), 5–33 (1996)
34. Zhang, Q., Luo, X., Bertagnolli, D., Maslennikov, S., Nubile, B.: PMU data validation at iso new england. In: 2013 IEEE Power and Energy Society General Meeting (PES), pp. 1–5. IEEE (2013)
35. Zhang, Y., Huang, T., Bompard, E.F.: Big data analytics in smart grids: a review. Energy Inform. **1**(1), 8 (2018)

Balancing Performance Measures in Classification Using Ensemble Learning Methods

Neeraj Bahl and Ajay Bansal[✉]

Arizona State University, Mesa, AZ 85296, USA
{nbahl,ajay.bansal}@asu.edu

Abstract. Ensemble learning methods have recently been widely used in various domains and applications owing to the improvements in computational efficiency and distributed computing advances. However, with the advent of wide variety of applications of machine learning techniques to class imbalance problems, further focus is needed to evaluate, improve and balance other performance measures such as sensitivity (true positive rate) and specificity (true negative rate) in classification. This paper demonstrates an approach to evaluate and balance the performance measures (specifically sensitivity and specificity) using ensemble learning methods for classification that can be especially useful in class imbalanced datasets. In this paper, ensemble learning methods (specifically bagging and boosting) are used to balance the performance measures (sensitivity and specificity) on a diabetes dataset to predict if a patient will be readmitted to the hospital based on various feature vectors. From the experiments conducted, it can be empirically concluded that, by using ensemble learning methods, although accuracy does improve to some margin, both sensitivity and specificity are balanced significantly and consistently over different cross validation approaches.

Keywords: Ensemble methods · Classification · Boosting · Balancing

1 Introduction

There are wide varieties of applications for machine learning algorithms. Also, there are various performance measures that need to be evaluated when an algorithm is applied to a specific problem. We know that accuracy is one of the most common performance measures to evaluate any algorithm or model. However, we also know that higher accuracy does not necessarily means good performance of the algorithm. The other common performance measures used to evaluate the machine learning algorithms are specificity, sensitivity, precision, F1 score, ROC curve, etc. The type of dataset governs the need to evaluate these performance measures. For example, in the application of machine learning to medical diagnostics or medical domain in general, failing to predict positive individuals may result in fatal cost, however, failing to predict negative instances is also serious in some applications, for example, information retrieval, email spam filtering and facial recognition [2]. Hence, it is imperative to balance sensitivity (which is a measure of the classification algorithm in avoiding false

© Springer Nature Switzerland AG 2019
W. Abramowicz and R. Corchuelo (Eds.): BIS 2019, LNBIP 354, pp. 311–324, 2019.
https://doi.org/10.1007/978-3-030-20482-2_25

negative) and specificity (which is a measure of the classification algorithm in avoiding false positive). For example, in the case of email spam filtering, false positive can lead to an important email to land up in the spam mailbox.

There has been research done on optimizing or balancing the F1 measure or the F1 score to achieve a compromise between the performance measures of precision and recall by tuning the parameters in the algorithms especially by tuning the parameters in Support Vector Machines. However, in this paper, a novel approach is used to evaluate and balance the sensitivity (recall) and specificity by using ensemble learning methods. It can be observed from the results that the balancing of the performance measures (sensitivity and specificity) with respect to each other is achieved by running the data over ensemble learners. The comparison of the performance measures is made with the base machine learning algorithms like the Logistic Regression, Naïve Bayes Classifier, k Nearest Neighbor, Decision Trees and Support Vector Machines.

The need for evaluation of these performance measures apart from accuracy and the need to balance these performance measures with respect to each other is imperative especially when we have class imbalanced datasets. Class imbalance data leads to class imbalance problem. When the data has imbalanced amount of two class labels, it is called as class imbalanced data. For example, in Fig. 1 below, we observe that the 80% of the actual target values are of class 0 (red color represents class 0) and 20% of actual target values are of class 1 (green color represents class 1).

Fig. 1. Class imbalance dataset example (Color figure online)

Fig. 2. Class imbalance dataset example (Color figure online)

Most of the datasets these days are class imbalanced datasets and hence the need to evaluate and balance the performance measures such as sensitivity, specificity, precision, F1 score, etc. are imperative for most of the datasets. Let us take an example of Fig. 2 above, for instance, this is a medical diagnostics dataset and we are predicting whether unhealthy patients need to be readmitted to the hospital. In a scenario where only accuracy is considered for performance valuation of the algorithm/model applied to the dataset. Suppose, if we are training a model on this dataset to predict if a patient should be admitted to the hospital or discharged from the hospital based on diagnostic feature vectors, where the actual class 0 (in red) represents unhealthy patients and actual class 1 (in green) represents healthy patients, it is possible that with an algorithm or model that gives us 80% accuracy, we may still have all the 20% of unhealthy patients discharged from the hospital, which can be fatal. Hence, with this example, we underscore our claim that the performance measures such as sensitivity and specificity need to be evaluated and balanced.

The objectives of this paper are to evaluate and balance the performance measures of sensitivity and specificity for at least two of the ensemble learning methods (i.e. bagging and boosting) and compare them with the base machine learning algorithms. The balancing is achieved with respect to the values of sensitivity and specificity with each other. While doing so, this paper achieves balanced sensitivity and specificity using ensemble learning methods as opposed to the previous methods that have focused on tuning the parameters of the base learners or creating framework. This paper focuses on sensitivity and specificity for the analysis of the performance measures, using one dataset. The dataset is divided into various sub-sets of data and two types of cross validation methods are employed to divide the data into test and training. The hold-out cross-validation methods employed are 80–20 and 70–30. The rest of the paper is organized as follows. We present related work in Sect. 2. Section 3 discusses the methodology followed by experimental setup. The results are presented in Sect. 5 followed by conclusions and future work.

2 Related Work

Researchers have used different performance measures for optimization, depending on the application and domain [8, 11]. Typically, we can see that the various performance measures such as F1 score, Sensitivity, Specificity, Precision, Recall, ROC, Area Under the Curve (AUC) etc. are based on the confusion matrix which helps us better analyze and evaluate the performance of the classifiers. These measures are very important, which directly and indirectly puts emphasis on approximating to correct prediction for each instance in the dataset in an ideal case. In this paper, however, the visualization and analysis of the results is done using sensitivity and specificity, by finding the relation with each other, and by visualizing the standard deviation for sensitivity, specificity and accuracy.

Studies have been done attempting to optimize the performance measures using various methods that include base machine learning tuning, usage of ensemble learning and other methods to form a framework, etc. [2] proposed a two-stage framework and a novel evaluation criterion, namely optimal specificity under perfect sensitivity (OSPS). They argued that for medical data classification, this criterion is more suitable than other conventional measures such as accuracy, f-score, or area-under-ROC curve. Here, they employ instance ranking strategy to optimize the specificity given perfect sensitivity using threshold for the perfect sensitivity.

Further, Support Vector Machine (SVM) parameter tuning has also shown to optimize F1 score in unbalanced data. F measure, which is the harmonic mean of precision and recall, is considered the relevant performance measure to be achieved especially when we have unbalanced data. SVMs have been traditionally used for classification of unbalanced data by "weighting more heavily the error contribution from the rare class" [4]. In this paper, the authors provided significant and new theoretical results that support this popular heuristic. Specifically, they demonstrated that with the right parameter settings SVMs approximately optimize F-measure in the same way that SVMs have already been known to approximately optimize accuracy.

Moreover Chai et al. [5] investigated the connections and theoretical justification of the two methods commonly used to optimize F score. The two algorithms for learning discussed to maximize F-measures are: the empirical utility maximization (EUM) approach that learns a classifier having optimal performance on training data, and the decision-theoretic approach (DTA) approach that "learns a probabilistic model and then predicts labels with maximum expected F-measure." [5]. In the paper, the authors further studied the conditions under which one approach is preferable to the other using synthetic and real datasets. They claim that their results suggested that the two approaches are asymptotically equivalent given large training and test sets. Empirically they also prove that the EUM approach appeared to be more robust against model misspecification and the decision-theoretic approach appears to be better for handling rare classes and a common domain adaptation scenario.

This paper proposes to optimize performance measures using ensemble learning methods. We know that the diversity in the ensemble learning methods help achieve better accuracy and other performance measures. Hence, we also need to understand if there is any relation between accuracy and diversity in ensemble learning methods. Zeng et al. [12] proposed a novel method to evaluate the quality of a classifier ensemble, which is inspired by the F measure used in information retrieval. They proposed a Weight Accuracy Diversity (WAD) measure to assess the quality of the classifier ensemble. They claim that WAD would help find a balance between accuracy and diversity which is needed for enhancing the predictive ability of an ensemble learner over unknown data as higher diversity not necessarily means better performance by the ensemble learning algorithm. This research would not only help us understand the relation between accuracy and diversity in classifying ensemble learning methods but would also help us qualitatively assess and compare the various ensemble learning methods using classification.

The work done by these researchers have significantly helped discuss and achieve significant success in performance measures and ensemble learning methods. However, no apparent focus has been made in direct usage of ensemble learning methods for improving or optimizing the performance measures i.e. sensitivity (recall) and specificity. In this paper, in the further chapters we will see how the ensemble learning methods are empirically proven to improve, optimize and balance the performance measures of sensitivity and specificity.

3 Methodology

The emphasis of this paper is on evaluating and balancing the performance measures using different base machine learning algorithms and the ensemble learning algorithms. Confusion matrix is a table of 2 * 2 that gives us the number of true positives, true negatives, false positives and false negatives. This table helps us to find the performance measures required to evaluate the machine learning algorithms in question. Sensitivity (or Recall) is the ratio of number of True Positive instances divided by the sum of number of True Positive and False Negative instances. Hence, higher the sensitivity, higher is the amount of class label predicted as 1 correctly. This performance measure is important in applications where we cannot afford to have any positive class label misclassified (Table 1).

Table 1. Confusion matrix

	Predicted as 1	Predicted as 0
Actual 1	True Positive	False Negative
Actual 0	False Positive	True Negative

$$Sensitivity = \frac{Number\ of\ True\ Positives}{Number\ of\ True\ Positives + Number\ of\ False\ Negatives}$$

Specificity is the ratio of number of True Negative instances divided by the sum of number of True Negative and False Positive instances. Hence, higher the specificity, higher is the amount of class label predicted as 0 correctly. This performance measure is important in applications where we cannot afford to have any negative class label classified as 1. For example, for information retrieval, in a search engine, having a false positive can be costly to the functionality and business of the search engine.

$$Specificity = \frac{Number\ of\ True\ Negatives}{Number\ of\ True\ Negatives + Number\ of\ False\ Positives}$$

Precision (or Positive Predictive value) is the ratio of number of True Positive instances divided by the sum of number of True positive and False positive instances. Here, higher the precision, higher is the amount of fraction of positive class labels predicted from the total positive class labels predicted.

$$Precision = \frac{Number\ of\ True\ Positives}{Number\ of\ True\ Positives + Number\ of\ False\ Positives}$$

F1 score is a measure of the classification test's accuracy. It considers both Sensitivity (Recall) and Precision. F1 score is the weighted harmonic mean of precision and recall (sensitivity). Ensemble learning is a method to combine multiple models such as classifiers to solve and predict a machine learning problem. It is typically used to improve predictive performance of a model, reduce the likelihood of selecting a poor learner and to assign to confidence to the decision made by the model. [3]. Commonly used ensemble learning algorithms are (1) Bagging: Bagging which stands for Bootstrap aggregation obtains diversity in classification by randomly sampling the data - with replacement - from the entire training dataset. Each training dataset is used to train a different classifier of the same type or different types. Simple majority vote is used to fuse the resultant models. (2) Boosting: Boosting is similar to the ensemble learning method of bagging in that it resamples the data, however, it strategically samples the subset of the training data to create several weak classifiers [10]. The classifiers are combined through a n-way majority vote. For example, if there are three classifiers, the first subset of the training data is selected randomly; the second subset is selected in an informative way as per the boosting algorithm. The third subset is sampled based on the instances where both the first and second classifier disagrees with each other. Hence, during fusion of data, a strong classifier is created.

3.1 High-Level Implementation

The Fig. 3 below shows the high-level flowchart of the experimental implementation set up. These six high-level steps are consistent with the details provided below. These steps were implemented and were common to all the learners (base and ensemble learners). First for each learner, the libraries from the R Comprehensive R Archive Network (CRAN) Package were imported. Then, the data was pre-preprocessed. This step of preprocessing included various important and critical sub-steps. Once the data pre-processing was completed which includes factoring, normalization, making the feature vectors binary wherever necessary, the data was split using hold out cross validation methods into training and test datasets. The respective base machine learning or ensemble learning model were then trained using the training data over these cross-validation methods. The test data is tested on the generated model using the predict methods. Finally, the predicted feature vector thus created is compared to the actual target feature vector with the help of confusion matrix. The base learning classifiers and ensemble learning methods were implemented using the R CRAN packages library. To generate the confusion matrix or the cross table and the performance measures, the respective R library methods were used. The large dataset was divided into 10 subsets of 10000 instances.

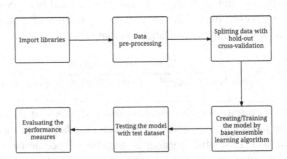

Fig. 3. High level implementation flowchart

3.2 Software Used

The following software have been used for implementation and experiments in this paper: (1) R version 3.2.3 and associated R CRAN packages [7] was used for implementing machine learning algorithms and evaluate performance measures. (2) R Studio 1.0.136 is a free and open source integrated development environment for R was used for the ease of implementation. (3) Microsoft Excel 2016 was used to generate tables and plot graphs and Lucidchart was used to create flowchart.

3.3 Dataset

In this paper, the UC Irvine (UCI) Diabetes 130-US hospitals for the years 1999–2008 dataset [1, 6] has been used. This dataset is multi-variate dataset, contains 100,000 instances and 55 features. "The dataset represents 10 years (1999–2008) of clinical care at 130 US hospitals and integrated delivery networks. It includes over 50 features representing patient and hospital outcomes. Information was extracted from the database for encounters that satisfied the following criteria: (1) It is an inpatient encounter (a hospital admission). (2) It is a diabetic encounter, that is, one during which any kind of diabetes was entered to the system as a diagnosis. (3) The length of stay was at least 1 day and at most 14 days. (4) Laboratory tests were performed during the encounter. (5) Medications were administered during the encounter. The data contains such attributes as patient number, race, gender, age, admission type, time in hospital, medical specialty of admitting physician, number of lab test performed, HbA1c test result, diagnosis, number of medication, diabetic medications, number of outpatient, inpatient, and emergency visits in the year before the hospitalization, etc." [1]. It is a three-class label dataset with the values being NO, <30, >30. Here, NO means that the patient was not readmitted after the discharge. <30 means that the patient was readmitted within 30 days of discharge. >30 means that the patient was readmitted after 30 days of discharge.

4 Experiment

4.1 Data Pre-processing

Before using this data in the implementation of machine learning algorithms to evaluate and compare the performance measures of ensemble learning methods to the base machine learning algorithms, the data is pre-processed. This was needed to ensure that all the classification algorithms and ensemble learners can be equally and fairly implemented on the dataset. First, the missing values in the dataset are converted into NA and in turn into 0. The following eight feature vectors were dropped from dataset as they do not actively represent the classification problem and do no contribute to the target class label values for classification: Encounter ID, Patient Number, Weight (this feature vector was dropped as 97% values are missing), Payer Code, Medical Specialty, Diagnosis 1, Diagnosis 2, Diagnosis 3. Then, the target feature vector "readmitted" was changed to a binary feature vector as 0 or 1 by converting the three class label dataset into two class label classification dataset as follows: If the target feature vector value is <30 or >30, it was changed to 1. If the target feature vector is NO, the target feature vector value was changed to 0. To ensure consistency for the implementation, the nominal feature vectors were changed to binary feature vectors. After changing the feature vectors, the dataset now contained 98 feature vectors. Further, the numeric data, which is now contained in the first eight feature vectors, is normalized to ensure that the evaluation and comparison is fairly made when using these feature vectors for classification. After the normalization, the data in these 8 numeric feature vectors lied in the range of 0 to 100. The normalization range was chosen to be from 0 to 100 to ensure that the wide variety of the values in the dataset are included without having any common values. For example, it would be statistically incorrect to bring 100000 instances with different nominal features in a range of 0 to 10 as most of the instances in that case would have common values. This would lose the consistency and integrity of data in the dataset. Now the dataset was made ready to be implemented for all the algorithms.

Now, the instances were separated in 10 blocks of 10000 instances of dataset to implement as individual separate datasets. This would give us wide range of results for effective comparison. The hold out cross-validation methods used were 80–20 and 70–30. These different methods of cross validation helped to demonstrate whether any changes in the cross-validation affects the performance measures, especially, the measures that are critical to the application domain i.e. accuracy, sensitivity and specificity. In this paper, only one subset of 10000 instances of the dataset was used.

4.2 Implementation and Experimental Setup

The base machine learning algorithms and the ensemble learning algorithms were implemented in the R version 3.2.3. As seen in the Table 2 below, the following algorithms were implemented and following are the corresponding R methods and packages used for the implementation. The training data was separated according to the hold out cross-validation methods described above and the model was trained on the training dataset. Once the model was trained, the test data was fed to the model using the respective prediction methods to determine the performance measures. In case of logistic regression, the threshold probabilities generated are converted into binary target values. When the probability is greater than 0.5, the binary target value is changed to 1 else 0. In the case of k Nearest Neighbor (kNN), the k value is selected as the optimal value, which is the square root of the number of training examples. This is dynamically calculated for each dataset and encoded in the code. The target value is converted into a factor wherever required. The Support Vector Machine (SVM) algorithm is implemented for both the radial basis function and the sigmoid basis function. For bagging, the treebag method (algorithm) is used which means the base learning algorithm for bagging was decision tree. Similarly, the boosting algorithm is implemented using Decision Trees and Gradient Boosting Model.

Table 2. Algorithms and corresponding R methods/packages

Algorithm	R Methods	R Packages
Logistic Regression	Glm	Multiple
Naïve Bayes Classifier	naiveBayes	Multiple
K Nearest Neighbor (kNN)	knn	Multiple
Decision Trees	C5.0	C50/Multiple
Support Vector Machine (sigmoid)	svm	e1071/Multiple
Support Vector Machine (radial)	svm	e1071/Multiple
Bagging (CART)	treebag	caretEnsemble/Multiple
Boosting (Decision Trees)	C5.0	caretEnsemble/Multiple
Boosting (Gradient Boosting Machine)	gbm	caretEnsemble/Multiple

The bagging and both the boosting ensemble methods were resampled over 30 iterations. This value was referred in the article by [9]. As part of future work, more experiments can be conducted by fine-tuning the number of classifiers/iterations for each ensemble learning method that is implemented. For bagging, boosting and boosting using gradient boosting model, decision tree was used as a base machine learning algorithm. Decision tree was chosen as a base learning algorithm for the ensemble learners as it did not balance both the performance measures as good as the other base learning algorithms on the same dataset, so to compare and evaluate if the ensemble learning algorithms perform better than the base learning algorithm, decision trees were used.

4.3 Evaluation Measures

The evaluation measures are generated and demonstrated by plotting the confusion matrix and by using the summary of the results using R packages. These results include various performance measures including the ones that are integral to this paper i.e. accuracy, sensitivity and specificity. The two R methods that were used to generate the evaluation measures are CrossTable() and confusionMatrix(). CrossTable generates a 2*2 confusion matrix of the True Positives, True Negatives, False Positives and False Negatives. confusionMatrix() provides a detailed summary including the confusion matrix and the associated performance measure values.

5 Results

Experimental results clearly show that the ensemble learning methods of Bagging using Decision Trees, Boosting using Decision Trees and Boosting using Gradient Boosting Model shows consistent improvement in the balancing of both sensitivity and specificity. It is also observed from the plots and the graphs that balancing is achieved without compromising the accuracy of the prediction. For example, the Sensitivity and Specificity are balanced and optimum consistently when bagging and both the boosting methods are used with 80–20 cross validation method. Further for the same dataset, in the column (bar) graph we can observe that both sensitivity and specificity almost have equal height. From Table 3 and Fig. 4 followed below, we can see that the same results are consistently achieved across the different cross-validation methods that are experimented. Although, Logistic regression, kNN and SVM with radial basis function also demonstrates balanced results as proved by the graphs below, we can see consistent balancing results are achieved with the help of ensemble learning algorithms such as Bagging, Boosting using CART (Decision Trees) and Boosting using Gradient Boosting Machine (GBM) using Decision Trees.

The results are arranged in the following order. First, the table of the results for each hold-out cross validation method is seen, Then, a graph of absolute values of sensitivity, specificity and accuracy are plotted as points that are connected with dotted lines. This graph is very helpful in visualizing whether a specific learner was able to achieve optimization/balancing of the performance measures. The x axis on this graph has the absolute names of the algorithms and y axis has the value of the specific learner. Further, a bar plot graph is plotted to compare only the height of sensitivity and specificity to better visualize the performance measures. Finally, in the end the standard deviation of each learner is calculated for accuracy, sensitivity and specificity. The lower the standard deviation, more the algorithm is balanced with respect to performance measures sensitivity and specificity.

Table 3. Performance measures for 80–20 cross validation

Algorithm	Sensitivity	Specificity	Accuracy
Logistic Regression	0.6683	0.5068	0.6125
Naive Bayes Classifier	0.9315	0.2632	0.584
K Nearest Neighbor	0.6091	0.5631	0.591
Decision Trees	0.7759	0.4221	0.5975
SVM (kernel = sigmoid)	0.91692	0.04496	0.489
SVM (kernel = radial)	0.6076	0.5491	0.5775
Bagging Decision Tree	0.5873	0.5867	0.587
Boosting Decision Tree	0.6631	0.5338	0.601
Boosting GBM	0.6781	0.56	0.62

First, for the 80–20 holdout cross validation method, we can observe from Table 3, that the performance measure values are diverse and lie in the range from 0.04 to 0.91. We have achieved wide variety of balancing of sensitivity and specificity with respect to each other. Hence, to visualize we observe from Fig. 4, that the bagging, boosting and boosting using GBM have consistently performed better in balancing both sensitivity and specificity. Although, we can observe that Logistic Regression, k Nearest Neighbor (kNN) and Support Vector Machine using Radial basis function has also performed better in balancing both sensitivity and specificity. However, we also need to ascertain whether the results hold consistency and confidence across all the hold-out cross validation methods. The worst performance has been shown by Support Vector Machine (SVM) using Sigmoid basis function.

In SVM using sigmoid, we see that Sensitivity is very high but specificity is very low. This performance may be accepted in medical domain but not in information retrieval where a higher specificity is accepted. This can be visualized from Fig. 5 as well. Further, we can observe from Table 4, Figs. 6 and 7 for 70–30 hold out cross validation method, largely most of the values are consistent with 80–20, however, there has been a visible change in performance of Logistic Regression, Naïve Bayes and Boosting using GBM.

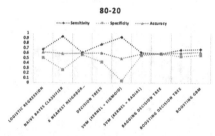

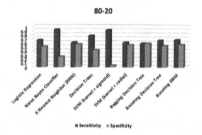

Fig. 4. Plot to visualize balancing of 80–20 cross validation

Fig. 5. Bar Plot for 80–20 cross validation

Table 4. Performance measures for 70–30 cross validation

Algorithm	Sensitivity	Specificity	Accuracy
Logistic Regression	0.7562	0.4156	0.5907
Naive Bayes Classifier	0.4514	0.6994	0.5743
K Nearest Neighbor (KNN)	0.6286	0.5425	0.585
Decision Trees	0.7886	0.4253	0.6087
SVM (kernel = sigmoid)	0.94184	0.02623	0.488
SVM (kernel = radial)	0.5435	0.5967	0.5693
Bagging Decision Tree	0.6077	0.5622	0.5847
Boosting Decision Tree	0.5878	0.5968	0.5922
Boosting GBM	0.6775	0.5331	0.6068

As per Fig. 6, we can observe that again the bagging, boosting and boosting using GBM methods have consistently performed better in balancing both sensitivity and specificity. In this case, we observe that only k Nearest Neighbor (kNN) and Support Vector Machine using Radial basis function has performed better in balancing both sensitivity and specificity as compared to other base learners. The same observation is made through the bar plot of the 70–30 cross validation method. The height of most of the bars for ensemble learning algorithms for both sensitivity and specificity are of mostly equal size. In this case, for Logistic Regression, we observe that although sensitivity has increased, however, specificity has reduced as compared to the 80–20 hold out cross validation method.

Standard Deviation for Sensitivity and Specificity

Next, we observed how spread out sensitivity and specificity across the cross-validation approaches are for the ensemble learning and base machine learning algorithms. These standard deviations are calculated using values of accuracy, specificity and sensitivity. As accuracy is approximately same or better across both experiments conducted, higher standard deviation would mean, more the sensitivity and specificity are spread out and lower the standard deviation would mean, more the two measures are balanced and close to each other. This measure helps us understand the balancing process if the above graphs are hard to visualize.

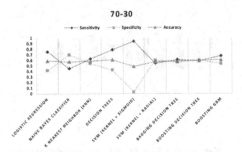

Fig. 6. Plot to visualize balancing of 70–30 cross validation

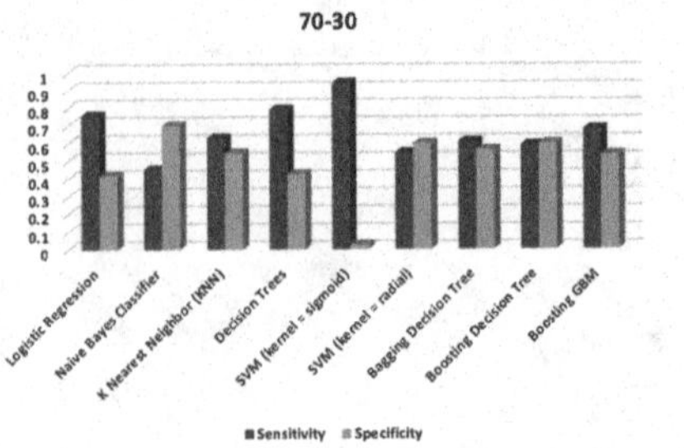

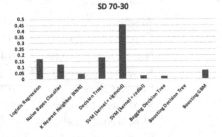

Fig. 7. Bar Plot for 70–30 cross validation

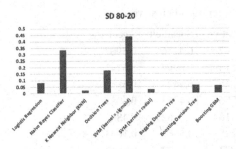

Fig. 8. SD for 80–20

Fig. 9. SD for 70–30

Here, in Fig. 8, as we have observed in earlier graphs, we can observe that the ensemble learning methods and three of the single classifier system viz., Logistic Regression, k Nearest Neighbor and SVM using Radial basis function has lower Standard Deviation. However, for 70–30 hold out cross validation method, as seen in Fig. 9, the standard deviation for Logistic Regression is seen to be increased and hence the Sensitivity and Specificity are spread out from each other.

6   Conclusions and Future Work

Given the evaluation and analysis done on the results for the experiments conducted on this dataset over different cross validation approaches, in this paper, it can be empirically concluded that by using ensemble learning methods such as bagging, boosting using decision trees and boosting – Gradient Boosting Modeling (GBM) using decision tress, sensitivity and specificity can be balanced with respect to each other without compromising accuracy. This usage of ensemble learning methods for balancing sensitivity and specificity would help apply this technique to various domains of applications. Although depending on the application and domain, the need for sensitivity and specificity differs, however, given this method, the avenues of balancing and nearly approximating the performance measures of sensitivity and specificity to 1 using ensemble learning methods are opened. The comparison made with the base machine learning algorithms helped underscore the claim.

As the results showed that the base learning algorithms such as the kNN, Logistic Regression and SVM using radial basis function performed as better as the ensemble learning methods, further research can be conducted to investigate the cause of the same. Also, the ensemble of these base learners as opposed to only Decision Trees can be implemented to study if there are any further improvements in the performance measures. Further, theoretical research can be done to assess, evaluate and validate these improvements in performance measures by ensemble learning methods. Finally, there is a scope to improve the computational efficiency of the ensemble learning methods as these take more computational resources as compared to the base learning algorithms.

References

1. Strack, B., et al.: Impact of HbA1c measurement on hospital readmission rates: analysis of 70,000 clinical database patient records. BioMed Research International, vol. 2014, Article ID 781670, 11 pages (2014)
2. Hsiao, J.C.-Y., Lo, H.-Y., Yin, T.-C., Lin, S.-D.: Optimizing specificity under perfect sensitivity for medical data classification. In: Proceedings of International Conference on Data Science and Advanced Analytics (DSAA), Shanghai, pp. 163–169 (2014). https://doi.org/10.1109/dsaa.2014.7058068
3. Polikar, R.: Ensemble learning. Scholarpedia 4(1), 2776 (2009)
4. Musicant, D., Kumar, V., Ozgur. A.: Optimizing f-measure with support vector machines. In: Proceedings of FLAIRS (2003)
5. Nan, Y., Chai, K.M., Lee, W.S., Chieu, H.L.: Optimizing F-measure: a tale of two approaches. In: Proceedings of International Conference on Machine Learning (ICML) (2012)
6. Diabetes 130-US hospitals for years 1999–2008 Data Set, UC Irvine (UCI) Machine Learning Repository. https://archive.ics.uci.edu/ml/datasets/Diabetes+130-US+hospitals+for+years+1999-2008. Accessed 2 Apr 2017
7. R CRAN Packages. https://cran.r-project.org/web/packages/available_packages_by_name.html. Accessed 2 Apr 2017
8. Mandal, I.: A novel approach for predicting DNA splice junctions using hybrid machine learning algorithms. Soft Comput. 19(12), 3431–3444 (2015). http://dx.doi.org/10.1007/s00500-014-1550-z
9. Brownlee, J.: How to Build an Ensemble of Machine Learning Algorithms in R (2016). http://machinelearningmastery.com/machine-learning-ensembles-with-r/. Accessed 20 Apr 2017

10. Amunategui, M.: Bagging/ Bootstrap Aggregation with R (2015). http://amunategui.github. io/bagging-in-R/index.html. Accessed 20 Apr 2017
11. Asmita, S., Shukla, K.K.: Review on the architecture, algorithm and fusion strategies in ensemble learning. Int. J. Comput. Appl. (0975 - 8887). **108**(8), December 2014
12. Zeng, X., Wong, D.F., Chao, L.S.: Constructing better classifier ensemble based on weighted accuracy and diversity measure. Sci. World J. vol. 2014, Article ID 961747, 12 pages (2014). https://doi.org/10.1155/2014/961747

Machine Learning for Engineering Processes

Christian Koch[(⌧)] [ID]

University of Technology, Construction Management Gothenburg,
42133 Gothenburg, Sweden
kochch@chalmers.se

Abstract. Buildings are realized through engineering processes in projects, that however tend to result in cost and/or time overrun. Therefore, a need is highlighted by the industry and the literature, to develop predictive models, that can aid in decision-making and guidance, especially in a preparation effort before production is initiated.

This study aims at investigating what are possible applications of machine learning in building engineering projects and how they impact on their performance?

First, a literature review about machine learning (ML) is done. The first case is drawing on a productivity survey of building projects in Sweden (n = 580). The most influential factors behind project performance are identified, to predict performance. Features that are strongly correlated with four performance indicators are identified: cost variance, time variance and client- and contractor satisfaction and a regression analysis is done. Human related factors predict success best, such as the client role, the architect performance and collaboration. But external factors and technical aspects of a building are also important.

The second case combines constructability and risk analysis on a basis on civil engineering project from several different countries and with very different character; a town square, a biogas plant, road bridges and sub projects from an airport. The data encompasses 30 projects. The development build on literature study, expert interview, unsupervised and supervised learning. The strength lies more in the conceptual work of risk sources enabled by ML. Human reasoning is needed in building projects. Also after the introduction of ML.

Keywords: Machine learning · Engineering · Hybrid learning ·
Project performance

1 Introduction

The digital transformation of business and society has recently taken a new turn where the mutually overlapping concepts of Artificial Intelligence (AI) and Machine learning has become revitalized [1, 2]. Ascribing intelligence and learning capabilities to computers has previously received skepticism [2] and is prone to renewed skepticism. Nevertheless, contemporary applications of AI and Machine learning do actually exhibit new capabilities especially the ability to process large amount of data and operate several if not numerous concerted rules. AI is in this sense a moving target and

© Springer Nature Switzerland AG 2019
W. Abramowicz and R. Corchuelo (Eds.): BIS 2019, LNBIP 354, pp. 325–336, 2019.
https://doi.org/10.1007/978-3-030-20482-2_26

exhibits an effect where ever new areas and capabilities are developed leading McCullock to propose the AI effect, where AI is redefined to "what machines have not been able to do yet" [2].

[2] similarly propose that the highest level of AI should be artificial super intelligence defined to be above human skills. The outset of this contribution is nevertheless a far more practical understanding of the many places building engineering project performance can be improved by using machine learning techniques. The aim of the present paper is therefore to investigate:

What are possible applications of machine learning in building engineering projects and how might they impact on their performance?

We review selected contributions of machine learning (ML) applications for building engineering projects published in 2018 or recently. Two cases of prediction systems are presented and discussed: a prototype system for predicting building project success and a prototype system for predicting constructability of projects. This material is used to draw out some implications of the present developments and their impact for building engineering projects.

2   Recent Machine Learning Contributions

Applications of machine learning within engineering in general and of buildings particularly is rapidly developing, both drawing on general machine learning literature, technologies and methods. The review of recent contributions is therefore structured according to the general scientific discourse [3, 4] and its categories exemplified by recent building engineering applications. There are at present a portfolio of promising prototype systems for application in various parts of building engineering and construction [5–11].

Jordan and Mitchell [2] define Machine Learning as "computer systems that automatically improve through experience" [2]: 255, [3, 12]). As with the notion of (machine) "learning", "experience" is here meaning feeding the system with new data from the domain it comes from. A central task for the machine learning development are to identify and verify the possible underlying statistical computational-information-theoretic laws that govern the learning systems in question [2]. In trying to do this task tools from data mining, statistics (linear regression) and optimization theory have been adopted [2].

As noted initially the position here is that AI and machine learning are overlapping fields, but other positions view machine learning as adjacent to AI and others again as a sub discipline to AI [2, 3]. Kaplan and Haenlein [1] define Artificial Intelligence as "a system's ability to correctly interpret external data, to learn from such data, and to use those learnings to achieve specific goals and tasks through flexible adaptation". With such a definition it is difficult to see the difference between AI and ML. Moreover if one attempt to focus AI as the use of artificial neural networks, fuzzy logic and genetic algorithms as a possible core of AI in contrast to ML it is quickly revealed that ML also use these algorithms [5]. This blurred delimitation is not further discussed here.

ML was originally often classified in three types: supervised machine learning, unsupervised machine learning and hybrids. Today hybrid systems have become the dominant both in terms of combining algorithms in the final system but often also with a test encompassing several algorithms. Supervised learning [13, 14] concerns using algorithms, that are provided with training data and correct answers, in a context where it is assumed that the reasoning of the application domain is known. The task of the ML algorithm is to learn based on the training data, and to apply the knowledge that was gained using real data [4]. Many different algorithms are used in supervised ML; Decision trees, decision forests, logistic regression, support vector machines, kernel machines, Bayesian classifiers and more [4]. More recently deep learning system have received attention. Deep learning systems make use of gradient-based optimization algorithms to adjust parameters throughout a multilayered network based on errors of its output [2].

Within building engineering one group of systems using supervised learning aims at supporting the tendering process of building projects [15, 16]. Le et al. [15] presents a system for analyzing texts in building contract extracting stated requirements. The system uses a supervised learning approach with Naïve Bayes algorithms. A test with some 1100 statements showed a very high accuracy according to [10]. Zhang et al. [26] develops a ML system for compliance checking vis a vis building code. The system uses semantic role labeling of the building code text. The method is combining multiple ML algorithms: first capturing the syntactic and semantic features of the building code sentences with natural language processing techniques. Second using external training data selected by data similarity for the ML system. Third by performing semantic role labeling using a conditional random field model [26]. The conceptual model was tested on a corpus of annotated text from the International Building Code encompassing 300 sentences, and achieved promising precision according to [26]. Yet the use of an external data set of linguistic origin for testing appears problematic comparing to building code. The two are quite different types of text. [25] presents a study of a compliance code checker prototype based on natural language processing and machine learning. As their system is tested on one semantic frame only, it is a very early prototype far from being applicable in bidding processes. Petrova et al. [8] using data collected from sensors installed in a finished building to make the prediction of outcomes in building engineering more accurate, by reducing the occurrence of design errors using a combination of data mining and semantic modelling techniques. [5] presents a crack detection system based on web images of cracks in roads. The proposed method incrementally retrieves and classify crack images. A weak convolutional Neural Network classifier first models and sorts data from a limited set of Web images, and then acts as a machine annotator and further labels a larger size of data. The proposed method was according to [5]: 553 "able to retrieve and label a set of images with 95% labeling recall".

Unsupervised machine learning can be defined as "the analysis of unlabeled data under assumptions about structural properties of the data" [3]. In unsupervised learning ML algorithms do not have a training set. The systems are presented with some data about a domain and have to develop models of relations from that data "on their own" by running internal procedures [4]. Unsupervised learning algorithms are mostly focused on finding hidden patterns in data. Counter to supervised learning there are not preset assumptions about internals laws of the dataset. Rather the machine learning systems are supposed to find these. Also, here statistical principles and laws are used.

In building examples are few, but we provide an example in the second case below. Hybrid types involves mixing two or more approaches or algorithms. This include semi-supervised [2, 3] and reinforcement learning [2, 3]. Many studies and prototypes use a combination of machine learning algorithms [3, 6] Amasyali and El-Gohary [6] test a series for their energy performance prediction system: Gaussian process regression, support vector regression, Artificial Neural Networks, and linear regression. It is not so clearly described in these studies what the hybridization implies for their usability in building engineering process. In the general literature on machine learning a large and quickly growing number of applications are developing socalled smart assistants and recommender systems [3, 24] propose a BIMbot as a smart assistant for a building engineering team using Building Information Modelling product design. BIMbot is intended to be a socalled cognitive assistant to the BIM team. i.e. the team of building engineers. Another area of applications are image recognition [4, 8, 28]. Tixier et al. [28] reviews several examples of applications of image analysis in building surveilling issues such as concrete quality and workers movement. Siddula et al. [8] presents a ML system for analyzing pictures of roofs to prevent occupational accidents. [4] presents a crack detection system for roads using images.

Also deep learning systems are emerging [2]. Large-scale deep learning systems have had a major effect in recent years in computer vision and speech recognition, where they have yielded major improvements in performance over previous approaches. [6] present a model for energy consumption prediction consisting of three base models: (1) a machine learning model that models the impact of outdoor weather conditions from simulation-generated data, (2) a machine learning model that learns the impact of occupant behavior from real data, and (3) an ensembler model that predicts cooling energy consumption based on weather-related and occupant behavior-related factors predicted by the first two models. In doing so several machine learning algorithms were tested: gaussian process regression, support vector regression, Artificial Neural Networks, and linear regression. The predicted energy consumption levels showed agreement with the actual levels. Nevertheless, it is unclear what the impact on the result is that some data are simulation generated. There is a risk to inhering weaknesses of the previous energy consumption model. Similarly [9] presents a proposal of a classification of office buildings energy consumption using multiple machine learning algorithms. The classification sorts by building characteristics, occupant behaviors, geographical and climate and classify the building in three energy consumption levels. To support this the ML algorithms used encompasses support vector regression models, naïve bayes, decision trees, and random forests [9]: 757. [5] develop an EEG-based stress recognition framework by applying deep learning algorithms and a neural net- work to recognize building workers' stress while performing different tasks at actual construction sites. They experiment with different number of layers to optimize the efficiency of the system. Yet the connection to the practical domain appear weak as alternative forms of coping with stress on a building site not evaluated along this proposed technical solution. [24] uses accident reports to categorize safety outcomes, with a natural language processing tool and two ML algorithms random forest and stochastic gradient tree boosting. The model is claimed by [24]: 102 to be able to predict injury type, energy type, and body part with "high skill". An important area of further development is the criteria for application. In their review [2] point to the need for "function approximation" when making ML systems. Attempts has been made to categorize ML algorithms based on the purpose for which they are designed [15, 21] however categorization by purpose does not assure "fit" with the domain. [3] identify another aspect of difficulties of accommodating machine learning algorithms to a context, namely the large number of algorithms described in the literature complicating the ML-designers choices. But as a general observation ML studies and prototypes are not well described when it comes to criteria for application.

Summarizing, the literature review has identified three main forms of machine learning; supervised, unsupervised and hybrid. Much current machine learning development involves hybrids of both these categories and other intermediate forms as well as elements from data mining, statistics and AI [2, 3, 6]. The hybrid category appears undertheorized at present. The combination of several algorithms and ability to derive relations (if not causalities) out of large data set are central characteristics [3]. It appears to be a common weakness that machine learning approaches are unprecise in understanding their possible applicability to particular practices or domain, characterized by a particular knowledge structure, decision practice and structure of reasoning. Not all proposed prototypes of systems are equally promising, and some appear to have a weak connection to the context they are intended to support. It is commonplace that researchers provide their own internal evaluation as basis for presenting estimated performance for the systems. These estimations of performance are usually quite promising, yet may not be trustworthy compared to practical domains.

3 Methodology

The overall approach adopted here is an interpretive sociological with a mixed method [17, 18]. The identification and selection of recent applications of ML to building engineering builds on a literature review consisting of several explorative searches on machine learning and AI in building carried out in spring, autumn and winter of 2018. This encompassed scientific publications as well as professional press coverage in the IT and the building press. All contributions on ML at the recent CIB W78 conference was reviewed [19]. The two cases presented are selected for convenience reasons. They represent current work on ML in a construction project management context. The first case draws on project performance data from a productivity survey of building projects in Sweden (n = 580 projects, author reference). The study that is analyzed in this research includes answers from 324 main contractor representatives and 256 clients that participated in the survey in 2014. A main property of the investigation was an ambition to measure productivity as more than cost per square meter. Processual and soft aspects are entered, looking at disturbances during the process and performance of the project organization members, i.e. the client, the consultants, the contractor, and the suppliers. The design of the questionnaire led to a set of questions, where most had pregiven categories for answers in liker scales. This include project technical complexity, such as preparation work i.e. the use of blasting work, amount of prefabrication, the structural engineering technology (concrete, steel, timber). But also, a series of project organization questions such as the clients and contractor's evaluation of the consulting engineers architect and supplier performance and collaboration. However, there were also a series of questions where facts and figures were demanded, as well as some open questions, relating to stated definitions, such as on client costs, and partnering. Finally, a few questions were open without definition, including questions on satisfaction, disturbances and learnings. The design and operation of data collection was done in autumn 2014. Access were provided to the data from the productivity investigation organized in six large excel sheets. The development was a master thesis.

The second case on constructability is developed within a ph.d. project by one of the authors. It encompassed an extensive literature study on constructability. Data was extracted on risk sources from literature and risk analysis was corroborated by expert inter- views. Thirty civil engineering projects were used for training the ML system. This include one biogas plant, two bridges (Greece), Four added classrooms for a school (Greece), reconstruction of a municipal road axis (Greece), sustainable public installations (including a square, Greece), 4 road projects from Estonia, 3 renewable technology projects, 4 electrical lighting projects (Greece) and 10 subcontracts of the Midfield Terminal, Abu Dhabi Airports. The main limitations of the present research are first a highly selective set of literature and second a lack of context appreciation in the two cases. For example, the ML case of building engineering project performance does not make use of the data materials distinction between project type, building type, geographical location, but aim at a general comparison between projects. It covers the Swedish building market as the applicability to other contexts such as other countries or other building industries is not known. Cost, time and satisfaction are selected and considered as key performance indicators, and other possible indicators disregarded. The other case encompasses limitations as well.

4   Case Building Production Project Performance Prediction

To be able to predict building engineering project performance is an aspiration by practitioners. Project performance can be measured by using key performance indicators. Performance are mainly constituted by two different elements: the general project conditions and the project organization. The general project conditions involves the type of contract, size of building and external factors encompasses market conditions, regulations and requirements. The second element, the project organization, is a process which is embodied by the characteristics and performance of the client, main contractor, engineers, subcontractors and consultants. Machine learning algorithms were used to extract and analyze project performance data from a productivity survey of building projects in Sweden (n = 580 projects [26]). The data covers the mentioned aspects and dynamics of project performance and it was chosen to focus on four key performance indicators; cost, time and client- and contractor satisfaction. The factors include project attributes, external factors, the project organization. The tool that is used to do the linear regression and attribute selection is Weka (Waikato Environment for Knowledge Analysis), which is a data mining software [19]. Weka contains a series of machine learning algorithms including tools for data processing, feeding data into schemes, regression analysis, classification, clustering, association rule mining and attribute selection [19]. Features strongly correlated with three performance indicators: cost variance, time variance and client- and contractor satisfaction, was extracted.

A regression analysis is done to develop a model for predicting project cost, time and satisfaction. For example, to generate insight in the client satisfaction performance indicator a process of cleaning and organizing the data about the client's perspective was extracted from the excel sheet covering the clients answers to questions about the building projects for. The client's satisfaction is here conceptualized to be the sum of the three questions regarding satisfaction, success and expectations of the results of the project. A correlation test was carried out to evaluate if the three questions are representing satisfaction. Forward and backward stepwise regression was carried out starting with the full set of data and systematically deleting irrelevant attributes, one by one, until reaching the final set of attributes. The final set of input variables is evaluated to have the highest capability to produce accurate predictions. An attribute is irrelevant when it's value does not change systematically with the output class. It is also important to remove redundant attributes that are characterized as being correlated with one or more than other attributes. Cross validation was used to test the ML system. The error rates show that the best prediction error is the one associated with predicting contractor cost variance of building (9.79%). The prediction for building cost based on the pre-construction predictors variables is also performing better than other models, with a root mean square error of 10.06%. In the results, project technical complexity, the use of blasting work or amount of prefabrication among other features that are important factors that affect project performance. Human related factors in the project life cycle are of high impact, such as the client role, the architect performance and collaboration throughout projects. These are also the factors that are most suitable for predicting if a building project will be successful. The conclusion is that, although external factors and technical aspects of a building are important, the most recurring factors behind project performance can be linked to human related aspects. The method proves to be successful to be used to test the shared assumptions in the industry. The study provides statistical evidence for what factors that are important for modelling prediction.

5 Case Prediction of Constructability and Risk

Two important aspects of preparing building engineering projects are management of constructability and risk. The need for such activities are more over enforced by the divide of engineering design and production. The present ML system aimed at supporting these preparation issues. Constructability was in this context understood as the optimal use of building knowledge and experience throughout the building project lifecycle until its delivery (e.g. planning, design, procurement, field operations etc.), so that its performance objectives (cost, time, quality, client satisfaction) can be optimized. Risk was conceptualized noting the prevalence of non-standard risk definitions, with different emphasis of the uncertainty of the culmination of a potentially harmful event, while in other cases it is on the magnitude of the consequences if that event happens and definitional discrepancy between risk, hazard, impact, defect, and other risk-related notions. This was tackled by discussing sources of risks, rather than risk itself. An unsupervised machine learning algorithm was used, which performs semantic processing and clustering of linguistic data organized in lists of aspects of risks which were then applied to the literature research found on risks, which constitutes the dataset. 3434 risk elements were initially identified. They were sorted alphabetically and according to notions and fields of application, as they are entered in the literature sources (due to the definitional discrepancy, similar elements may be differently defined, resulting in the iteration of linguistically same, but differently designated, elements).

Modelling was then carried out aiming to develop and algorithm: extraction of a concise list of general building project risk sources from the constructed database. This was done by first doing semantic processing to find the "roots" of the risk elements (i.e. the risk sources) and then through clustering for the robustification of the list of the semantically processed elements. The tools used was "stop word removal" by using suitable libraries, word suffix stemming by using the Porter stemmer, including the reduction of inflected and derived words to their basic word stem or root form, and linguistic clustering tool k-means: Unsupervised machine learning algorithm, performing vector quantization and then clustering of linguistic elements. This lead to first a clustering in ten overall themes and then a second level clustering into 129 centers. The ten overall clusters are: Technical design and drawings, productivity in building, economy, cost and finances, time and schedule, building process, environment, site safety and accidents, project management, contracts and procurement and sociopolitical factors. The second main step in the development was then the development and toolification of an algorithm using supervised machine learning for the derivation of a classification equation, which will characterize the constructability class of a new building project when given the values of the identified general risk sources affecting it. This ML system was trained and validated with the use of real risk- and constructability class-related data from civil engineering projects using WEKA [19]. The training was done as a soft-margin support vector machine, with an optimized version of the sequential minimal optimization process, and simultaneous validation with n-fold cross-validation. The project data came from thirty projects ranging from a town square, a biogas electric power plant station, a prestressed road bridge, to parts of an airport. Two main results of this development are an extensive if not fully defined analysis and systematization of risk sources and the design of a risk source-based constructability ML based prototype that is intended to assess and predict the constructability of a project based on an assessment of risk sources.

6 Analysis

The literature review showed that here are many elements of building engineering projects where ML can improve performance. It was also highlighted that hybrid ML systems, systems using more than one ML algorithm are recurrent and possibly also in need of a further conceptualisation as combinations of algorithms might have unclear impact of the generated predictions. The two cases are both prototypes developed in a university context, they are still to be tested in practice processes. This also means that little can be said about their impact on building project performance. Nevertheless, the two systems address some recurrent and important factors in building projects. The first case develops some interesting insights in factors that affect Swedish building project, such as for example the contract form. The data set behind is large compared to its domain, Swedish building project in 2014, it covers geographical, size, company, time and other types of typical variance. When human related factors in the project, such as the client role, the architect performance and collaboration, stands out as being of high impact, this is probably a general result that has bearing beyond the Swedish context. Nevertheless, part of this result can also be interpreted as a resonance with a Swedish building culture of mutual trust and willingness to cooperate. The second case combines constructability and risk analysis on a basis on civil engineering project from several different countries and with very different character; a town square, a biogas plant, a roadbridge and an airport. The data is restricted to 30 projects. The strength thus lies more in the conceptual work of risk sources enabled by ML than of the systems coverage of context. Both presented systems the testing is tested on limited data sets from specific domains. Nevertheless, they built on an assumption of generalizability; building engineering project success and risk and constructability are a result of a set of generic parameters. The contributions does not elaborate on the limitations of ML Therefore, it makes sense to draw on the more general literature available on the limitations. At the basis of ML lies a quantification of certain parts of a domain. There is risk of missing an organic complexity in issues such as "collaboration" and "satisfaction" through quantifying them [24]. Stilgoe [27] point to that designer of contemporary ML systems loose control over the content of what the ML systems learns, i.e. the generation of relations are too many and too complex to be matched even by their designers. [16] problematizes the willingness to generalized and analyze four main ML techniques; logistic regression, Naïve Bayes, K nearest neighbor and decision trees and claims that their commonality s an unsolicited assumption about stable and distinct categories. [16] point to that ML data is also images, videos, and sensor signals. The critical question according to [16] is how the combination in ML comes about. Prediction comes about through approximation, and ML tends to attempt to overcome the messy, the unforeseen and substitute it with order. Returning to building projects performance, a human touch in ML is therefore both needed in the content of the ML such as including soft parameter in the modelling as it is done in two case examples, but also a using ML require reflection and reasoning in the subsequent decision making in building engineering project processes.

7 Conclusion

We position this paper on the many places building project performance might be improved by using present machine learning techniques inquiring into what the possible applications of machine learning in building engineering projects could be and how might they impact on building engineering project performance. A review of recently proposed ML solutions for building projects were carried out. It showed that here are many elements in building engineering projects where ML can improve performance and many more that haven't been explored yet. ML is not about using a single algorithm (any more). It is rather systems using more than one ML algorithm that are recurrent. These types of systems should be further researched and conceptualized. Combinations of algorithms might have unclear impact of the generated predictions and their relation to the application domain's body of knowledge and reasoning is under illuminated. Two cases of ML prototypes are discussed: a ML based system for predicting building project success and a system to evaluate constructability of project. They are not tested in practice yet. Their effect on building project performance are still unknown. Both presented systems built on an assumption of generalizability; it is posited that building engineering project success and risk and constructability are a result of a set of generic parameters. This assumption should be addressed in the future. The human role in future preparation activities in building engineering project is possibly even more important after the introduction of ML. It is relevant, yet not enough to introduce human aspects such as satisfaction and risk perception in the content of the ML. Human reflection and reasoning should also be strengthened in the decision making in project processes.

References

1. Abramowicz, W., Paschke, A. (eds.): BIS 2018. LNBIP, vol. 320. Springer, Cham (2018). https://doi.org/10.1007/978-3-319-93931-5
2. Kaplan, A., Haenlein, M.: Siri, Siri in my Hand, who's the fairest in the land? On the interpretations, illustrations and implications of artificial intelligence. Bus. Horiz. **62**(1), 15–25 (2019)
3. McCorduck, P.: Machines Who Think: A Personal Inquiry into the History and Prospects of Artificial Intelligence, 2nd edn. A.K. Peters, Natick (2004)
4. Jordan, M.I., Mitchell, T.M.: Machine learning: trends, perspectives, and prospects. Science **349**(6245), 255–260 (2015)
5. Portugal, I., Alencar, P., Cowan, D.: The use of machine learning algorithms in recommender systems: a systematic review. Expert Syst. Appl. **97**(1), 205–227 (2018)
6. Liu, P.C.-Y., El-Gohary, N.: Automatic annotation of web images for domain-specific crack classification. In: Mutis, I., Hartmann, T. (eds.) Advances in Informatics and Computing in Civil and Construction Engineering, pp. 553–560. Springer, Cham (2019). https://doi.org/10.1007/978-3-030-00220-6_66
7. Jebelli, H., Khalili, M.M., Lee, S.H.: Mobile EEG-based workers' stress recognition by applying deep neural network. In: Mutis, I., Hartmann, T. (eds.) Advances in Informatics and Computing in Civil and Construction Engineering, pp. 173–180. Springer, Cham (2019). https://doi.org/10.1007/978-3-030-00220-6_21
8. Amasyali, K., El-Gohary, N.: Predicting energy consumption of office buildings: a hybrid machine learning-based approach. In: Mutis, I., Hartmann, T. (eds.) Advances in Informatics and Computing in Civil and Construction Engineering, pp. 695–700. Springer, Cham (2019). https://doi.org/10.1007/978-3-030-00220-6_83
9. Petrova, E., Pauwels, P., Svidt, K., Jensen, R.L.: In search of sustainable design patterns: combining data mining and semantic data modelling on disparate building data. In: Mutis, I., Hartmann, T. (eds.) Advances in Informatics and Computing in Civil and Construction Engineering, pp. 19–26. Springer, Cham (2019). https://doi.org/10.1007/978-3-030-00220-6_3
10. Siddula, M., Dai, F., Ye, Y., Fan, J.: Classifying construction site photos for roof detection: a machine-learning method towards automated measurement of safety performance on roof sites. Constr. Innov. **16**(3), 368–389 (2016)

11. Wang, L., El-Gohary, N.M.: Machine-learning-based model for supporting energy performance benchmarking for office buildings. In: Mutis, I., Hartmann, T. (eds.) Advances in Informatics and Computing in Civil and Construction Engineering, pp. 757–764. Springer, Cham (2019). https://doi.org/10.1007/978-3-030-00220-6_91

12. Sarkar, D., Bali, R., Sharma, T.: Practical Machine Learning with Python. Apress, New York (2013)

13. Kotsiantis, S.B.: Supervised machine learning: a review of classification techniques. Informatica **31**(3), 249–268 (2007)

14. Habrich, T., Wagner, C., Hellingrath, B.: Qualitative assessment of machine learning techniques in the context of fault diagnostics. In: Abramowicz, W., Paschke, A. (eds.) BIS 2018. LNBIP, vol. 320, pp. 359–370. Springer, Cham (2018). https://doi.org/10.1007/978-3-319-93931-5_26

15. Le, T., Le, C., David Jeong, H., Gilbert, Stephen B., Chukharev-Hudilainen, E.: Requirement text detection from contract packages to support project definition determination. In: Mutis, I., Hartmann, T. (eds.) Advances in Informatics and Computing in Civil and Construction Engineering, pp. 569–576. Springer, Cham (2019). https://doi.org/10.1007/978-3-030-00220-6_68

16. Shalev-Shwartz, S., Ben-David, S.: Understanding Machine Learning: From Theory to Algorithms. Cambridge University Press, Cambridge (2013)

17. Bryman, A., Bell, E.: Business Research Methods. Oxford University Press, Oxford (2011)

18. Creswell, J.W., Clark, V.L.P.: Designing and Conducting Mixed Methods Research, 2nd edn. SAGE Publications, Los Angeles (2011)

19. Mutis, I., Hartmann, T. (eds.): Advances in informatics and computing in civil and construction engineering. Springer, Cham (2019). https://doi.org/10.1007/978-3-030-00220-6

20. Frank, E., Hall, M.A., Witten, I.H.: WEKA Workbench. Online Appendix for Data Mining: Practical Machine Learning Tools and Techniques, 4th edn. Morgan Kaufmann, Burlington (2016)

21. Kulkarni, S.: Machine Learning Algorithms for Problem Solving in Computational Applications: Intelligent Techniques. IGI Global, Hershey (2012)

22. Mackinsey, A.: The production of prediction: what does machine learning want. Eur. J. Cult. Stud. **18**(4–5), 429–445 (2015)

23. Min, K., Jung, M., Kim, J., Chi, S.: Sound event recognition-based classification model for automated emergency detection in indoor environment. In: Mutis, I., Hartmann, T. (eds.) Advances in Informatics and Computing in Civil and Construction Engineering, pp. 529–535. Springer, Cham (2019). https://doi.org/10.1007/978-3-030-00220-6_63

24. Mutis, I., Ramachandran, A., Martinez, M.G.: The BIMbot: a cognitive assistant in the BIM room. In: Mutis, I., Hartmann, T. (eds.) Advances in Informatics and Computing in Civil and Construction Engineering, pp. 155–163. Springer, Cham (2019). https://doi.org/10.1007/978-3-030-00220-6_19

25. Power, M.: Counting, control and calculation: reflections on measuring and management. Hum. Relat. **57**(6), 765–783 (2004)

26. Koch, C., och Lundholm, M.: Produktivitetsläget i svenskt byggande 2014. Lokaler, Grupphus, och anläggning. Chalmers Tekniska Högskola, Prolog, SBUF. Göteborg (2018)

27. Stilgoe, J.: Machine learning, social learning and the governance of self-driving cars. Soc. Stud. Sci. **48**(1), 25–56 (2018)

28. Tixier, A.J.-P., Hallowell, M.R., Rajagopalan, B., Bowman, D.: Application of machine learning to construction injury prediction. Autom. Const. **69**, 102–114 (2016)

29. Xu, X., Cai, H.: Semantic frame-based information extraction from utility regulatory documents to support compliance checking. In: Mutis, I., Hartmann, T. (eds.) Advances in Informatics and Computing in Civil and Construction Engineering, pp. 223–230. Springer, Cham (2019). https://doi.org/10.1007/978-3-030-00220-6_27

30. Zhang, R., El-Gohary, N.: A machine learning approach for compliance checking-specific semantic role labeling of building code sentences. In: Mutis, I., Hartmann, T. (eds.) Advances in Informatics and Computing in Civil and Construction Engineering, pp. 561–568. Springer, Cham (2019). https://doi.org/10.1007/978-3-030-00220-6_67

Author Index

Printed in the United States
By Bookmasters